Intermediate
Counting and Probability

David Patrick
Art of Problem Solving

Art of Problem Solving

Books • Online Classes • Videos • Interactive Resources

www.artofproblemsolving.com

Published by: AoPS Incorporated
                 P.O. Box 2185
                 Alpine, CA 91903-2185
                 (619) 659-1612
                 books@artofproblemsolving.com

ISBN-13: 978-1-934124-06-2

Visit the Art of Problem Solving website at http://www.artofproblemsolving.com

 Scan this code with your mobile device to visit the Art of Problem Solving website, to view our other books, our free videos and interactive resources, our online community, and our online school.

Cover image designed by Vanessa Rusczyk using KaleidoTile software.

Printed in the United States of America.

Printed in 2012.

How to Use This Book

Learn by Solving Problems

We believe that the best way to learn mathematics is by solving problems. Lots and lots of problems. In fact, we believe that the best way to learn mathematics is to try to solve problems that you don't know how to do. When you discover something on your own, you'll understand it much better than if someone had just told you.

Most of the sections of this book begin with several problems. The solutions to these problems will be covered in the text, but try to solve the problems *before* reading the section. If you can't solve some of the problems, that's OK, because they will all be fully solved as you read the section. Even if you can solve all of the problems, it's still important to read the section, both to make sure that your solutions are correct, and also because you may find that the book's solutions are simpler or easier to understand than your own.

Explanation of Icons

Throughout the book, you will see various shaded boxes and icons.

Concept: This will be a general problem-solving technique or strategy. These are the "keys" to becoming a better problem solver!

Important: This will be something important that you should learn. It might be a formula, a solution technique, or a caution.

WARNING!! Beware if you see this box! This will point out a common mistake or pitfall.

> **Sidenote:** This box will contain material which, although interesting, is not part of the main material of the text. It's OK to skip over these boxes, but if you read them, you might learn something interesting!

> **Bogus Solution:** Just like the impossible cube shown to the left, there's something wrong with any "solution" that appears in this box.

Exercises, Review Problems, and Challenge Problems

Most sections end with several **Exercises**. These will test your understanding of the material that was covered in the section. You should try to solve *all* of the exercises. Exercises marked with a ★ are more difficult.

Most chapters conclude with several **Review Problems**. These are problems that test your basic understanding of the material covered in the chapter. Your goal should be to solve most or all of the Review Problems for every chapter—if you're unable to do this, it means that you haven't yet mastered the material, and you should probably go back and read the chapter again.

All of the chapters (except for Chapter 1) end with **Challenge Problems**. These problems are generally more difficult than the other problems in the book, and will really test your mastery of the material. Some of them are very, very hard—the hardest ones are marked with a ★. Don't expect to be able to solve all of the Challenge Problems on your first try—these are difficult problems even for experienced problem solvers. If you are able to solve a large number of Challenge Problems, then congratulations, you are on your way to becoming an expert problem solver!

Hints

Many problems come with one or more hints. You can look up the hints in the Hints section in the back of the book. The hints are numbered in random order, so that when you're looking up a hint to a problem you don't accidentally glance at the hint to the next problem at the same time.

It is very important that you first try to solve each problem without resorting to the hints. Only after you've seriously thought about a problem and are stuck should you seek a hint. Also, for problems which have multiple hints, use the hints one at a time; don't go to the second hint until you've thought about the first one.

Solutions

The solutions to all of the Exercises, Review Problems, and Challenge Problems are in the separate solutions book. If you are using this textbook in a regular school class, then your teacher may decide not to make this solutions book available to you, and instead present the solutions him/herself. However, if you are using this book to learn on your own, then you probably have a copy of the solutions book, in which case there are some very important things to keep in mind:

1. Make a serious attempt to solve each problem before looking at the solution. Don't use the solutions book as a crutch to avoid really thinking about the problem first. You should think *hard* about a problem before deciding to look at the solution. On the other hand, after you've thought hard about a problem, don't feel bad about looking at the solution if you're really stuck.

2. After you solve a problem, it's usually a good idea to read the solution, even if you think you know how to solve the problem. The solutions book might show you a quicker or more concise way to solve the problem, or it might have a completely different solution method than yours.

3. If you have to look at the solution in order to solve a problem, make a note of that problem. Come back to it in a week or two to make sure that you are able to solve it on your own, without resorting to the solution.

Resources

Here are some other good resources for you to further pursue your study of mathematics:

- *The Art of Problem Solving* books, by Sandor Lehoczky and Richard Rusczyk. Whereas the book that you're reading right now will go into great detail of one specific subject area—counting and probability—the *Art of Problem Solving* books cover a wide range of problem solving topics across many different areas of mathematics.

- The www.artofproblemsolving.com website. The publishers of this book are also the webmasters of the Art of Problem Solving website, which contains many resources for students:
  - a discussion forum
  - online classes
  - resource lists of books, contests, and other websites
  - a LATEX tutorial
  - a math and problem solving Wiki
  - and much more!

- You can hone your problem solving skills (and perhaps win prizes!) by participating in various math contests. For middle school students in the United States, the major contests are MOEMS, MATHCOUNTS, and the AMC 8. For U.S. high school students, some of the best-known contests are the AMC/AIME/USAMO series of contests (which are used to choose the U.S. team for the International Mathematical Olympiad), the American Regions Math League (ARML), the Mandelbrot Competition, the Harvard-MIT Mathematics Tournament, and the USA Mathematical Talent Search. More details about these contests are on page vii, and links to these and many other contests are available on the Art of Problem Solving website.

A Note to Teachers

We believe students learn best when they are challenged with hard problems that at first they may not know how to do. This is the motivating philosophy behind this book.

Rather than first introducing new material and then giving students exercises, we present problems at the start of each section that students should try to solve *before* new material is presented. The goal is to get students to discover the new material on their own. Often, complicated problems are broken into smaller parts, so that students can discover new techniques one piece at a time. After the problems, new material is formally presented in the text, and full solutions to each problem are explained, along with problem-solving strategies.

We hope that teachers will find that their stronger students will discover most of the material in this book on their own by working through the problems. Other students may learn better from a more traditional approach of first seeing the new material, then working the problems. Teachers have the flexibility to use either approach when teaching from this book.

The book is linear in coverage. Generally, students and teachers should progress straight through the book in order, without skipping chapters. Sections marked with a ★ contain supplementary material that may be safely skipped. In general, chapters are not equal in length, so different chapters may take different amounts of classroom time.

Links

The Art of Problem Solving website has a page containing links to websites with content relating to material in this book, as well as an errata list for the book. This page can be found at:

http://www.artofproblemsolving.com/BookLinks/IntermCounting/links.php

---

Extra! Occasionally, you'll see a box like this at the bottom of a page. This is an "Extra!" and
▪▶▪▶▪▶▪▶ might be a quote, some biographical or historical background, or perhaps an interesting
idea to think about.

Acknowledgements

Contests

We would like to thank the following contests for allowing us to use a selection of their problems in this book:

- The **American Mathematics Competitions**, a series of contests for U.S. middle and high school students. The **AMC 8**, **AMC 10**, and **AMC 12** contests are multiple-choice tests that are taken by over 400,000 students every year. Top scorers on the AMC 10 and AMC 12 are invited to take the **American Invitational Mathematics Examination (AIME)**, which is a more difficult, short-answer contest. Approximately 10,000 students every year participate in the AIME. Then, based on the results of the AMC and AIME contests, about 500 students are invited to participate in the **USA Mathematical Olympiad (USAMO)** and the **USA Junior Mathematical Olympiad (USAJMO)**, a 2-day, 9-hour examination in which each student must show all of his or her work. Results from the USAMO and USAJMO are used to invite students to the Math Olympiad Summer Program, and to choose the U.S. team for the **International Mathematical Olympiad (IMO)**. More information about the AMC contests can be found on the AMC website at http://amc.maa.org.

- The **Mandelbrot Competition**, which was founded in 1990 by Sandor Lehoczky, Richard Rusczyk, and Sam Vandervelde. The aim of the Mandelbrot Competition is to provide a challenging, engaging mathematical experience that is both competitive and educational. Students compete both as individuals and in teams. The Mandelbrot Competition is offered at the national level for more advanced students and at the regional level for less experienced problem solvers. More information can be found at www.mandelbrot.org.

- The **USA Mathematical Talent Search (USAMTS)**, which was founded in 1989 by Professor George Berzsenyi. The USAMTS is a free mathematics competition open to all United States middle and high school students. As opposed to most mathematics competitions, the USAMTS allows students a full month to work out their solutions. Carefully written justifications are required for each problem. More information is available at www.usamts.org.

- The **American Regions Math League (ARML)**, which was founded in 1976. The annual ARML competition brings together nearly 2,000 of the nation's finest students. They meet, compete against, and socialize with one another, forming friendships and sharpening their mathematical skills. The contest is written for high school students, although some exceptional junior high students attend each year. The competition consists of several events, which include a team round, a power question (in which a team solves proof-oriented questions), an individual round, and two relay rounds. More information is available at www.arml.com.

- The **Harvard-MIT Mathematics Tournament**, which is an annual math tournament for high school students, held at MIT and at Harvard in alternating years. It is run exclusively by MIT and Harvard students, most of whom themselves participated in math contests in high school. More information is available at web.mit.edu/hmmt/.

Some other sources for problems in the book are listed below.

MATHCOUNTS MATHCOUNTS, the premier national math contest for U.S. middle school students.

Putnam The William Lowell Putnam Competition, an annual math competition for undergraduate students in North America.

PSS *Problem-Solving Strategies* by Arthur Engel. (See the references on page 366 for the complete citation.)

ToT The International Mathematics Tournament of Towns, a mathematical problem solving competition for students throughout the world.

We have also included problems from various countries' national math Olympiad contests. These problems are cited by country name in the text.

How We Wrote This Book

This book was written using the LaTeX document processing system. We thank the authors of the various LaTeX packages that we used while preparing this book, and also the authors of *The LaTeX Companion* for writing a reference book that is not only thorough but also very readable. The diagrams were prepared using METAPOST.

About Us

This book is a collaborative effort of the staff of the Art of Problem Solving. Richard Rusczyk wrote the original lecture notes for the online course that was the inspiration for this book; he also read several drafts of this book and made many, many helpful suggestions. Extensive proofreading of the manuscript was done by Greg Brockman, Chris Chang, Larry Evans, Sean Markan, Jeff Nanney, Adrian Sanborn, Nathan Savir, and Valentin Vornicu. Special thanks to Hyun Soo Kim for his assistance with selecting many of the problems. Vanessa Rusczyk designed the cover.

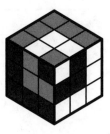

Contents

Ernő Rubik (1944–present) and his Magic Cube

The diagrams at the start of each chapter depict a **Rubik's Cube**.

Ernő Rubik was an inventor, sculptor, and architecture professor living in Budapest, Hungary, when in 1974 he invented a nifty little puzzle that would go on to sell hundreds of millions of copies. It was originally called the "Magic Cube," but when it became available worldwide in 1980, it was known simply as "Rubik's Cube."

The Cube itself is about 2 inches long on each side and is divided into a $3 \times 3 \times 3$ array of 27 smaller cubes. Each of the 6 faces can rotate independently of each other, which changes the configuration of the colors on the outside of the Cube. Initially, the Cube starts with a solid color on each of its 6 faces, but rotating the sides will cause the colors to become mixed.

There are exactly 43,252,003,274,489,856,000 different configurations of Rubik's Cube! This is a *very* difficult number to count: for those of you keeping score at home, it is $8! \times 12! \times 3^7 \times 2^{10}$. The goal of the puzzle is to first put the Cube into a random configuration, and then to try to return it to its original configuration, with each side a solid color.

Rubik's Cube became incredibly popular worldwide in the early 1980s. According to the official Rubik's Cube website at www.rubiks.com, over 100,000,000 cubes were sold in the period from 1980 to 1982 alone. In 1983, the immense popularity of the Cube even led to a Saturday-morning cartoon called *Rubik, The Amazing Cube*.

As if solving the Cube is not challenging enough, many people enjoy the further challenge of trying to solve it as fast as possible, in some cases while blindfolded! The World Cube Association is the official keeper of Rubik's Cube speed-solving records. The current world record (as of May 2012) is held by Feliks Zemdegs, who solved a randomly-configured Cube in 5.66 seconds. A detailed list of records is on the WCA's website, which is on the links page referenced on page vi.

If you don't have a Rubik's Cube at home, there's a Java applet allowing you to play with a virtual Cube; this applet is also available via our links page.

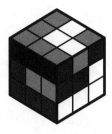

A bad review is like baking a cake with all the best ingredients and having someone sit on it. – Danielle Steele

CHAPTER 1

Review of Counting & Probability Basics

1.1 Introduction

Before beginning with new material, we'll spend this chapter making sure that you have a good grasp of the basics of counting and probability. This chapter is basically a one-chapter review of the contents of *Introduction to Counting & Probability*. If a lot of the material in this chapter is unfamiliar to you, then you should probably start with the *Introduction* book, before tackling this book.

The chapter will consist of some problems selected from *Introduction to Counting & Probability* together with their solutions.

> **Important:** You should feel confident that you are able to solve most or all of the problems in this chapter before continuing on to the rest of the book.

It is important that basic counting and probability computations of the type that we do in this chapter are as natural and familiar to you as arithmetic and basic algebra are. As a child, you wouldn't try to learn how to multiply three-digit numbers together without having absolutely mastered multiplication of 1-digit numbers. In the same way, you shouldn't plunge into the more advanced counting techniques in this book without having the basics mastered.

Also, there is one brief section at the end of this chapter (namely, Section 1.6) that includes new material that is not in the *Introduction* book. This section discusses **summation notation**, an example of which is the left side of the equation below:

$$\sum_{i=1}^{n} i = \frac{n(n+1)}{2}.$$

If you are not familiar with this notation, please make sure that you read this section before continuing on with the book.

1.2 Basic Counting Techniques

 **Problems**

Problem 1.1: My city is running a lottery. In the lottery, 25 balls numbered 1 through 25 are placed in a bin. Four balls are drawn one at a time and their numbers are recorded. The winning combination consists of the four selected numbers in the order they are selected. How many winning combinations are there, if:

(a) each ball is discarded after it is removed?

(b) each ball is replaced in the bin after it is removed and before the next ball is drawn?

Problem 1.2: On the island of Mumble, the Mumblian alphabet has only 5 letters, and every word in the Mumblian language has no more than 3 letters in it. How many words are possible? (A word can use a letter more than once, but 0 letters does not count as a word.)

Problem 1.3: The Smith family has 4 sons and 3 daughters. In how many ways can they be seated in a row of 7 chairs, such that at least 2 boys are next to each other?

Problem 1.4: How many 3-digit numbers have exactly one zero?

Problem 1.5: Our math club has 20 members and 3 officers: President, Vice President, and Treasurer. However, one member, Ali, hates another member, Brenda. In how many ways can we fill the offices if Ali refuses to serve as an officer if Brenda is also an officer?

Problem 1.6: How many possible distinct arrangements are there of the letters in the word BALL?

Problem 1.7: In how many different ways can 6 people be seated at a round table? Two seating arrangements are considered the same if, for each person, the person to his or her left is the same in both arrangements.

Problem 1.8: Consider a club that has n people. What is the number of ways to form an r-person committee from the total of n people?

Problem 1.9: Each block on the grid shown at right is 1 unit by 1 unit. Suppose we wish to walk from A to B via a 7 unit path, but we have to stay on the grid—no cutting across blocks. How many different paths can we take?

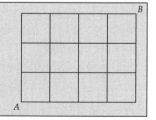

Problem 1.10: In how many ways can a dog breeder separate his 10 puppies into a group of 4 and a group of 6 if he has to keep Biter and Nipper, two of the puppies, in separate groups?

Problem 1.1: My city is running a lottery. In the lottery, 25 balls numbered 1 through 25 are placed in a bin. Four balls are drawn one at a time and their numbers are recorded. The winning combination consists of the four selected numbers in the order they are selected. How many winning combinations are there, if:

(a) each ball is discarded after it is removed?

(b) each ball is replaced in the bin after it is removed and before the next ball is drawn?

Solution for Problem 1.1:   (a) We have 25 choices for the first ball. After the first ball is drawn and discarded, there are 24 balls left in the bin, so there are 24 choices for the second ball. Similarly, there are 23 choices for the third ball and 22 choices for the fourth ball. Hence the number of winning combinations is $25 \times 24 \times 23 \times 22 = 303{,}600$.

(b) We have 25 choices for each of the four balls, since when each ball is drawn, all 25 balls are in the bin. Hence there are $25 \times 25 \times 25 \times 25 = 25^4 = 390{,}625$ winning combinations. □

Concept: In an event with multiple stages, we *multiply* the number of choices at each stage to get the total number of choices.

Concept: It is important to distinguish selecting *without* replacement from selecting *with* replacement. When choosing k items, in order, from a group of n, with replacement, we have n^k choices. When choosing k items, in order, from a group of n, without replacement, we have

$$n(n-1)(n-2)\cdots(n-k+1)$$

choices. Selecting k items from a group of n items, without replacement, where order matters, is called a **permutation**. The number of permutations of k items from a group of n items is sometimes denoted $P(n,k)$ or $_nP_k$.

Problem 1.2: On the island of Mumble, the Mumblian alphabet has only 5 letters, and every word in the Mumblian language has no more than 3 letters in it. How many words are possible? (A word can use a letter more than once, but 0 letters does not count as a word.)

Solution for Problem 1.2: For this problem, it makes sense to divide the words into cases, based on the number of letters in each word.

Case 1: 1-letter words
There are 5 1-letter words (each of the 5 letters is itself a 1-letter word).

Case 2: 2-letter words
To form a 2-letter word, we have 5 choices for our first letter, and 5 choices for our second letter. Thus there are $5 \times 5 = 25$ 2-letter words possible.

Case 3: 3-letter words
To form a 3-letter word, we have 5 choices for our first letter, 5 choices for our second letter, and 5 choices for our third letter. Thus there are $5 \times 5 \times 5 = 125$ 3-letter words possible.

To get the total number of words in the language, we add the number of words from each of our cases. (We need to make sure that the cases are exclusive, meaning they don't overlap. But that's clear in this solution, since, for example, a word can't be both a 2-letter word and a 3-letter word at the same time.)

Therefore, there are 5 + 25 + 125 = 155 words possible on Mumble. □

> **Concept:** **Casework** is the general technique of breaking up the possibilities into two or more categories. We can then *add* the possibilities of the various cases to get the total number of possibilities.

> **Concept:** When using casework, it is important that the cases be *exclusive*, meaning that they don't overlap. Otherwise, you'll end up counting some outcomes multiple times. (Although, later in this book, we'll see some techniques for dealing with overlapping cases.)

Problem 1.3: The Smith family has 4 sons and 3 daughters. In how many ways can they be seated in a row of 7 chairs, such that at least 2 boys are next to each other?

Solution for Problem 1.3: It will be fairly difficult to try to count this directly with casework, since there are lots of possible cases (for example, two possibilities are BBBBGGG and BGGBBGB, where B is a boy and G is a girl). But there is only one way to assign genders to the seating so that no two boys are next to each other, and that is BGBGBGB. So we use **complementary counting**: we count the items that we *don't* want.

If we seat the children as BGBGBGB, then there are 4! orderings for the 4 boys, and 3! orderings for the 3 girls, giving a total of 4! × 3! = 144 seatings for the 7 children.

These are the seatings that we *don't* want, so to count the seatings that we *do* want, we need to subtract these seatings from the total number of seatings without any restrictions. Since there are 7 kids, there are 7! ways to seat them.

Therefore, the answer is 7! − (4! × 3!) = 5040 − 144 = 4896. □

> **Concept:** Often, complicated casework means that you should think about trying **complementary counting**: that is, counting what we don't want and subtracting this count from the number of possibilities (without restriction). If it's hard to count all the cases that we want, then it may be relatively easy to count what we don't want.

> **Concept:** When a problem asks "How many are not?", we might think instead to count "How many are?" When a problem asks "How many have at least one?", we might think instead to count "How many have none?"

Problem 1.4: How many 3-digit numbers have exactly one zero?

Solution for Problem 1.4: We could do this by casework.

Case 1: 3-digit numbers with 0 as the middle digit
There are 9 choices for the first digit and 9 choices for the last digit, for a total of $9 \times 9 = 81$ numbers in this case.

Case 2: 3-digit numbers with 0 as the final digit
There are 9 choices for the first digit and 9 choices for the middle digit, for a total of $9 \times 9 = 81$ numbers in this case.

Therefore, there are $81 + 81 = 162$ such numbers.

On the other hand, we can also solve the problem directly, by thinking about the steps necessary to construct such a 3-digit number, and counting the number of choices that we have at each step.

When doing such a construction, we usually want to start by dealing with the most severe restriction in the construction. In this problem, the restriction is that the number must have exactly one zero.

So when attempting to construct such a 3-digit number, our first choice should be: where do we put the 0? We have 2 choices—the zero can go in the middle (tens) digit or in the rightmost (units) digit. Note that a number cannot begin with a 0, so we can't put the zero in the leftmost (hundreds) digit.

Now that we've placed the zero, we need to choose the other two digits. Each of the other two digits can be any digit from 1 through 9 (but cannot be 0). So we have 9 choices for each of the other two digits.

Therefore, there are $2 \times 9 \times 9 = 162$ such 3-digit numbers. □

> **Concept:** This general idea—thinking about how to construct the items that we wish to count, and then keeping track of the choices that we have to make during the construction—is known as **constructive counting**.

> **Concept:** Deal with the restriction first. Considering the restriction first usually helps when solving constructive counting problems.

Problem 1.5: Our math club has 20 members and 3 officers: President, Vice President, and Treasurer. However, one member, Ali, hates another member, Brenda. In how many ways can we fill the offices if Ali refuses to serve as an officer if Brenda is also an officer?

Solution for Problem 1.5: The best way to approach this problem is to use complementary counting.

We know that there are $20 \times 19 \times 18$ ways to choose the 3 officers if we ignore the restriction about Ali and Brenda. So now we want to count the number of outcomes that we don't want: both Ali and Brenda serving as officers.

To count the outcomes we don't want, we will use constructive counting. We need to pick an office for Ali, then pick an office for Brenda, then put someone in the last office.

We have 3 choices for an office for Ali, either President, VP, or Treasurer.

Once we pick an office for Ali, we have 2 offices left from which to choose an office for Brenda.

Once we have both Ali and Brenda installed in offices, we have 18 members left in the club to pick from for the remaining vacant office.

So there are $3 \times 2 \times 18$ ways to pick officers such that Ali and Brenda are both in an office. Remember that these are the cases that we want to exclude, so to finish the problem we subtract these cases from the total number of cases. Hence the answer is:

$$(20 \times 19 \times 18) - (3 \times 2 \times 18) = ((20 \times 19) - 6) \times 18 = 374 \times 18 = 6732.$$

□

> **Concept:** Many problems will require you to use more than one counting method. In the previous problem, we used both complementary counting and constructive counting.

> **Problem 1.6:** How many possible distinct arrangements are there of the letters in the word BALL?

Solution for Problem 1.6: First, note that the answer is not simply 4!:

> **Bogus Solution:** We have 4 ways to pick the first letter, 3 ways to pick the second, and so on, for a total of 4! possibilities.

This method overcounts. The reason for this is that two of our letters are the same.

Let's pretend that the two L's are different, and call them L_1 and L_2. So our word BALL is now really BAL_1L_2. In making the expected 4! arrangements, we make both BAL_1L_2 and BAL_2L_1. But when we remove the numbers, we have BALL and BALL, which are the same.

With 4!, we've overcounted, and we need to correct for this.

Every possible arrangement of BALL is counted twice among our arrangements of BAL_1L_2. For example, LLAB is counted as both L_1L_2AB and L_2L_1AB, LABL is counted as both L_1ABL_2 and L_2ABL_1, and so on for every possible arrangement of BALL. We can see this in Figure 1.1:

$$
\begin{array}{llll}
BAL_1L_2, BAL_2L_1 & \Rightarrow & BALL \qquad & BL_2AL_1, BL_1AL_2 \Rightarrow BLAL \\
BL_1L_2A, BL_2L_1A & \Rightarrow & BLLA & ABL_1L_2, ABL_2L_1 \Rightarrow ABLL \\
L_1BAL_2, L_2BAL_1 & \Rightarrow & LBAL & L_1BL_2A, L_2BL_1A \Rightarrow LBLA \\
AL_1BL_2, AL_2BL_1 & \Rightarrow & ALBL & L_1ABL_2, L_2ABL_1 \Rightarrow LABL \\
L_1L_2BA, L_2L_1BA & \Rightarrow & LLBA & AL_1L_2B, AL_2L_1B \Rightarrow ALLB \\
L_1AL_2B, L_2AL_1B & \Rightarrow & LALB & L_1L_2AB, L_2L_1AB \Rightarrow LLAB \\
\end{array}
$$

Figure 1.1: BALL's with different L's

Thus, the number of arrangements of BAL_1L_2 is exactly twice the number of arrangements of BALL. So to get the number of arrangements of BALL, we have to divide the number of arrangements of BAL_1L_2 by 2.

Therefore, the number of arrangements of BALL is $4!/2 = 12$. □

> **Concept:** Often we can count outcomes by strategically **overcounting** the number of outcomes, and then correcting for the overcounting. One common use of this is in counting permutations with repeated items.

> **Problem 1.7:** In how many different ways can 6 people be seated at a round table? Two seating arrangements are considered the same if, for each person, the person to his or her left is the same in both arrangements.

Solution for Problem 1.7: If the 6 people were sitting in a row, rather than around a table, there would be 6! arrangements. But this is clearly an overcounting, since several different row arrangements correspond to the same round table arrangement. For example, suppose we take a row of 6 people and sit them at the round table by putting the first person in the "top" seat and proceeding counterclockwise. Six row arrangements and their corresponding circular arrangements are shown in Figure 1.2 below.

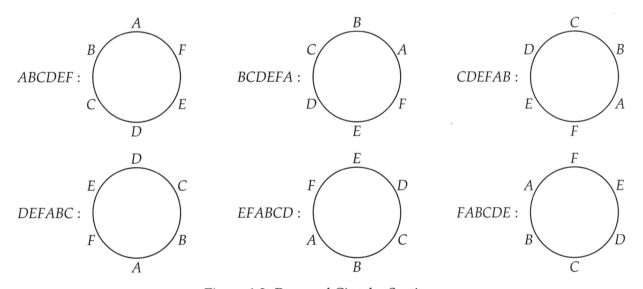

Figure 1.2: Row and Circular Seatings

All six arrangements of the people A through F shown in the diagram correspond to the same arrangement of the people seated around a round table.

The reason for this is that our problem has **symmetry**: each arrangement can be rotated 6 ways (one of them being no rotation) to make other arrangements. Hence, when we put our 6! arrangements of $ABCDEF$ in a circle, we count each circular arrangement 6 times, once for each rotation.

Therefore, we must divide our initial overcount of 6! by 6, to account for the fact that there are 6 row arrangements corresponding to each circular arrangement. So our answer is that there are

$6!/6 = 5! = 120$ ways to arrange the 6 people around the table. $\square$

> **Concept:** Counting outcomes with **symmetry** usually means some sort of strategic overcounting.

> **Problem 1.8:** Consider a club that has n people. What is the number of ways to form an r-person committee from the total of n people?

Solution for Problem 1.8: We start by counting the number of ways to choose r people if order matters. There are n choices for the first person, $n-1$ choices for the second person, $n-2$ choices for the third person, and so on, up to $n-r+1$ choices for the rth person.

So there are

$$n \times (n-1) \times (n-2) \times \cdots \times (n-r+1) \tag{1.1}$$

ways to choose r people from a total of n people *if order matters*. (Recall that this is the quantity that is often denoted by $P(n,r)$.)

But we know that there are $r!$ ways to order r people. Therefore, each *unordered* committee of r people will correspond to $r!$ *ordered* choices of r people. So we need to divide our count in equation (1.1) by $r!$ to correct for the overcounting.

Therefore our answer is

$$\binom{n}{r} = \frac{n \times (n-1) \times (n-2) \times \cdots \times (n-r+1)}{r!} = \frac{n!}{r!(n-r)!}.$$

$\square$

> **Concept:** If we don't care about the order when choosing r items from a set of n items—for example, when choosing a committee—we have a **combination**. The number of ways to choose r items from a set of n items, without regard to order, is $\binom{n}{r}$, pronounced "n choose r." (Note: some sources denote this as $C(n,r)$ or as $_nC_r$.)

> **Important:** The formula for combinations is:
>
> $$\binom{n}{r} = \frac{n!}{r!(n-r)!} = \frac{n(n-1)(n-2)\cdots(n-r+1)}{r!}.$$
>
> The first formula is more typically used in algebraic proofs involving combinations, whereas the latter formula is the one that we most often use to actually compute combinations. For example:
>
> $$\binom{11}{3} = \frac{11 \times 10 \times 9}{3 \times 2 \times 1} = 165.$$

Problem 1.9: Each block on the grid shown below is 1 unit by 1 unit. Suppose we wish to walk from A to B via a 7 unit path, but we have to stay on the grid—no cutting across blocks. How many different paths can we take?

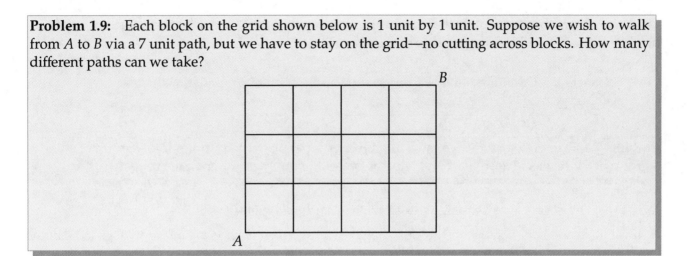

Solution for Problem 1.9: We know that we must take a 7 unit path. If we look at the grid a little more carefully, we can see that our path must consist of 4 steps to the right and 3 steps up, and we can take those steps in any order. So in order to specify a path, we must choose 3 of our 7 steps to be "up" (and the other 4 steps will thus be "right"). Hence the number of paths is

$$\binom{7}{3} = \frac{7 \times 6 \times 5}{3 \times 2 \times 1} = 35.$$

If you are not convinced, we can describe a path by writing "r" for a right step and "u" for an up step. For every arrangement of 4 r's and 3 u's, we get a path from A to B, as shown in Figure 1.3.

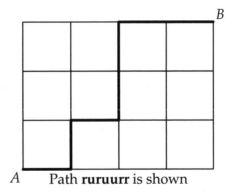

A Path **ruruurr** is shown

Figure 1.3: Example path from A to B

Therefore, to count the number of paths from A to B, we need only to count the number of arrangements of "rrrruuu." The number of arrangements is

$$\frac{7!}{4!3!} = \binom{7}{3} = 35.$$

Remember how we get the above expression: there are 7! arrangements of "$r_1 r_2 r_3 r_4 u_1 u_2 u_3$," where we think of each r and each u as different. However, we want to count the number of arrangements of "rrrruuu," where the r's and u's are all the same, so we must divide by the 4! possible arrangements of the r's and the 3! possible arrangements of the u's. □

> **Concept:** Counting paths on a grid is one application of combinations.

Problem 1.10: In how many ways can a dog breeder separate his 10 puppies into a group of 4 and a group of 6 if he has to keep Biter and Nipper, two of the puppies, in separate groups?

Solution for Problem 1.10: Let's first do the problem by complementary counting.

If we have no restrictions on the groups, then we simply need to choose 4 of the 10 dogs to be in the smaller group, and the rest of the dogs will make up the larger group. There are $\binom{10}{4}$ ways to do this.

But we can't have Biter and Nipper in the same group. So we have to subtract the number of ways that we can form the two groups with Biter and Nipper in the same group. This calls for a little bit of casework.

Case 1: Biter and Nipper are both in the smaller group.
If they are both in the smaller group, then we have to choose 2 more dogs from the 8 remaining to complete the smaller group, and we can do this in $\binom{8}{2}$ ways.

> **WARNING!!** Don't mistakenly count the possibilities in Case 1 as $\binom{8}{2}\binom{8}{6}$, by reasoning that we must choose 2 out of 8 dogs for the smaller group, and 6 out of 8 dogs for the larger group. These choices are not independent! Once we pick the 2 dogs for the smaller group, then we have no choice but to put the remaining 6 dogs into the larger group.

Case 2: Biter and Nipper are both in the larger group.
If they are both in the larger group, then we have to choose 4 dogs from the 8 remaining to compose the smaller group, and we can do this in $\binom{8}{4}$ ways.

So to get the number of ways to form groups such that Biter and Nipper are both in the same group, we add the counts from our two cases, to get $\binom{8}{2} + \binom{8}{4}$.

But remember that these are the cases that we *don't* want, so to solve the problem, we subtract this count from the number of ways to form the two groups without restrictions. Thus, our answer is

$$\binom{10}{4} - \left(\binom{8}{2} + \binom{8}{4}\right) = 210 - (28 + 70) = 112.$$

The other way that we could solve this problem is by direct casework. There are two cases of possible groupings.

Case 1: Biter is in the smaller group, Nipper is in the larger group.
To complete the smaller group, we need to choose 3 more dogs from the 8 remaining dogs. We can do this in $\binom{8}{3}$ ways.

Case 2: Nipper is in the smaller group, Biter is in the larger group.
Again, to complete the smaller group, we need to choose 3 more dogs from the 8 remaining. We can do this in $\binom{8}{3}$ ways.

To count the total number of groupings, we add the counts from our two cases, to get $\binom{8}{3} + \binom{8}{3} = 56 + 56 = 112$ as our final answer. $\square$

> **Concept:** Many counting problems can be solved in more than one way. If it's easy to do, solving problems in more than one way is both good practice and a good way to check your answer.

1.3 Basic Probability Techniques

 **Problems**

Problem 1.11: What is the probability that when a fair 6-sided die is rolled, a prime number faces up?

Problem 1.12: A standard deck of cards has 52 cards divided into 4 suits, each of which has 13 cards. Two of the suits ($\heartsuit$ and $\diamondsuit$, called "hearts" and "diamonds") are red, the other two ($\spadesuit$ and $\clubsuit$, called "spades" and "clubs") are black. The cards in the deck are placed in random order (usually by a process called "shuffling"). What is the probability that the first two cards are both red?

Problem 1.13: We have a standard deck of 52 cards, with 4 cards in each of 13 ranks. We call a 5-card poker hand a *full house* if the hand has 3 cards of one rank and 2 cards of another rank (such as 33355 or AAAKK). What is the probability that five cards chosen at random form a full house?

Problem 1.14: A bag has 3 red and k white marbles, where k is an (unknown) positive integer. Two of the marbles are chosen at random from the bag. Given that the probability that the two marbles are the same color is $\frac{1}{2}$, find k.

Problem 1.15: Mary and James each sit in a row of 7 chairs. They choose their seats at random. What is the probability that they don't sit next to each other?

Problem 1.16: The Grunters play the Screamers 4 times. The Grunters are the much better team, and are 75% likely to win any given game. What is the probability that the Grunters will win all 4 games?

Problem 1.17: A bag has 4 red and 6 blue marbles. A marble is selected and not replaced, then a second is selected. What is the probability that both are the same color?

Problem 1.18: Point C is chosen at random atop a 5 foot by 5 foot square table. A circular disk with a radius of 1 foot is placed on the table with its center directly on point C. What is the probability that the entire disk is on top of the table (in other words, none of the disk hangs over an edge of the table)?

Problem 1.11: What is the probability that when a fair 6-sided die is rolled, a prime number faces up?

Solution for Problem 1.11: There are 6 equally likely outcomes. Three of those outcomes are successful: ⚀, ⚁, or ⚄. Therefore, the probability is $\frac{3}{6} = \frac{1}{2}$. □

Concept: If all outcomes are **equally likely**, then the probability of success is

$$P(\text{success}) = \frac{\text{Number of successful outcomes}}{\text{Number of possible outcomes}}.$$

Problem 1.12: A standard deck of cards has 52 cards divided into 4 suits, each of which has 13 cards. Two of the suits ($\heartsuit$ and $\diamond$, called "hearts" and "diamonds") are red, the other two ($\spadesuit$ and $\clubsuit$, called "spades" and "clubs") are black. The cards in the deck are placed in random order (usually by a process called "shuffling"). What is the probability that the first two cards are both red?

Solution for Problem 1.12:

Method 1: For the total number of possibilities, there are 52 ways to pick the first card, then 51 ways to pick the second card, for a total of 52×51 possibilities.

For the number of successful possibilities, there are 26 ways to pick a red card first (since there are 26 total red cards), then there are 25 ways to pick a second red card (since there are 25 red cards remaining after we've chosen the first card). Thus, there are a total of 26×25 successful possibilities.

Therefore, the probability is

$$P(\text{first two cards are red}) = \frac{\text{Number of successful outcomes}}{\text{Number of possible outcomes}} = \frac{26 \times 25}{52 \times 51} = \frac{25}{102}.$$

Method 2: For the total number of possibilities, there are $\binom{52}{2} = 1326$ ways to pick two cards (without regard to order).

For the number of successful possibilities, there are $\binom{26}{2} = 325$ ways to pick two red cards (without regard to order).

Therefore, the probability is

$$P(\text{first two cards are red}) = \frac{\text{Number of successful outcomes}}{\text{Number of possible outcomes}} = \frac{325}{1326} = \frac{25}{102}.$$

Method 3: The probability that the first card is red is $\frac{26}{52} = \frac{1}{2}$. If the first card is red, then the probability that the second card is red is $\frac{25}{51}$. Therefore:

$$P(\text{first two cards are red}) = P(\text{first card red}) \times P(\text{second card red}) = \frac{1}{2} \times \frac{25}{51} = \frac{25}{102}.$$

□

> **Concept:** There are often many ways to approach the counting within probability problems. However you choose to approach a problem, make sure that you are consistent! For example, if you count possible outcomes without regard to order, don't count successful outcomes *with* regard to order. In other words, don't compare apples and oranges!

Problem 1.13: We have a standard deck of 52 cards, with 4 cards in each of 13 ranks. We call a 5-card poker hand a *full house* if the hand has 3 cards of one rank and 2 cards of another rank (such as 33355 or AAAKK). What is the probability that five cards chosen at random form a full house?

Solution for Problem 1.13: The total number of outcomes is just the number of ways to choose 5 cards from a set of 52, which is $\binom{52}{5} = 2,598,960$. Notice that in this count, we don't care about the order in which the cards are chosen. (Remember—apples and oranges!)

To count the number of successful outcomes, we turn to constructive counting, thinking about how we'd construct a full house.

To form a full house, we have to choose:

(a) A rank for the 3 cards. This can be done in 13 ways.

(b) 3 of the 4 cards of that rank. This can be done in $\binom{4}{3} = 4$ ways.

(c) A rank for the other 2 cards. This can be done in 12 ways (since we can't choose the rank that we chose in (a)).

(d) 2 of the 4 cards of that rank. This can be done in $\binom{4}{2} = 6$ ways.

Again, note that in each of the steps in our constructive count, we don't care about the order in which the cards are chosen.

So there are $13 \times 4 \times 12 \times 6 = 3,744$ full houses. Thus, the probability is

$$\frac{3,744}{2,598,960} = \frac{6}{4165}.$$

□

> **Concept:** You will often have to use your toolbox of counting techniques (in this case, constructive counting) to solve probability problems.

Problem 1.14: A bag has 3 red and k white marbles, where k is an (unknown) positive integer. Two of the marbles are chosen at random from the bag. Given that the probability that the two marbles are the same color is $\frac{1}{2}$, find k.

Solution for Problem 1.14: Here we compute the probability in terms of k that the two marbles are the same color, and then set that probability equal to $\frac{1}{2}$.

To calculate the probability, we simply use our usual method of counting both the total number

---

of outcomes and the number of successful outcomes. There are $k + 3$ total marbles in the bag, so the number of ways to choose 2 of them, without regard to order, is $\binom{k+3}{2}$.

One type of successful outcome is to choose two red marbles. This can be done in $\binom{3}{2} = 3$ ways. The other successful outcome is to choose two white marbles. This can be done in $\binom{k}{2}$ ways. Since these two cases are mutually exclusive, we can add our counts. So

$$P(\text{both marbles are the same color}) = \frac{3 + \binom{k}{2}}{\binom{k+3}{2}}.$$

We set this equal to the given probability of $\frac{1}{2}$, and then solve for k. Our equation is

$$\frac{3 + \binom{k}{2}}{\binom{k+3}{2}} = \frac{1}{2}.$$

In order to solve this, we'll need to first write out the combinations:

$$\frac{3 + \frac{k(k-1)}{2}}{\frac{(k+3)(k+2)}{2}} = \frac{1}{2}.$$

We can get rid of the 2's in the denominators by multiplying the numerator and denominator of the left side by 2:

$$\frac{6 + k(k - 1)}{(k + 3)(k + 2)} = \frac{1}{2},$$

and then we cross-multiply to get rid of the fractions:

$$12 + 2k(k - 1) = (k + 3)(k + 2).$$

Multiplying out, we get

$$2k^2 - 2k + 12 = k^2 + 5k + 6,$$

so $k^2 - 7k + 6 = 0$. This factors as $(k - 6)(k - 1) = 0$, so either $k = 6$ or $k = 1$. Both solutions work. $\square$

> **Concept:** Many probability problems will require some algebraic manipulation in order to solve them. Don't be afraid to use algebra (in particular, to use variables) if you need to!

> **Problem 1.15:** Mary and James each sit in a row of 7 chairs. They choose their seats at random. What is the probability that they don't sit next to each other?

Solution for Problem 1.15: There are $\binom{7}{2} = 21$ ways in which Mary and James can choose 2 chairs, if we don't worry about the order in which they sit.

Although we can use casework to count the number of ways they can choose chairs that are not next to each other, it is easier to use complementary counting. If we number the chairs #1, #2, . . . , #7 in order, then there are 6 ways Mary and James can choose chairs next to each other: they can sit in the first two chairs, or chairs #2 and #3, or chairs #3 and #4, etc., up to chairs #6 and #7. Therefore

$$P(\text{they sit next to each other}) = \frac{6}{21} = \frac{2}{7},$$

and therefore

$$P(\text{they don't sit next to each other}) = 1 - \frac{2}{7} = \frac{5}{7}.$$

☐

> **Concept:** Just as with counting, sometimes it's easier to calculate the probability of an event *not* occurring than it is to calculate the probability of the event occurring.

> **Problem 1.16:** The Grunters play the Screamers 4 times. The Grunters are the much better team, and are 75% likely to win any given game. What is the probability that the Grunters will win all 4 games?

Solution for Problem 1.16: The following solution is incorrect:

> **Bogus Solution:** Since each game has 2 possible outcomes, there are $2^4 = 16$ possible outcomes for the series. Only 1 of these outcomes is what we want, namely the Grunters winning all 4 games. Therefore the probability is $\frac{1}{16}$.

One clue that this is an incorrect solution is that we never used the "75%" information that was presented in the problem. The reason this solution is incorrect is that the outcomes described in the bogus solution are not all equally likely! We can only use this counting approach to probability when we have equally likely outcomes.

So instead, we'll use multiplication of probabilities of independent events. (The games are independent, because the outcome of each game does not depend on what happened in the earlier games.) Each of the 4 games is independent of the others, and in each game, the Grunters have probability $\frac{3}{4}$ of winning. Therefore, to get the probability that the Grunters will win all 4 games, we multiply the probabilities that the Grunters win each individual game. This gives:

$$
\begin{aligned}
P(\text{Grunters win all 4 games}) &= P(\text{Grunters win Game 1}) \times \cdots \times P(\text{Grunters win Game 4}) \\
&= \frac{3}{4} \times \frac{3}{4} \times \frac{3}{4} \times \frac{3}{4} \\
&= \left(\frac{3}{4}\right)^4 = \frac{81}{256}.
\end{aligned}
$$

☐

Using the same logic, we can show that $P(\text{Screamers win all 4 games}) = \left(\frac{1}{4}\right)^4 = \frac{1}{256}$.

> **Concept:** When computing probability by counting outcomes and using
>
> $$P(\text{success}) = \frac{\text{Number of successful outcomes}}{\text{Number of possible outcomes}},$$
>
> this approach will work only if the outcomes are equally likely.

> **Concept:** In an event that is made up of multiple, independent, sequential sub-events, we multiply the probabilities of the sub-events to get the probability of the overall event.

> **Problem 1.17:** A bag has 4 red and 6 blue marbles. A marble is selected and not replaced, then a second is selected. What is the probability that both are the same color?

Solution for Problem 1.17: We could solve this problem by counting the outcomes, but let's instead solve it by multiplying probabilities.

The probability that both marbles are red is given by:

$$P(\text{both red}) = P(\text{first red}) \times P(\text{second red } \textbf{after} \text{ first red is drawn}).$$

The probability that the first marble is red is $\frac{4}{10}$. After drawing a red marble, there are 3 red marbles and 9 marbles total left in the bag, so the probability that the second marble is also red is $\frac{3}{9}$. Therefore

$$P(\text{both red}) = \frac{4}{10} \times \frac{3}{9} = \frac{2}{15}.$$

Similarly, the probability that both marbles are blue is given by:

$$P(\text{both blue}) = P(\text{first blue}) \times P(\text{second blue } \textbf{after} \text{ first blue is drawn}).$$

The probability that the first marble is blue is $\frac{6}{10}$. After drawing a blue marble, there are 5 blue marbles and 9 marbles total left in the bag, so the probability that the second marble is also blue is $\frac{5}{9}$. Therefore

$$P(\text{both blue}) = \frac{6}{10} \times \frac{5}{9} = \frac{1}{3}.$$

Since drawing two red marbles and drawing two blue marbles are exclusive events, we add the individual probabilities to get the probability of one or the other occurring. Therefore:

$$P(\text{both same color}) = P(\text{both red}) + P(\text{both blue}) = \frac{2}{15} + \frac{1}{3} = \frac{7}{15}.$$

□

> **Concept:** When multiplying probabilities of **dependent** events, be sure to take the prior events into account when computing the probabilities of later events.

Problem 1.18: Point C is chosen at random atop a 5 foot by 5 foot square table. A circular disk with a radius of 1 foot is placed on the table with its center directly on point C. What is the probability that the entire disk is on top of the table (in other words, none of the disk hangs over an edge of the table)?

Solution for Problem 1.18: The "total outcomes" region is easy: it's just the surface of the table, so it's a square region with side length 5.

The "successful outcomes" region is trickier. We can draw a diagram as in Figure 1.4.

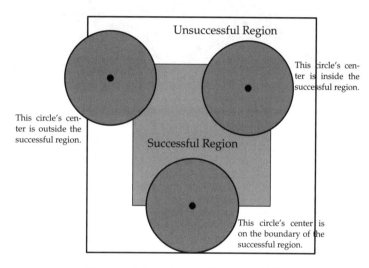

Figure 1.4: Table and some disks

We can see that the disk will be entirely on the table if and only if C is at least 1 foot away from each edge of the table. Therefore, C must be within a central square region of side length 3, as shown in Figure 1.4.

So, now we can compute the probability:

$$P(\text{success}) = \frac{P(\text{Area of successful outcomes region})}{P(\text{Area of possible outcomes region})} = \frac{3 \times 3}{5 \times 5} = \frac{9}{25}.$$

□

Concept: In geometry problems, or in other problems in which the possible outcomes are **continuous** (as opposed to **discrete**, meaning that they can be counted), we need to use geometry to compute the probability:

$$P(\text{success}) = \frac{P(\text{Size of successful outcomes region})}{P(\text{Size of possible outcomes region})}.$$

Extra! *Twice and thrice over, as they say, good is it to repeat and review what is good. – Plato*
⟶⟶⟶⟶

1.4 Expected Value

Problems

Problem 1.19: Suppose you have a weighted coin in which heads comes up with probability $\frac{3}{4}$ and tails with probability $\frac{1}{4}$. If you flip heads, you win $2, but if you flip tails, you lose $1. What is the expected value of a coin flip?

Problem 1.20: In an urn, I have 20 marbles: 2 red, 3 yellow, 4 blue, 5 green, and 6 black. I select one marble at random from the urn, and I win money based on the following chart:

Color	Red	Yellow	Blue	Green	Black
Amount won	$10	$5	$2	$1	$0

What is the expected value of my winnings?

Problem 1.21: Suppose I have a bag with 12 slips of paper in it. Some of the slips have a 2 on them, and the rest have a 7 on them. If the expected value of the number shown on a slip randomly drawn from the bag is 3.25, then how many slips have a 2?

Recall that **expected value** is the notion of a "weighted average," where the possible outcomes of an event are weighted by their respective probabilities. We can state this more precisely:

Concept: Suppose that we have an event with a list of possible values of the outcomes: $x_1, x_2, \ldots, x_n$. Value x_1 occurs with probability p_1, value x_2 occurs with probability p_2, and so on. Note that

$$p_1 + p_2 + \cdots + p_n = 1,$$

since the probabilities must total to 1. Then the **expected value** of the outcome is defined as the sum of the probabilities of the outcomes times the value of the outcomes:

$$E = p_1 x_1 + p_2 x_2 + \cdots + p_n x_n.$$

Problem 1.19: Suppose you have a weighted coin in which heads comes up with probability $\frac{3}{4}$ and tails with probability $\frac{1}{4}$. If you flip heads, you win $2, but if you flip tails, you lose $1. What is the expected value of a coin flip?

Solution for Problem 1.19: We multiply the outcomes by their respective probabilities, and add them up:

$$E = \frac{3}{4}(+\$2) + \frac{1}{4}(-\$1) = \$1.50 - \$0.25 = \$1.25.$$

$\square$

Concept: Another way to think of the expected value in Problem 1.19 is to imagine flipping the coin 1000 times. Based on the probabilities, we would expect to flip heads 750 times and to flip tails 250 times. We would then win $1500 from our heads but lose $250 from our tails, for a net profit of $1250. Since this occurs over the course of 1000 flips, our average profit per flip is

$$\frac{\$1250}{1000} = \$1.25.$$

Problem 1.20: In an urn, I have 20 marbles: 2 red, 3 yellow, 4 blue, 5 green, and 6 black. I select one marble at random from the urn, and I win money based on the following chart:

Color	Red	Yellow	Blue	Green	Black
Amount won	$10	$5	$2	$1	$0

What is the expected value of my winnings?

Solution for Problem 1.20: I draw a red marble with probability $\frac{2}{20}$, a yellow marble with probability $\frac{3}{20}$, and so on. Therefore, the expected value is

$$\frac{2}{20}(\$10) + \frac{3}{20}(\$5) + \frac{4}{20}(\$2) + \frac{5}{20}(\$1) + \frac{6}{20}(\$0) = \frac{\$48}{20} = \$2.40.$$

$\square$

Concept: One common use of the expected value in a problem like Problem 1.20 is to determine a **fair price** to play the game. The fair price to play is the expected winnings, which is $2.40. In the long run, if I were charging people the fair price to play the game, I would expect to break even, neither making nor losing money. If I were running this game at a carnival, I could charge carnival-goers $2.50 to play the game, and I would expect to make a 10-cent profit, on average, from each person that plays.

Problem 1.21: Suppose I have a bag with 12 slips of paper in it. Some of the slips have a 2 on them, and the rest have a 7 on them. If the expected value of the number shown on a slip randomly drawn from the bag is 3.25, then how many slips have a 2?

Solution for Problem 1.21: We let x denote the number of slips with a 2 written on them. (This is the usual tactic of letting a variable denote what we're trying to solve for in the problem.) Then there are $12 - x$ slips with a 7 on them.

The probability of drawing a 2 is $\frac{x}{12}$ and the probability of drawing a 7 is $\frac{12-x}{12}$, so the expected value of the number drawn is

$$E = \frac{x}{12}(2) + \frac{12-x}{12}(7) = \frac{84-5x}{12}.$$

But we are given that $E = 3.25$, so we have the equation

$$3.25 = \frac{84-5x}{12}.$$

This simplifies to $39 = 84 - 5x$, which means that $x = 9$. Thus 9 of the 12 slips have a 2 written on them.
□

> **Concept:** We said this before about probability, and it's true here too: don't be afraid to use algebra in solving expected value problems.

1.5 Pascal's Triangle and the Binomial Theorem

Problem 1.22: Suppose that we consider Pascal's Triangle to be a grid of dots (in other words, everywhere there's a number, we're just going to place a dot). We can count the number of paths from the top dot to any of the lower dots, where each step of the path is from a dot to one of the two dots immediately below it.

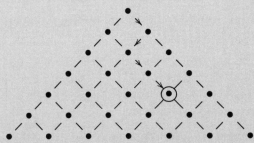

For example, the above diagram shows a path to the circled dot. How many paths are there from the top dot to the circled dot in the above picture?

Problem 1.23: Prove that
$$\binom{n-1}{r-1} + \binom{n-1}{r} = \binom{n}{r}.$$

Problem 1.24: Expand $(x + y)^n$.

Problem 1.25: What is the coefficient of the term of $(x + 2y^2)^6$ with a y^8 in it?

Problem 1.26: Prove that for any n,
$$\binom{n}{0} + \binom{n}{1} + \cdots + \binom{n}{n} = 2^n.$$

> **Extra!** *All men's miseries derive from not being able to sit in a quiet room alone.* – Blaise Pascal

Problem 1.22: Suppose that we consider Pascal's Triangle to be a grid of dots (in other words, everywhere there's a number, we're just going to place a dot). We can count the number of paths from the top dot to any of the lower dots, where each step of the path is from a dot to one of the two dots immediately below it.

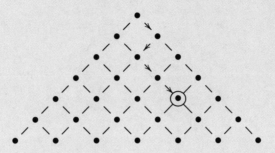

For example, the above diagram shows a path to the circled dot. How many paths are there from the top dot to the circled dot in the above picture?

Solution for Problem 1.22: We must take a total of 4 steps, since the circled dot is 4 rows below the top dot. Of those 4 steps, 3 of the steps must be down and to the right, and the other step is down and to the left. We can take the 4 steps in any order, so long as 3 are to the right and 1 is to the left. Therefore, the number of paths to the circled dot is the same as the number of ways to arrange 3 steps to the right and 1 step to the left. We must choose 3 of the 4 steps to be to the right, and we know that we can do this is $\binom{4}{3}$ ways. So there are $\binom{4}{3}$ paths from the top of the triangle to the circled dot. $\square$

Important: More generally, the number of paths to the rth dot of Row n (where the top dot is considered to be Row 0) is $\binom{n}{r}$.

Important: This is **Pascal's Triangle**:

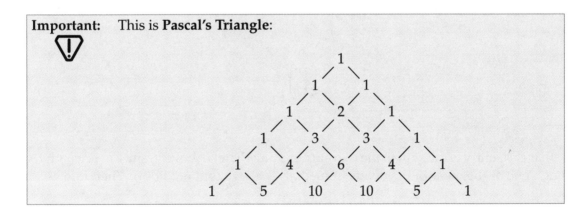

While knowing the numbers is important, it's not nearly as important as knowing what the numbers *mean*—and what they mean is combinations!

Concept: Most experienced counters think of Pascal's Triangle as:

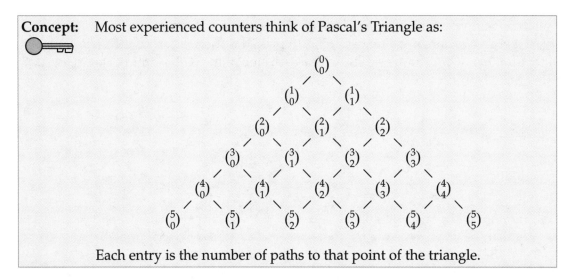

Each entry is the number of paths to that point of the triangle.

Problem 1.23: Prove that
$$\binom{n-1}{r-1} + \binom{n-1}{r} = \binom{n}{r}.$$

Solution for Problem 1.23:

Proof by block-walking on Pascal's Triangle: Let's look at a piece somewhere in the middle of Pascal's Triangle:

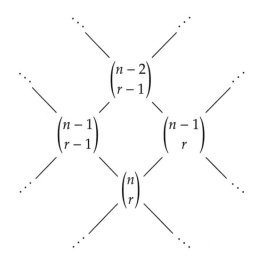

We know that each entry is the sum of the two entries immediately above it, since every path to $\binom{n}{r}$ must pass through one or the other of the points immediately above (but not both). Therefore, we conclude that
$$\binom{n-1}{r-1} + \binom{n-1}{r} = \binom{n}{r}.$$

Proof by algebra: We start by writing out the algebraic definition of the combinations:
$$\binom{n-1}{r-1} + \binom{n-1}{r} = \frac{(n-1)!}{(r-1)!((n-1)-(r-1))!} + \frac{(n-1)!}{r!(n-1-r)!}.$$

We now can now simplify the denominators, and put both fractions over a common denominator:

$$\binom{n-1}{r-1} + \binom{n-1}{r} = \frac{(n-1)!}{(r-1)!(n-r)!} + \frac{(n-1)!}{r!(n-1-r)!} = \frac{r(n-1)!}{r!(n-r)!} + \frac{(n-r)(n-1)!}{r!(n-r)!}.$$

Now we add the fractions and factor the numerator:

$$\binom{n-1}{r-1} + \binom{n-1}{r} = \frac{r(n-1)! + (n-r)(n-1)!}{r!(n-r)!} = \frac{n(n-1)!}{r!(n-r)!} = \frac{n!}{r!(n-r)!} = \binom{n}{r}.$$

Proof by committee-forming: Given a club with n members, we know that we can form an r-person committee in $\binom{n}{r}$ ways. This is just our basic definition of combinations.

Imagine that one of the club members is named Bob. If Bob is on the committee, then we must choose $r-1$ remaining committee members from the $n-1$ remaining club members. So there are $\binom{n-1}{r-1}$ committees with Bob on them.

However, if Bob is not on the committee, then we must choose all r committee members from the $n-1$ remaining club members. So there are $\binom{n-1}{r}$ committees with Bob not on them.

The two situations above are exclusive cases, so to get the total number of committees, we add the number of committees with Bob and the number of committees without Bob. Thus, the number of r-person committees is $\binom{n-1}{r-1} + \binom{n-1}{r}$. But we already know that the number of r-person committees is $\binom{n}{r}$, so equating these two counts, we again get

$$\binom{n-1}{r-1} + \binom{n-1}{r} = \binom{n}{r}.$$

$\square$

Important: **Pascal's identity:**

$$\binom{n-1}{r-1} + \binom{n-1}{r} = \binom{n}{r}.$$

Concept: Many combinatorial identities can be proved in one or more of the following ways:

- By a committee-forming argument

- By a block-walking argument

- By algebra

Problem 1.24: Expand $(x+y)^n$.

Solution for Problem 1.24: Let's look at how this works for $(x+y)^2$:

$$(x+y)^2 = (x+y)(x+y)$$
$$= xx + xy + yx + yy$$
$$= x^2 + 2xy + y^2.$$

On the second line above, we get four terms: each term corresponds to choosing x or y from the first term, then choosing x or y from the second term, and multiplying them together. After simplification, we get our usual result on the third line above.

Just to make sure that it's clear what's happening, let's also look at how this works for $(x + y)^3$:

$$(x + y)^3 = (x + y)(x + y)(x + y)$$
$$= xxx + xxy + xyx + xyy + yxx + yxy + yyx + yyy$$
$$= x^3 + 3x^2y + 3xy^2 + y^3.$$

As before, in the second line above, each term is a product of x or y from the first term, times x or y from the second term, times x or y from the third term. We take these products for every possible choice of x or y from each term, and then add them up. When we simplify, we get our usual expression for $(x + y)^3$.

Now we can see where the coefficients come from. For example, the coefficient of the x^2y term in $(x+y)^3$ is 3. That's because there are 3 ways to choose two x's and one y: xxy, xyx, and yxx. Alternatively, we can think of this as the number of ways to arrange two x's and one y to make a three-letter "word". And we know how to count this: it's

$$\frac{3!}{2!1!} = \binom{3}{1}.$$

We can also think of this as the number of ways to choose one "slot" for the y in our 3-letter word.

Now let's go back to the general case of $(x + y)^n$. Every term in the product results from choosing an x or y from each of the n different $(x + y)$ terms in the product. So if, for example, we choose k y's, the other $n - k$ choices will be x's, and we'll get an $x^{n-k}y^k$ term in the product. But k y's can be chosen from n terms in $\binom{n}{k}$ ways. Therefore, the coefficient of $x^{n-k}y^k$ is $\binom{n}{k}$. □

> **Important:** The **Binomial Theorem**: for any positive integer n, the coefficient of the
> $x^{n-k}y^k$ term of $(x + y)^n$ is $\binom{n}{k}$. In other words,
>
> $$(x + y)^n = \binom{n}{0}x^n + \binom{n}{1}x^{n-1}y + \cdots + \binom{n}{k}x^{n-k}y^k + \cdots + \binom{n}{n}y^n.$$

Problem 1.25: What is the coefficient of the term of $(x + 2y^2)^6$ with a y^8 in it?

Solution for Problem 1.25: How do we get a term in the expansion of $(x + 2y^2)^6$ with a y^8 in it? The term with a y^8 in it is the term that has a $(2y^2)^4$ in it. We know that when expanding $(x + 2y^2)^6$, we have to choose 4 copies of $2y^2$ from the six $(x+2y^2)$ terms in order to get a term with y^8 in it. This can be done in $\binom{6}{4}$ ways. We then take an x from each of the remaining two $(x + 2y^2)$ terms that didn't contribute a $2y^2$. Therefore, the relevant term in the expansion of $(x + 2y^2)^6$ is $\binom{6}{4}x^2(2y^2)^4 = 240x^2y^8$. The answer is 240. □

> **Concept:** When dealing with the Binomial Theorem, as with most theorems or formulas, it is important to not merely memorize the formula, but to *understand* it well enough to be able to use it flexibly.

Problem 1.26: Prove that for any n,

$$\binom{n}{0} + \binom{n}{1} + \cdots + \binom{n}{n} = 2^n.$$

Solution for Problem 1.26: The sum of combinations should definitely remind us of the Binomial Theorem:

$$(x + y)^n = \binom{n}{0}x^n + \binom{n}{1}x^{n-1}y + \cdots + \binom{n}{n}y^n.$$

Plug in $x = 1$ and $y = 1$ into the Binomial Theorem. This makes all the x's and y's go away!

After we plug in $x = 1$ and $y = 1$, we're left with:

$$(1 + 1)^n = 2^n = \binom{n}{0} + \binom{n}{1} + \cdots + \binom{n}{n}.$$

That's what we wanted to prove, so we're done! □

Concept: Often we can solve a problem by taking a general expression (which is stated in terms of variables) and plugging in some nice values for the variables to make them go away.

1.6 Summation Notation

You've probably already seen expressions like

$$1 + 2 + \cdots + n = \frac{n(n + 1)}{2}$$

or

$$a + ar + ar^2 + \cdots = \frac{a}{1 - r}.$$

These expressions use "$\cdots$" to denote missing terms in a sum. Although writing sums like this may be intuitively clear, it is not very mathematically precise. For some expressions, the pattern of the missing terms might not be sufficiently well-defined. For example, if we were to write

$$1 + 3 + \cdots + 81,$$

are we summing the first 41 positive odd integers, or are we summing the first 5 positive powers of 3? Also, in an expression like

$$1 + 2 + \cdots + n,$$

we run into a bit of ambiguity if $n \leq 2$: what are the "missing" terms?

To avoid many of these difficulties, we have a convenient notation that we use to write sums, called **summation notation**. We use the symbol $\sum$, the capital Greek letter sigma, which stands for "sum." For example, we would write

$$1 + 2 + \cdots + n = \sum_{k=1}^{n} k.$$

The symbol $\displaystyle\sum_{k=1}^{n}$ means to take the sum where the different terms are given by plugging in $k = 1$, $k = 2$, and so on up through $k = n$. The letter k is not important, and can be any letter not in the sum's terms. The variable k is sometimes called a **dummy variable**.

To take another simple example,

$$\sum_{i=5}^{11} i^2 = 5^2 + 6^2 + 7^2 + 8^2 + 9^2 + 10^2 + 11^2 = 25 + 36 + 49 + 64 + 81 + 100 + 121 = 476.$$

Often we will use summation notation to write identities more concisely. For instance, we can write the Binomial Theorem as

$$(x + y)^n = \sum_{k=0}^{n} \binom{n}{k} x^{n-k} y^k.$$

If the sum is infinite, then we can use the symbol ∞ to indicate this. For example, if $|r| < 1$, then

$$\sum_{j=0}^{\infty} ar^j = \frac{a}{1 - r}.$$

We will discuss more subtle aspects of summation notation throughout the book as we need them.

1.7 Summary

This chapter is a review of the concepts and techniques that you should know before proceeding further with this book. The rest of the book will assume that you know and are comfortable with these techniques. If many of the solutions to the problems in this chapter did not come easily to you, you may want to review more introductory-level material before continuing on with this book.

The book *Introduction to Counting & Probability* covers all of the topics in this chapter (except for summation notation), and does so in much greater detail than we did here.

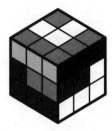

Logic is the beginning of wisdom, not the end. – Mr. Spock

CHAPTER **2**

Sets and Logic

2.1 Introduction

In order to discuss some of the more advanced concepts in this book, we'll first need to discuss two very fundamental areas of mathematics: **sets** and **logic**.

Some of these concepts may be familiar to you, but you should nonetheless read through this chapter carefully—there may be some subtleties that you're not already familiar with. Also, you'll see that there are relatively few problems in the text in this chapter. The concepts in this chapter are tools that we use to describe and solve problems later in the book, but these tools don't necessarily give rise to many interesting problems in their own right.

The notation and terminology might seem a bit heavy in this chapter. Try not to worry too much about all the new notation and terminology, especially in the logic sections. It's much more important to try to keep the big picture in mind: sets and logic are tools that we use to formulate and solve more complicated ideas and problems. (Although, to be fair, set theory and the theory of formal logic are rich branches of mathematics in their own right, and many mathematicians have made careers studying them.) As you use sets and logic in more and more problems, the terminology will become more and more natural to you.

Also, because a lot of the exercises in this chapter depend on knowing the appropriate definitions and notation, we're not using our usual style of putting practice problems at the start of every section, because without knowing the terminology or notation, you wouldn't be able to make any headway. Instead, for this chapter only, we'll just be introducing problems as we go through the text. When you reach a problem in the text, try to solve it before reading the solution; then, whether you think you've solved it or not, carefully read the solution presented.

2.2 Sets

Sets are the building blocks of mathematics. Like many other extremely fundamental mathematical concepts (such as "point" or "number"), sets are difficult to precisely define, and we're not going to try to be precise here.

Roughly speaking, a **set** is a collection of objects. The objects can be essentially anything: numbers, functions, other sets, any combination of these, or nothing at all. The order of the objects in the set is unimportant. All that matters is what objects are in the set. There might only be a **finite** number of objects in the set (meaning basically that we could count them if we liked), in which case the set is called (big surprise) a **finite set**. Otherwise we call it an **infinite set**. The objects in the set are called the **elements** or **members** of the set.

There are two basic ways that we can describe a set. The first is to simply list its elements. For example:

$$A = \{2, 9, 22\}.$$

This is a set with three elements, namely 2, 9, and 22. This is the most basic way to define or describe a set: We list the elements inside of curly braces, and separate the different elements by commas. As we said above, the order of the elements doesn't matter, so $A = \{9, 22, 2\}$ is exactly the same set as $A = \{2, 9, 22\}$. Also, each element can only be in the set once, so for example $B = \{3, 6, 3\}$ is not a legal set (or, alternatively, we can think of the second "3" in $\{3, 6, 3\}$ as being redundant and write $\{3, 6, 3\} = \{3, 6\}$).

Sometimes it's impractical to list a big set, so we use ellipses if the pattern of the elements in the set is clear. For example, we feel pretty safe describing a set as $\{1, 2, 3, \ldots, 99, 100\}$ and knowing that this is the set of the first 100 positive integers.

If a set is infinite, then we obviously have no hope of being able to list all the elements, since such a list would go on forever! But if it is clear which elements are in the set, then we can list the elements using ellipses. For example, the set of all positive integers can be written as $\{1, 2, 3, \ldots\}$, because the pattern is clear. As another example, we can be pretty sure that $\{1, 2, 4, 8, 16, 32, \ldots\}$, without any further description, is the set of all nonnegative powers of 2. Be careful though: you should only do this if your pattern is absolutely clear. Listing a set as $\{1, 2, 4, \ldots\}$ is pretty ambiguous: is it the set of all nonnegative powers of 2, or the set of all positive integers not divisible by 3, or something else that we didn't think of? It's not at all clear, so we need more sample elements to make the pattern clear, or some words describing the set, or we can define the set via the properties of its elements, as we're about to see.

Aside from listing the elements, the other basic way to describe a set is to provide a property that precisely defines the elements of the set. For example:

$$B = \{x \mid x \text{ is an integer}\}.$$

In this example, the set B consists of all the integers. Some people use a colon (:) instead of the vertical bar (|); in either case, you should read the symbol as "such that." For example, we would read our set B above as "the set of all x such that x is an integer." Another common example is an interval on the real line; for example,

$$\{x \mid x \text{ is a real number and } 2 < x \leq 3\}$$

is the interval of all real numbers that are greater than 2 and less than or equal to 3. One of the major strengths of this way of describing a set is that we can use this even if we don't know explicitly what the elements are. For example,

$$\{y \mid y \text{ is a real number and } 2y^4 - y^3 + 6y^2 - 11y + 12 = 0\}$$

is perfectly valid, even if we don't necessarily know at first glance exactly what values of y are in the set, or even if there are any elements in the set.

Most of the time, we will be using sets whose elements are numbers or other mathematical objects. However, a set can contain pretty much anything as elements, as long as the elements are precisely defined. For example, we could define a set

$$S = \{x \mid x \text{ is a state in the United States}\}.$$

Then $S = \{\text{Alabama, Alaska, } \ldots\}$. We can even "mix and match" different types of elements to our heart's content. For example,

$$T = \{5, \{2, \pi\}, \text{Boston}\}$$

is a set with three elements: a number (5), a set (with elements 2 and π), and a city (Boston). If an object x is an element of S, we write this as $x \in S$. If x is not an element of S, we write $x \notin S$. For any object x and any set S, either $x \in S$ or $x \notin S$, but (of course) never both. Indeed, this is the whole point of sets: a set is a collection of the objects that belong to it, and everything else does not belong.

There is a very special set called the **empty set**, denoted by $\emptyset$. This is the set with no elements at all. For example,

$$\{x \mid x \text{ is a real number and } x^2 < 0\} = \emptyset,$$

because there is no real number satisfying the property that its square is less than 0. Note that $x \notin \emptyset$ for any x. We sometimes also write the empty set as a list: $\emptyset = \{\}$. Of course, it's an empty list, since the empty set has no elements.

If S is a finite set, then we let $\#(S)$ denote the number of elements of S. We say that $\#(S)$ is the **cardinality** of S. (Note that many sources use the notation $|S|$ in place of $\#(S)$.) For example, if $S = \{2, 4, 9, 11\}$ then $\#(S) = 4$, since S has four elements. Note that $\#(\emptyset) = 0$, since $\emptyset$ has zero elements. If S is an infinite set (such as the set of all integers), then we cannot define $\#(S)$ without resorting to so-called *transfinite cardinal numbers*, which are beyond the scope of this book.

A set A is called a **subset** of a set B if every element of A is also an element of B. We think of A as a smaller set that is made up of some of the elements of B. More informally, we think of A as sitting "inside" B. The notation that we use is $A \subseteq B$. For example,

$$\{3, 8, 11\} \subseteq \{2, 3, 8, 10, 11, 14, 16\}$$

and

$$\{x \mid x \text{ is an even integer}\} \subseteq \{x \mid x \text{ is an integer}\}.$$

If A is a subset of B, we also say that B is a **superset** of A.

Sometimes it is convenient to have a notation for when a set is a **proper subset** of another set, meaning that it is a subset but not equal to the larger set. For example, $\{1, 2, 3\}$ is a proper subset of $\{1, 2, 3, 4\}$, but $\{1, 2, 3, 4\}$ is not a proper subset of $\{1, 2, 3, 4\}$, although it is a subset.

We use the notation $A \subset B$ to denote that A is a proper subset of B. (This is very similar to the notations $<$ for "less than" and $\leq$ for "less than or equal to": $3 < 4$ and $3 \leq 4$ and $4 \leq 4$, but $4 \not< 4$.) Another way to think of this is that $A \subset B$ means that A is a subset of B, but there is some element of B that is not in A. For example, $\{1, 2, 3\} \subset \{1, 2, 3, 4\}$ because 4 is not in $\{1, 2, 3\}$.

WARNING!! ☢	Unfortunately, this notation is not universally agreed upon. Many authors use $A \subset B$ to mean that A is any subset of B, not necessarily proper. Some of these authors then use the notation $A \subsetneq B$ or $A \subsetneqq B$ to mean that A is a proper subset of B. However, in this book, we will always use $A \subseteq B$ to mean that A is a subset of B, possibly equal, and use $A \subset B$ to mean that A is a proper subset of B.

WARNING!! ☢	The concepts and notations can get a bit confusing, and it takes a little bit of practice to use them properly. For example, if $A = \{1, 2, 3\}$, then it is correct to say that 1 is an element of A and that $\{1\}$ is a subset of A. In notation, we would say $$1 \in A \quad \text{and} \quad \{1\} \subseteq A.$$ But it is *not* correct to say that $\{1\}$ is an element of A.

Let's practice with some basic exercises involving elements and subsets:

Problem 2.1: Consider the following sets:

$$A = \{1, 2, 3, 4, 5\}, \quad B = \{2, 3, 4\}, \quad C = \{3, \{4, 5\}\}.$$

(a) Is $A \subseteq A$? Is $A \subset A$?

(b) Is $B \subseteq A$? Is $B \subset A$?

(c) Is $C \subseteq A$? Is $C \subset A$?

(d) Is $4 \in B$? Is $4 \in C$?

(e) List all of the subsets of B. How many are there?

Solution for Problem 2.1:

(a) The elements of A are 1, 2, 3, 4, and 5. All of these elements are elements of A, so $A \subseteq A$. However, $A = A$, so A is not a proper subset of A. Therefore, $A \not\subset A$.

(b) All of the elements of B are also elements of A, so $B \subseteq A$. Further, A contains elements (namely, 1 and 5) that are not in B, so B is a proper subset of A; that is, $B \subset A$.

(c) One of the elements of C is $\{4, 5\}$. This is not an element of A (it is a subset of A, which is not the same thing). So $C \not\subseteq A$ and by the same reasoning $C \not\subset A$.

(d) 4 is an element of B, so $4 \in B$. However, 4 is not an element of C. The set C has two elements, the number 3 and the set $\{4, 5\}$. So $4 \notin C$.

(e) We can list the subsets of B:

$$\emptyset, \{2\}, \{3\}, \{4\}, \{2,3\}, \{2,4\}, \{3,4\}, \{2,3,4\}$$

(Don't forget that $\emptyset$ and B itself are subsets of B.) There are 8 subsets of B.

$\square$

Note some special properties of subsets:

- Every set is a subset of itself; that is, $A \subseteq A$ for any set A.

- The empty set is a subset of any set; that is, $\emptyset \subseteq A$ for any set A.

- If A, B are two sets such that $A \subseteq B$ and $B \subseteq A$, then $A = B$.

- If A and B are any two sets, then we cannot have both $A \subset B$ and $B \subset A$.

Let's explain one of these properties now, and you'll be asked to explain the others in the exercises.

Problem 2.2: Show that the empty set is a subset of any set.

Solution for Problem 2.2:   By the definition of subset, we know that $\emptyset \subseteq A$ if every element of $\emptyset$ is an element of A. But $\emptyset$ has no elements, so every element of $\emptyset$ (there aren't any of them!) is in A (there's nothing to check!). $\square$

Every set S has a special associated set called its **power set**, denoted by $\mathcal{P}(S)$. This is a set whose elements are all of the subsets of S. More precisely,

$$\mathcal{P}(S) = \{T \mid T \subseteq S\}.$$

For example, if $S = \{1,3,4\}$, then

$$\mathcal{P}(S) = \{\emptyset, \{1\}, \{3\}, \{4\}, \{1,3\}, \{1,4\}, \{3,4\}, \{1,3,4\}\}.$$

Notice that the elements of $\mathcal{P}(S)$ are *sets*, not numbers. This is perfectly legal—remember that "set" is a very general concept, and that elements of sets can be anything! Also notice that $\mathcal{P}(\emptyset) = \{\emptyset\}$, which is not the empty set: $\{\emptyset\}$ is a set with one element, and that element is the set $\emptyset$.

This can get a bit confusing, so let's look at an example.

Problem 2.3: Let $A = \{1,2,3,6\}$ be the set of positive divisors of 6. Define

$$B = \{S \in \mathcal{P}(A) \mid 1 \notin S\}.$$

List the elements of B.

Solution for Problem 2.3:   This problem is most easily solved by trying to get past all the notation and thinking about it in words. The elements of B are exactly the subsets of A that don't contain 1 as an element. So we can list them:

$$B = \{\emptyset, \{2\}, \{3\}, \{6\}, \{2,3\}, \{2,6\}, \{3,6\}, \{2,3,6\}\}.$$

Note also that $B = \mathcal{P}(\{2,3,6\})$. $\square$

> **Sidenote:**
> **Some special sets**
> Some sets are so common that they have special names. (We've already seen one such set, namely ∅.) Some others are:
>
> - $\mathbb{Z}$, the set of integers
>
> - $\mathbb{Q}$, the set of rational numbers
>
> - $\mathbb{R}$, the set of real numbers
>
> - $\mathbb{C}$, the set of complex numbers
>
> Note that $\mathbb{Z} \subset \mathbb{Q} \subset \mathbb{R} \subset \mathbb{C}$. (These sets are usually written, as they are here, in a font called "blackboard bold," which consists of regular capital letters with an extra line in them.)

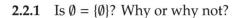
Exercises

2.2.1 Is $\emptyset = \{\emptyset\}$? Why or why not?

2.2.2 For each part, determine if the statement is true or false, and explain why or why not.

(a) $3 \in \{1, 3, 5, 9\}$

(b) $\{2, 6\} = \{6, 2, 2\}$

(c) $\{5\} \in \{3, 5, 9\}$

(d) $4 \in \{\{4\}\}$

(e) $\emptyset \subset \{1, 2, 9\}$

(f) $\emptyset \in \{\emptyset, \{1\}, 82\}$

(g) $\{x \mid x \text{ is an even integer}\} \subset \mathbb{Z}$

2.2.3 Explain why $A \subseteq A$ for any set A.

2.2.4 Explain why if $A \subseteq B$ and $B \subseteq A$, then $A = B$.

2.2.5 Is the subset relationship *transitive*? In other words, if $A \subseteq B$ and $B \subseteq C$, can we conclude that $A \subseteq C$? Why or why not? How about for proper subsets?

2.2.6 Suppose that B is a set such that $B \subseteq \emptyset$. What can we conclude about B?

2.2.7 Explain why it is true that if A and B are finite sets and $A \subseteq B$, then $\#(A) \leq \#(B)$. What does it mean if $A \subseteq B$ and $\#(A) = \#(B)$? If A and B are finite sets such that $A \subset B$, what can we conclude about $\#(A)$ and $\#(B)$?

2.2.8 Explain why it is impossible for two sets A and B simultaneously to satisfy $A \subset B$ and $B \subset A$.

2.2.9★ What is $\mathcal{P}(\mathcal{P}(\emptyset))$? **Hints:** 145

2.2.10★ Prove that if S is a finite set and $\#(S) = n$, then $\#(\mathcal{P}(S)) = 2^n$. **Hints:** 316, 276

2.3 Operations on Sets

Just as we can perform operations on numbers, such as addition and multiplication, to get new numbers, we can perform operations on sets to get new sets. We start by defining the two basic operations on sets.

> **Definition:** The **union** $A \cup B$ of two sets A and B is the set of all objects that are elements of at least one of A and B.

We can write this more formally as:

$$A \cup B = \{x \mid x \in A \text{ or } x \in B\}.$$

Note that our use of the word "or" in the line above does *not* mean "one or the other, but not both." Instead, we use the word "or" to mean "one or the other, or possibly both." (This is the way that logicians define "or," which we'll see in the next section of this chapter.) Elements that are in both A and B are also in their union.

Here are some examples:

- $\{2, 3, 8\} \cup \{1, 7, 11, 13\} = \{1, 2, 3, 7, 8, 11, 13\}$.

- $\{4, 8, 9\} \cup \{2, 4, 9, 11\} = \{2, 4, 8, 9, 11\}$. (We don't list 4 or 9 twice, even though they appear in both sets, since elements are not duplicated in a set.)

- $\{1, 2, 4\} \cup \{1, 2, 4, 8\} = \{1, 2, 4, 8\}$.

- $\{x \mid x \text{ is an even integer}\} \cup \{x \mid x \text{ is an odd integer}\} = \{x \mid x \text{ is an integer}\}$.

> **Definition:** The **intersection** $A \cap B$ of two sets A and B is the set of all objects that are elements of both of A and B.

We can write this more formally as:

$$A \cap B = \{x \mid x \in A \text{ and } x \in B\}.$$

Some examples:

- $\{3, 5, 9, 11\} \cap \{2, 5, 8, 11, 13\} = \{5, 11\}$.

- $\{2, 6, 9, 11\} \cap \{2, 9\} = \{2, 9\}$.

- $\{4, 9, 11, 16\} \cap \{2, 8, 10, 14\} = \emptyset$, since these two sets have no elements in common.

- $\{x \mid x \text{ is an even integer}\} \cap \{x \mid x \text{ is an odd integer}\} = \emptyset$.

- $\{x \mid x \text{ is an even positive integer}\} \cap \{x \mid x \text{ is a prime number}\} = \{2\}$.

A special case of union and intersection is shown in the following problem.

Problem 2.4: Suppose A and B are sets such that $A \subseteq B$.

(a) What is $A \cup B$?

(b) What is $A \cap B$?

Solution for Problem 2.4:

(a) If $A \subseteq B$, then every element of A is also an element of B. This means that any element in A or B must be in B. Therefore, $A \cup B \subseteq B$. On the other hand, every element of B is also an element of $A \cup B$, so $B \subseteq A \cup B$. Hence, $A \cup B = B$.

(b) Any element in A and B must be in A, so $A \cap B \subseteq A$. On the other hand, every element of A is also an element of B, and thus also an element of $A \cap B$, so $A \subseteq A \cap B$. Therefore, $A \cap B = A$.

$\square$

> **Important:** In order to show that two sets A and B are equal, we have to show that every element of A is also an element of B, and we have to show that every element of B is also an element of A.
>
> If we only do one of these but not both, all we're showing is that one set is a subset of the other. For example, if we show that every element of A is also an element of B, then we've shown that $A \subseteq B$. In order to show that they're equal, we have to show the reverse as well.

It is useful to have a word describing when two sets have no elements in common:

Definition: If $A \cap B = \emptyset$, then we say that A and B are **disjoint**.

We can visualize the operations of union and intersection using Venn diagrams. In particular, if A are B are sets, represented by circles in a Venn diagram, then $A \cup B$ is the region inside of the two circles combined, and $A \cap B$ is the region inside of both circles:

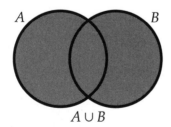

$A \cup B$

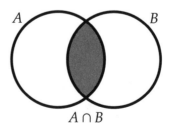

$A \cap B$

These two operations—union and intersection—are distributive with respect to each other. We'll prove one of the distributive laws and leave the other as an exercise.

> **Extra!** *Logic (and therefore probability as a branch of logic) is not concerned with what men do believe, but what they ought to believe, if they are to believe correctly.* – John Venn

Problem 2.5: Prove that if A, B, and C are sets, then

$$A \cap (B \cup C) = (A \cap B) \cup (A \cap C).$$

Solution for Problem 2.5:   To prove that two sets are equal, we must show that they have the same elements. This means that we must show that every element in one set is also in the other set, and vice versa.

So we start by letting x be an element of $A \cap (B \cup C)$. This means that $x \in A$ and $x \in (B \cup C)$, which means that

$$x \in A \text{ and } ((x \in B) \text{ or } (x \in C)).$$

This can be rewritten as

$$(x \in A \text{ and } x \in B) \text{ or } (x \in A \text{ and } x \in C).$$

Now, rewriting our statement using union and intersection of sets, we have

$$x \in ((A \cap B) \cup (A \cap C)).$$

So all elements of $A \cap (B \cup C)$ are also elements of $(A \cap B) \cup (A \cap C)$, which means that

$$A \cap (B \cup C) \subseteq (A \cap B) \cup (A \cap C). \tag{$*$}$$

We're not done! We have to show the reverse as well. To do so, let y be an element of $(A \cap B) \cup (A \cap C)$. Then

$$(y \in A \text{ and } y \in B) \text{ or } (y \in A \text{ and } y \in C).$$

This can be written as

$$y \in A \text{ and } (y \in B \text{ or } y \in C),$$

which means that $y \in A \cap (B \cup C)$. Thus,

$$(A \cap B) \cup (A \cap C) \subseteq A \cap (B \cup C). \tag{$**$}$$

Combining (*) and (**), we see that $A \cap (B \cup C)$ and $(A \cap B) \cup (A \cap C)$ have the same elements, so they are equal. $\square$

The logic steps in the middle of the above argument might be a bit unclear—we'll cover more details of the logic involved in the next section—but we can also visualize the result using Venn diagrams.

The first diagram, at right, shows A filled with diagonal lines in one direction, and $B \cup C$ filled with diagonal lines in the other direction. Their overlap, filled with the crossed diagonal lines, is the region that is in both sets; that is, it is the region $A \cap (B \cup C)$.

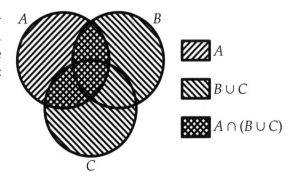

On the other hand, the diagrams below show $A \cap B$ shaded in the first picture, and $A \cap C$ shaded in the second picture. When we combine the pictures, the shaded region is the region that is in at least one of the intersections, so the shaded region in the right picture is $(A \cap B) \cup (A \cap C)$.

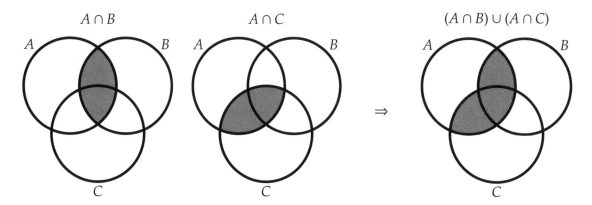

This is visual evidence that

$$A \cap (B \cup C) = (A \cap B) \cup (A \cap C).$$

Note that using Venn diagrams in this way is not a proof, but it does allow us to see visually why the identity in question is true.

There are a couple more set-theory concepts that come up occasionally. One is the idea of a **universal set**, which is a set that contains all possible elements in a given problem or situation. For example, in an introductory algebra setting, the universal set might be the set $\mathbb{R}$ of real numbers, so that, for example,

$$\{x \mid x^2 + 1 = 0\} = \emptyset,$$

because it is assumed that we are only considering elements $x \in \mathbb{R}$. On the other hand, if the universal set were the complex numbers, then

$$\{x \mid x^2 + 1 = 0\} = \{i, -i\}.$$

Note that the universal set may vary from problem to problem, and it is not always explicitly stated.

In a Venn Diagram, often a universal set is denoted by a large box surrounding the diagram, as shown to the right. In any context in which we have a universal set, we have the implicit assumption that every set under discussion is automatically a subset of the universal set. In other words, we're not allowed to go "outside the box" of the universal set.

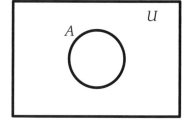

Generally, we prefer not to use universal sets unless the context is absolutely clear. For example, as discussed above, without a clear universal set, the set

$$\{x \mid x^2 + 1 = 0\}$$

is unclear. If the universal set is $\mathbb{R}$, then this set is empty, but if the universal set is $\mathbb{C}$, then this set is $\{i, -i\}$. It's usually much better to be upfront with our assumptions. For example, if we write

$$\{x \in \mathbb{R} \mid x^2 + 1 = 0\},$$

then it's clear that this is the empty set.

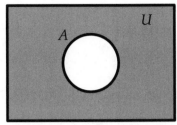

We also talk about the **complement** of a set A with respect to a universal set U. This is the set of all elements (in U) that are *not* in A. For example, if $U = \mathbb{Z}$, the set of integers, and A is the set of all odd integers, then the complement of A is the set of all even integers. The complement of A is sometimes denoted A^C or $U - A$ or $U \setminus A$. In a Venn Diagram, the complement of A is the shaded region in the diagram at left.

We can also define the **set-theoretic difference** of two sets A and B. This set, denoted either $A - B$ or $A \setminus B$, is the set of all elements of A that are not elements of B. For example, if

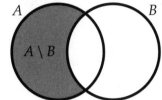

$$A = \{1, 4, 5, 9, 11\} \quad \text{and} \quad B = \{2, 4, 5, 8, 10\},$$

then

$$A \setminus B = \{1, 9, 11\},$$

since those are the elements of A that are not elements of B. The "extra" elements of B (2, 8, and 10) are irrelevant. The set-theoretic difference $A \setminus B$ is represented by the shaded area in the Venn Diagram above to the right.

Exercises

2.3.1 For any set A, what are $A \cup \emptyset$ and $A \cap \emptyset$?

2.3.2 Prove that $A \cup A = A \cap A = A$.

2.3.3 If $x \in S$, what are $S \cup \{x\}$ and $S \cap \{x\}$?

2.3.4 Show that $A \cup (B \cap C) = (A \cup B) \cap (A \cup C)$.

2.3.5

(a) Suppose that $A \cup B = A$. What can we conclude about A and B?

(b) Suppose that $A \cap B = A$. What can we conclude about A and B?

2.3.6★ If S and T are sets, describe $\mathcal{P}(S \cap T)$ in terms of $\mathcal{P}(S)$ and $\mathcal{P}(T)$. Can you similarly describe $\mathcal{P}(S \cup T)$? **Hints:** 103

2.4 Truth and Logic

Informally, for our purposes in this book, **logic** is the way that we interpret statements of fact and use them to prove new statements of fact. (What a professional mathematician means by "logic" is something far different, much more abstract, and well beyond the scope of this book.) In most cases, you will find that logic is just an extension of your natural common sense, but in this section we will be careful to try to explicitly state all of our assumptions and conclusions.

To start with, a **statement** (or **proposition**) is an assertion that is either **true** or **false** (but never both). "False" is the opposite of "true." We are not able to define what "true" or "false" means, but you should use your common sense. (Philosophers, mathematicians, and logicians have been debating for centuries about the meaning of "truth," and we're certainly not going to enter that debate here.)

Some examples of statements are:

- Paris is the capital of France. (This is a true statement.)

- New York is the capital of the United States. (This is a false statement.)

- $0 = 1$. (False.)

- $3 < 8$. (True.)

- All even integers are divisible by 2. (True. The word "all" means that this is an example of a **quantified** statement, which we'll talk more about in the next section of the chapter.)

- There exists an integer whose square is 7. (False. The phrase "there exists" is another example of a quantified statement.)

- There exists a real number whose square is 7. (True.)

Some examples of things that are *not* statements are:

- Is Madrid the capital of Spain? (This is a question, not a statement.)

- Pizza is good. (This is an opinion, not a statement of fact.)

- Kaflooy is the capital of Garglbox. (This is just nonsense; it's not true nor false since the words have no meaning.)

- $2 + 9$. (This is a value.)

- There exists a number whose square is 7. (This is not a statement unless we more carefully define what "number" means. If "number" means "integer," then it's false. If "number" means "real number," then it's true.)

- This statement is false. (This cannot be true or false without contradicting itself, so it is not a statement.)

There are three basic operations that we can perform on statements to get new statements, sometimes called **compound** statements.

- The **negation** of a statement p, denoted $\neg p$, is the statement that is true when p is false and is false when p is true. Stated more simply, $\neg p$ is the opposite of p. For example, if p is the statement "Paris is the capital of France," then $\neg p$ is the statement "Paris is not the capital of France."

- The **conjunction** of two statements p and q, denoted $p \wedge q$, is the statement "p and q". For example, if p is "Paris is the capital of France" and q is "Madrid is the capital of Spain," then $p \wedge q$ is "Paris is the capital of France and Madrid is the capital of Spain." The statement $p \wedge q$ is true provided that both p is true and q is true; otherwise $p \wedge q$ is false.

- The **disjunction** of two statements p and q, denoted $p \vee q$, is the statement "p or q". As above, if p is "Paris is the capital of France" and q is "Madrid is the capital of Spain," then $p \vee q$ is "Paris is the capital of France or Madrid is the capital of Spain." It is true provided that at least one of p and q is true (perhaps both); otherwise $p \vee q$ is false.

> **WARNING!!** "p or q" is one place where mathematical logic might differ from your everyday use of the English word "or." The statement "p or q" means that *at least* one of the statements p and q are true. It does *not* mean that *exactly* one of them is true.

> **Concept:** Don't worry about learning or memorizing the names (like "conjunction") or the symbols (like "$\wedge$"). What's important is that you know what the terms "and", "or", and "not" mean. We'll almost never use the names, and we'll only rarely use the symbols.

Sometimes we analyze more complicated statements by means of a **truth table**. A truth table simply lists all of the possible values of a compound statement, given the possible values of its component parts. For example, we show the truth tables of our basic operations in Figure 2.1 below. In these tables, T represents "true" and F represents "false." For example, the table on the right in Figure 2.1 lists the truth value of "p or q" for each possible pair of values for p and for q.

p	not p
T	F
F	T

p	q	p and q
T	T	T
T	F	F
F	T	F
F	F	F

p	q	p or q
T	T	T
T	F	T
F	T	T
F	F	F

Figure 2.1: Truth tables for the basic logic operations

> **Concept:** You should get comfortable enough with "and", "or", and "not" so that these truth tables are as automatic to you as an addition table.

> **Important:** In order to prove that a statement of the form "p and q" is true, you must prove that *both* p and q are true. In order to prove that a statement of the form "p or q" is true, you must prove that *at least one* of p and q are true.

Let's do a simple example of proving the truth value of a more complicated statement.

Problem 2.6: Prove that, for any statement p, the statement "p and (not p)" is always false.

Solution for Problem 2.6: First, we know that this statement must be false, just by common sense. There's no way that a statement and its opposite can both be true: in fact, we always have that one is true and the other is false.

But to formally establish this, we can resort to a truth table. The statement p can itself be either true or false. So we make a table listing the possible values of p, and the resulting values of "p and (not p)."

p	not p	p and (not p)
T	F	F
F	T	F

As we can see from the table, any input value of p makes the statement false. □

Here is a slightly more complicated example. We'll write it using symbols because it's a bit complicated to write it in words.

Problem 2.7: Prove that, for any statements p and q, the statements $p \lor q$ and $\neg((\neg p) \land (\neg q))$ are either both true or both false.

Solution for Problem 2.7: This one is a bit too complicated for us to use "common sense," so let's go straight to the truth table. We list the truth table for all possible values of p and q. Note that we show the truth value of all the intermediate terms, to make the results clear.

p	q	$p \lor q$	$\neg p$	$\neg q$	$(\neg p) \land (\neg q)$	$\neg((\neg p) \land (\neg q))$
T	T	**T**	F	F	F	**T**
T	F	**T**	F	T	F	**T**
F	T	**T**	T	F	F	**T**
F	F	**F**	T	T	T	**F**

Note that the 3rd and last columns are identical, proving the assertion. □

By the way, the statement of Problem 2.7 is known as one of **DeMorgan's Laws**. In English, we would say that "p or q" is equivalent to the statement "not((not p) and (not q))." Think about, in English, why that makes sense. You'll get a chance to prove a version of another one of DeMorgan's Laws in the exercises.

The other very important operation that we use to make compound statements out of simple statements is that of **implication**. This is the operation given by the statement "If p, then q" and is denoted by $p \Rightarrow q$. We may also read $p \Rightarrow q$ as "p implies q." This is a true statement unless p is true and q is false, in which case it is false. For example, if p is the statement "4 is an odd number" and q is the statement "10 is prime," then $p \Rightarrow q$ is the statement "If 4 is an odd number, then 10 is prime." This is an example of a true statement, although some people find this a bit counterintuitive, and draw the following incorrect conclusion:

Bogus Solution: 4 is an even number, not an odd number, and 10 is composite, not prime, so the statement "If 4 is an odd number, then 10 is prime" is a false statement.

This is incorrect logic. 4 is an even number, so "If 4 is an odd number, then 10 is prime" is true. If the part following the "If" in an implication is false, then the implication is automatically true. For example, "If pigs can fly, then I am a billionaire" is a true statement: pigs can't fly, so we can put any statement we like after the "then" and get a true statement (even though, sadly, I am not a billionaire).

We can also list the truth table of $p \Rightarrow q$:

p	q	$p \Rightarrow q$
T	T	T
T	F	F
F	T	T
F	F	T

The only time that $p \Rightarrow q$ is false is if p is true and q is false. For example, "If 9 is odd, then 28 is a perfect square" is false.

We also have the two-way implication $p \Leftrightarrow q$, which is a shorthand for $(p \Rightarrow q) \wedge (q \Rightarrow p)$. In words, we would write "p if and only if q." Some people abbreviate this to "p iff q," but this should be considered shorthand only, and not correct mathematical writing. Here's the truth table:

p	q	$p \Rightarrow q$	$q \Rightarrow p$	$p \Leftrightarrow q$
T	T	T	T	T
T	F	F	T	F
F	T	T	F	F
F	F	T	T	T

Note that $p \Leftrightarrow q$ is true whenever p and q have the same truth value (they're either both true or both false), and $p \Leftrightarrow q$ is false whenever p and q have different truth values (one is true and the other is false). So we could have rewritten Problem 2.7 as:

Problem 2.7: Prove that, for any statements p and q, the statement

$$(p \vee q) \Leftrightarrow (\neg((\neg p) \wedge (\neg q)))$$

is always true.

There are two important logical operations that we can do to an implication. The first is to take its **converse**.

Definition: The **converse** of the statement $p \Rightarrow q$ is the statement $q \Rightarrow p$.

In other words, we swap the "if" and "then" parts.

Note that the truth or falsehood of an inference gives us no information about the truth or falsehood of its converse. For example, the statement "If a man lives in Los Angeles, then he lives in California" is true, but its converse "If a man lives in California, then he lives in Los Angeles" is false (since, for example, he might live in San Diego).

We can also list a truth table to see that the truth values of $p \Rightarrow q$ and $q \Rightarrow p$ are not necessarily the same:

p	q	$p \Rightarrow q$	$q \Rightarrow p$
T	T	T	T
T	F	F	T
F	T	T	F
F	F	T	T

> **Important:** A very common logical mistake that students often make when working proof-style problems is to prove the *converse* of what the problem is asking. As we have seen, the converse of an implication can be true even if the original implication is false.

The other common operation that we can do to an implication is to take its **contrapositive**.

> **Definition:** The **contrapositive** of the statement $p \Rightarrow q$ is the statement $\neg q \Rightarrow \neg p$.

In other words, we swap the "if" and "then" parts and take the negation of both.

In contrast to the converse, the contrapositive of an implication has the same truth value as the original implication. For example, the statement "If a man lives in Los Angeles, then he lives in California" is true, and its contrapositive "If a man does not live in California, then he does not live in Los Angeles" is also true. We can also see this in a truth table:

p	q	$p \Rightarrow q$	$\neg p$	$\neg q$	$\neg q \Rightarrow \neg p$
T	T	T	F	F	T
T	F	F	F	T	F
F	T	T	T	F	T
F	F	T	T	T	T

> **Concept:** Sometimes in proof-style problems, it's easier to prove the contrapositive of what the problem is asking than it is to prove the original problem statement. As we've just seen, this is logically valid: an implication is logically equivalent to its contrapositive.

 Exercises

2.4.1 Which of the following are statements? (For any that are not statements, briefly explain why not.)

(a) 2 is an odd number.

(b) The author of this book has brown eyes.

(c) Infinity is really cool.

(d) George Washington was the first president of the United States if and only if ((a touchdown in football is worth 11 points) and ($8 + 3^2 = 2^4 + 1$)).

(e) Can you drive a car?

(f) $x^2 + x - 3$

2.4.2 Prove that "p or (not p)" is always true for any statement p.

2.4.3 Prove that, for any statements p and q, the statements $\neg(p \wedge q)$ and $(\neg p) \vee (\neg q)$ are either both true or both false. (This is another of DeMorgan's Laws.)

2.4.4 Prove that, for any statements p and q, the statement $(p \Rightarrow q) \Leftrightarrow ((\neg p) \vee q)$ is always true.

2.4.5 For each of the following implications, state whether it is true or false. Then state the converse and contrapositive, and state whether they are true or false.

(a) If $x = 3$, then $x + 2 = 5$.

(b) If $y^2 - 3y + 2 = 0$, then $y = 1$.

(c) If $A \subseteq B$, then $A \cap B = A$.

(d) If Paris is the capital of France and London is the capital of Japan, then Washington is the capital of the United States.

(e) If WXY is an equilateral triangle, then $WX = XY = YW$.

2.4.6

(a) Prove that, for any statements p, q, and r, the statement

$$(p \vee (q \wedge r)) \Leftrightarrow ((p \vee q) \wedge (p \vee r))$$

is always true. (This is known as the **distributive property** of $\vee$ over $\wedge$.)

(b)★ State and prove the distributive property of $\wedge$ over $\vee$. **Hints:** 58

2.4.7★ Define a new operation ↑ using the following truth table:

p	q	$p \uparrow q$
T	T	F
T	F	T
F	T	T
F	F	T

Write a statement, using only the operation ↑, that is equivalent to $p \vee q$. (The operation ↑ is called a **Sheffer stroke**.) **Hints:** 139

2.5 Quantifiers

Most statements that we work with depend on one or more **quantifiers**. For example, we wish to make statements like:

- All even integers are divisible by 2.

- There exists an integer x such that $x^2 - 5x + 6 = 0$.

- For every positive real number x, there exists a positive real number y such that $x = y^2$.

The words "all," "there exists," and "every" in the above statements are examples of quantifiers. When we start making statements involving variables or other undetermined quantities, we will usually need to include quantifiers.

There are two types of quantifiers:

- The **universal quantifier** is the equivalent of the words "for all" or "every." It is sometimes denoted by an upside-down "A", like so: $\forall$. For example, we can write the (false) statement "All integers are divisible by 2" as

$$\forall x \in \mathbb{Z}, \, 2 | x.$$

In English, we would read this as "For all elements x of the set of integers, 2 divides x." This means, naturally, that 2 divides *every* element of $\mathbb{Z}$, the integers. Of course, this is a false statement, but it is a legal statement nonetheless.

- The **existential quantifier** is the equivalent of the words "there exists." It is sometimes denoted by a backwards "E", like so: $\exists$. For example, we can write the statement "There exists an integer x such that $x^2 - 5x + 6 = 0$" as

$$\exists x \in \mathbb{Z}, \, x^2 - 5x + 6 = 0.$$

In English, this reads "There exists an element x in the set of integers such that $x^2 - 5x + 6 = 0$." This means that there is *some* element of $\mathbb{Z}$ that makes the statement true.

Here is the general description of the two quantifiers:

Concept: If S is a set and $p(x)$ is a statement that depends on an element x of S, then:

- $\forall x \in S, p(x)$ means that $p(x)$ is true for *all* elements x of S.

- $\exists x \in S, p(x)$ means that $p(x)$ is true for *at least one* element x of S.

Important: The symbols are not that important, and you won't see them that often. The *meaning* of the quantifiers is what's important. It is very important to understand the difference between a statement being true for *all* elements of a set, and merely that *there exists* an element of the set that makes the statement true.

For example, the statement "All triangles are equilateral" is clearly false. But the statement "There exists an equilateral triangle" is clearly true.

We can combine quantifiers in the same statement; for example, the statement:

For every positive real number x, there exists a positive real number y such that $x = y^2$.

This statement can be written symbolically as

$$\forall x \in \mathbb{R}, x > 0 \Rightarrow (\exists y \in \mathbb{R}, (y > 0) \wedge (x = y^2)).$$

I said that the statement *can* be written symbolically, but we almost never do so, because as you can see, it's pretty unreadable. Just use English.

Note that the order of quantifiers is important! The following is wrong:

Bogus Solution: There exists some positive real number y such that for every positive real number x, we have $x = y^2$.

Clearly this is not true: this would mean that every real number has the same square root!

> **Important:** Make sure your quantifiers are in the correct order!

When we want to negate a statement that involves a quantifier, we have to be careful to also *reverse* the quantifier. For example, the negation of the statement "All cars have four wheels" is the statement "There exists a car that does not have four wheels." In the first statement, *all* cars have four wheels, so to get the opposite of that statement, we don't need that all cars don't have four wheels, we just need *some* car without four wheels. Another example is "There exists a person from Canada who plays hockey." The negation of this statement is "All people from Canada do not play hockey."

Symbolically, if $p(x)$ is a statement that depends on a variable x that is an element of a set S, then the opposite of the statement

$$\text{For all } x \in S, p(x) \text{ is true}$$

is

$$\text{There exists } x \in S \text{ such that } p(x) \text{ is false.}$$

> **Important:** To prove a universally-quantified statement—a statement that is true "for all $x \in S$"—you must prove that the statement is true for *every* x in the set S. On the other hand, to prove an existentially-quantified statement—a statement for which "there exists $x \in S$" that makes it true—you only have to find a single x in the set S that makes the statement true.

Exercises

2.5.1 Explain why the statements

$$\text{There exists } x \in S \text{ such that } p(x) \text{ is true}$$

and

$$\text{For all } x \in S, p(x) \text{ is false}$$

are opposites.

2.5.2 Find an example of a statement $p(x, y)$ in terms of two integers x and y such that the statement

$$\exists x \in \mathbb{Z}, \forall y \in \mathbb{Z}, p(x, y)$$

is true, whereas the statement

$$\forall x \in \mathbb{Z}, \exists y \in \mathbb{Z}, p(x, y)$$

is false.

2.6 Summary

In this section, we discussed some of the basic building blocks of higher mathematics: sets and logic. The point of this chapter is not to study these topics very deeply in and of themselves, but rather to make sure that we have the necessary tools to construct valid proofs. The symbolism in this chapter is not really that important.

> **Important:** Normally, we do not write the logical symbols. We prefer to use words, because words convey the meaning of our logical arguments better than the symbols do.

What's important is to understand the concepts that underlie set theory and basic logic. Some of these concepts are:

➤ What sets are and how they can be described

➤ How sets are related to their subsets

➤ The empty set $\emptyset$

➤ The power set $\mathcal{P}(S)$ of a set S

➤ How to combine sets by union and intersection

➤ The basics of truth and logic: statements are either true or false

➤ The meanings of "and", "or", and "not"

➤ The meaning of an "if...then..." inference

➤ The converse and contrapositive of an inference

➤ The meanings of, and difference between, "for all" and "there exists"

REVIEW PROBLEMS

2.8 Let

$$S = \{1, 2, 3, \ldots, 30\},$$
$$D = \{n \in S \mid n \geq 0 \text{ and the units digit of } n \text{ is } 9\},$$
$$P = \{p \in S \mid p \text{ is a positive prime}\},$$
$$U = \{n \in S \mid n \equiv 1 \pmod 4\} = \{n \in S \mid \exists k \in \mathbb{Z}, n = 4k + 1\},$$
$$V = \{n \in S \mid n \equiv 3 \pmod 4\} = \{n \in S \mid \exists k \in \mathbb{Z}, n = 4k + 3\},$$
$$W = \{n \in S \mid n \equiv 2 \pmod 4\} = \{n \in S \mid \exists k \in \mathbb{Z}, n = 4k + 2\}.$$

(a) Find the sets $D, P, U, V, W, D \cap U, P \cap W, S \setminus P$, and $P \setminus (D \cup V)$, by explicitly listing the elements of each set.

(b) Which is greater, $\#(P \cap U)$ or $\#(P \cap V)$?

2.9 Let $T = \{-2, -1, 0, 1, 2, 3, 4, 5\}$.

(a) What is $\{x^2 \mid x \in T\}$?

(b) What is $\{x \in \mathbb{Z} \mid x^2 \in T\}$?

2.10 The sets A, B, and C satisfy $A \cap B = B \cap C = \emptyset$. Is it necessarily true that $A \cap C = \emptyset$?

2.11 Are the converses of the statements in Problem 2.4 true? In other words:

(a) If A and B are sets such that $A \cup B = B$, must $A \subseteq B$?

(b) If A and B are sets such that $A \cap B = A$, must $A \subseteq B$?

If they are true, explain why. If they are not true, show an example where they fail.

2.12 If $A \setminus B = A$, then what can we say about sets A and B?

2.13 The finite sets A and B satisfy $\#(A \setminus B) = \#(B \setminus A)$. Does it follow that $\#(A) = \#(B)$?

2.14 Show that for any statements p, q, and r, the statement

$$((p \Rightarrow q) \wedge (q \Rightarrow r)) \Rightarrow (p \Rightarrow r)$$

is always true.

2.15 Let A be the set of mathematicians, let B be the set of snobs, let C be the set of scientists, let D be the set of people who have been to Mars, let E be the set of people who have drunk coffee, and let U be the set of all people. Let $p(x, y)$ denote the statement "x has taken a photo of y". Express the following statements in plain English:

(a) $B \subset U$.

(b) $A \subseteq C$.

(c) $(A \cap D) \subseteq E$.

(d) $U = C \cup E$.

(e) $\forall x \in U, (p(x, x) \Rightarrow x \in B)$.

(f) $\forall x \in C, \exists y \in E, p(x, y)$.

(g) $\exists y \in E, \forall x \in C, p(x, y)$.

2.16 Using the same notation as in the last problem, express the following statements using set and logic notation.

(a) Only snobs drink coffee.

(b) Not all scientists are snobs.

(c) Some mathematician has drunk a coffee.

(d) If everyone has been to Mars, then every scientist has drunk a coffee.

(e) Every mathematician has taken a photo of a snob.

(f) Everyone who has been to Mars is a snob, except for scientists.

(g) The only people who have taken a photo of everyone who has drunk a coffee are scientists.

2.17 Let $p(x)$ be a statement that depends on the element x.

(a) What can you say about the truth of the statement "$\forall x \in \emptyset, p(x)$"?

(b) What can you say about the truth of the statement "$\exists x \in \emptyset, p(x)$"?

Challenge Problems

2.18 Show that for any sets A, B, and C, we have $C \subseteq A$ if and only if $(A \cap B) \cup C = A \cap (B \cup C)$.

2.19 A set S of real numbers is called an *interval* if for all $x, y \in S$, we have

$$\{z \in \mathbb{R} \mid x \le z \le y\} \subseteq S.$$

For example, the sets $[0,1]$, $(-10,5)$, and $[3, \infty)$ are all intervals.

(a) Is the union of two intervals always an interval?

(b) Is the intersection of two intervals always an interval?

2.20 The **exclusive or** operation, denoted by $\oplus$, is defined so that $p \oplus q$ is true if and only if p is true or q is true, but not both.

(a) Show that for any statements p, q, and r, the statements $(p \oplus q) \oplus r$ and $p \oplus (q \oplus r)$ are equivalent.

(b) Show that for any statements p, q, and r, the statements $p \wedge (q \oplus r)$ and $(p \wedge q) \oplus (p \wedge r)$ are equivalent.

2.21 Which of the following statements are equivalent to $\neg(\forall x \in S, \exists y \in T, p(x,y))$? (More than one may be equivalent.)

(i) $\forall x \in S, \exists y \in T, \neg p(x,y)$.

(ii) $\exists x \in S, \forall y \in T, \neg p(x,y)$.

(iii) $\forall x \in S, \neg(\exists y \in T, p(x,y))$.

(iv) $\exists x \in S, \neg(\exists y \in T, p(x,y))$.

(v) $\exists x \in S, \neg(\forall y \in T, p(x,y))$.

2.22 The **symmetric difference** of two sets A and B is defined as $(A \setminus B) \cup (B \setminus A)$, and is denoted by $A \ominus B$.

(a) Let $A = \{1,2,3,4,7,8\}$, $B = \{2,4,5,7,9,10\}$, and $C = \{3,6,7,8,9\}$. Find $A \ominus B$, $(A \ominus B) \ominus C$, $B \ominus C$, and $A \ominus (B \ominus C)$.

(b) Use part (a) of Problem 2.20 to show that $(A \ominus B) \ominus C = A \ominus (B \ominus C)$ for all sets A, B, and C.

(c) What identity of set theory does part (b) of Problem 2.20 give rise to?

If you want to make an apple pie from scratch, you must first create the universe. – Carl Sagan

CHAPTER **3**

A Piece of PIE

3.1 Introduction

In this chapter we'll have some PIE.

PIE is the abbreviation for the Principle of Inclusion and Exclusion. Despite the fancy name, PIE is actually pretty simple. In fact, for the first few problems, PIE is really just an extension of your common sense, but as we'll see later in this chapter, PIE can be a powerful tool for solving a wide variety of counting problems.

PIE is essentially a special application of one of our most fundamental counting techniques: strategic overcounting. In a typical PIE computation, we will repeatedly overcount and undercount until, at the end of the process, we arrive at exactly the correct count. It may seem a bit confusing, but trust us, it's not as complicated as it sounds. It does, however, require some thought to use PIE correctly; in particular, memorizing a "formula" for PIE is a really bad idea.

3.2 PIE With 2 Properties

 **Problems**

Problem 3.1: If 20 girls are on my school's soccer team, 25 girls are on my school's hockey team, and 11 girls play both sports, then how many girls play soccer or hockey?

Problem 3.2: If A and B are sets, write an expression for the number of elements in $A \cup B$ in terms of the number of elements in A, B, and $A \cap B$.

Problem 3.3: At my school, the only foreign languages offered are Spanish and French, and there are 40 students enrolled in at least one of the classes. If 28 students are in the Spanish class and 23 students are in the French class, then how many students are taking both languages?

Problem 3.4:

(a) How many 6-digit numbers start with an even digit?

(b) How many 6-digit numbers end with an even digit?

(c) How many 6-digit numbers start and end with an even digit?

(d) How many 6-digit numbers start or end with an even digit?

Suppose that we have two sets A and B. We'd like to count the number of elements in their union $A \cup B$, which is to say, we want to count the elements that are in A or B.

What's wrong with the following argument?

Bogus Solution: The number of elements in $A \cup B$ is the number of elements in A plus the number of elements in B. Or, as an equation,

$$\#(A \cup B) = \#(A) + \#(B).$$

This argument doesn't take into account the fact that some elements might be in both sets. To take a simple example, suppose that $A = \{1, 2, 3\}$ and $B = \{3, 4, 5\}$. Then $A \cup B = \{1, 2, 3, 4, 5\}$, and we see that

$$\#(A \cup B) = 5 \quad \text{but} \quad \#(A) + \#(B) = 3 + 3 = 6.$$

Let's see a basic example of this phenomenon in a problem setting.

Problem 3.1: If 20 girls are on my school's soccer team, 25 girls are on my school's hockey team, and 11 girls play both sports, then how many girls play soccer or hockey?

Solution for Problem 3.1: A bogus solution to this would be:

Bogus Solution: There are 20 girls on the soccer team and 25 girls on the hockey team, so there are $20 + 25 = 45$ girls on either team.

This doesn't work since there are 11 girls on both teams. If we simply count $20 + 25$, we've counted these 11 girls twice, once on the soccer team and once on the hockey team.

Therefore, we must subtract these girls from our count, since we've counted them twice and we only want to count them once.

So the number of girls playing soccer or hockey is $20 + 25 - 11 = 34$.

We can also see this by using a Venn Diagram.

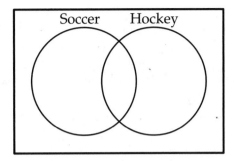

In order to fill in the numbers, we work from the inside out, as this is usually the best way to proceed with Venn diagrams. So, we start by placing the 11 girls who are on both teams into the center of the diagram.

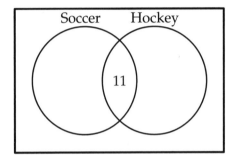

Then we can place $20 - 11 = 9$ girls in the "soccer only" part of the diagram, and $25 - 11 = 14$ girls in the "hockey only" part of the diagram.

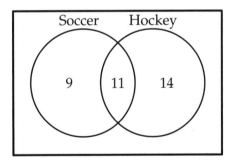

We can now see that there are $9 + 11 + 14 = 34$ girls on the two teams. $\square$

Here's the general idea of this basic version of PIE:

We want to count the number of elements in the union of two sets A and B. For example, in Problem 3.1, set A is "girls on the soccer team" and set B is "girls on the hockey team." However, if sets A and B overlap, we cannot merely sum the elements in each set, since elements in both sets would get counted twice. In order to count the elements in $A \cup B$, we need to **include** the elements of A and the elements of B, and then **exclude** the elements in both. This is why we call this method the Principle of Inclusion and Exclusion.

Let's see the general statement of PIE for counting the number of elements in the union of two sets:

Problem 3.2: If A and B are sets, write an expression for the number of elements in $A \cup B$ in terms of the number of elements in A, B, and $A \cap B$.

Solution for Problem 3.2: We can write PIE for 2 sets as a formula:

Important:

$$\#(A \cup B) = \#(A) + \#(B) - \#(A \cap B),$$

where $\#(S)$ denotes the number of elements in set S.

Going back to our first example, Problem 3.1, we can write this in words as

Number of girls playing soccer or hockey = Number of girls playing soccer
+ Number of girls playing hockey
− Number of girls playing both soccer and hockey.

$\square$

Problem 3.3: At my school, the only foreign languages offered are Spanish and French, and there are 40 students enrolled in at least one of the classes. If 28 students are in the Spanish class and 23 students are in the French class, then how many students are taking both languages?

Solution for Problem 3.3: This problem is slightly different from Problem 3.1. Here, we're told not only the sizes of the two sets, but also the size of their union; we want to find the size of their intersection.

Let x be the number of students in both language classes. (Note that we're using the very common problem solving technique of letting a variable denote what we want to find.) Then PIE tells us that

$$40 = 28 + 23 - x,$$

which we can solve to get $x = 11$.

As a quick check, we can draw the Venn Diagram for this problem, with all of the numbers filled in:

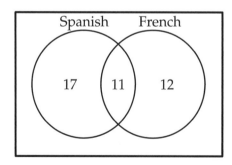

It is easy to verify from the diagram that there are 28 students in Spanish, 23 students in French, and 40 students total, and thus there are 11 students in both languages. $\square$

Problems 3.1 and 3.3 were pretty transparent—it was clear that we needed to count the elements in two overlapping sets, and that PIE was the tool to use. Most PIE problems are not quite so simply stated. Here's an example:

Problem 3.4: How many 6-digit numbers start or end with an even digit?

Solution for Problem 3.4: The word "or" suggests that we are counting elements in two sets. In this problem, we want to count the 6-digit numbers that are in the union of those that start with an even digit and those that end with an even digit.

> **Concept:** If we want to count things that satisfy some property **or** some other property, and the two properties overlap, then we probably want to use PIE.

Thus, our strategy is going to be

#(6-digit numbers that start or end with an even digit) =
#(6-digit numbers that start with an even digit)
+ #(6-digit numbers that end with an even digit)
− #(6-digit numbers that start and end with an even digit)

Each of these terms is fairly easy to count.

6-digit numbers that start with an even digit: There are 4 choices for the first digit (2, 4, 6, or 8), and 10 choices for each remaining digit, so there are $4 \times 10^5 = 400,000$ of these numbers.

6-digit numbers that end with an even digit: There are 9 choices for the first digit (since it can't be 0), 10 choices for each of the four middle digits, and 5 choices for the last digit (2, 4, 6, 8, or 0), so there are $9 \times 10^4 \times 5 = 450,000$ of these numbers.

6-digit numbers that start and end with an even digit: There are 4 choices for the first digit, 10 choices for each of the four middle digits, and 5 choices for the last digit, so there are $4 \times 10^4 \times 5 = 200,000$ of these numbers.

Therefore, the number of 6-digit numbers that start or end with an even digit is

$$400,000 + 450,000 - 200,000 = 650,000.$$

We also could have counted these numbers using complementary counting. We know there are 900,000 6-digit numbers (since there are 9 choices for the first digit and 10 choices for each of the other 5 digits). We also know that there are 250,000 6-digit numbers for which both the first and last digits are odd (since there are 5 choices for the first and last digits, and 10 choices for each of the 4 middle digits). Therefore, there are $900,000 - 250,000 = 650,000$ 6-digit numbers for which the first and last digits are not both odd, meaning that at least one of them is even. □

> **Concept:** When faced with counting things that are "at least" something or "one or more" of something, our two main tools are PIE and complementary counting.

Exercises

3.2.1 There are 47 dogs at the pound. All of them are big or very hairy. 30 are big. 42 are very hairy. How many of the dogs are big, very hairy dogs?

3.2.2 How many 10-digit binary numbers start with 2 ones or end with 2 ones (or both)?

3.2.3 Suppose that 80% of U.S. households own a DVD player and that 70% of U.S. households own a computer. What is the range of possible percentages of U.S. households that own both? **Hints:** 25

3.2.4 How many positive integers less than 100,000 are neither squares nor cubes? **Hints:** 62

3.2.5 Of the 85 teachers at my school, 25 have no children, 50 have a son, and 45 have a daughter. How many have a son and a daughter?

3.2.6 How many 9-digit numbers have the property that the product of their first and last digits is even?

3.2.7★ A school with 100 students offers French and Spanish as its language courses. Twice as many students are in the French class as the Spanish class. Three times as many students are in both classes as are in neither class. The number of students in both classes is even, and fewer than 10 students are in neither class. How many students are taking Spanish? **Hints:** 309, 141

3.3 PIE With 3 Properties

> **Problem 3.5:** Now my school offers 3 foreign languages: Spanish, French, and Chinese. There are 57 students enrolled in at least one of the classes. If 29 are in the Spanish class, 34 are in the French class, 33 are in the Chinese class, 15 are taking both French and Spanish, 16 are taking both French and Chinese, and 12 are taking both Spanish and Chinese, then how many students are taking all three languages?

> **Problem 3.6:** If A, B, and C are three sets, find an expression for the number of elements of $A \cup B \cup C$.

> **Problem 3.7:** How many positive integers less than 1000 are divisible by neither 2, 3, nor 5?

We've seen how to use PIE to count the number of elements in the union of two sets. It seems reasonable to ask whether PIE can be used to count the number of elements in the union of a bunch of sets. We'll explore this general question in Section 3.5, but in this section we'll look at the example of counting elements that are in one or more of three different sets. Let's start with a concrete example.

> **Problem 3.5:** Now my school offers 3 foreign languages: Spanish, French, and Chinese. There are 57 students enrolled in at least one of the classes. If 29 are in the Spanish class, 34 are in the French class, 33 are in the Chinese class, 15 are taking both French and Spanish, 16 are taking both French and Chinese, and 12 are taking both Spanish and Chinese, then how many students are taking all three languages?

Solution for Problem 3.5: Suppose that we try to count the number of students in at least one of the classes. We start with

$$(29 \text{ Spanish}) + (34 \text{ French}) + (33 \text{ Chinese}) = 96.$$

Clearly, though, this is way too big, since it overcounts those students who are in more than one class. Also, we already know that there should be 57 students total, because the problem told us so.

Our first attempt at correcting this is to subtract the number of students who are taking 2 classes: we know those students have been counted twice, so we need to subtract to make sure that they are only counted once. Now our count is.

$$(29 + 34 + 33) - (15 + 16 + 12) = 96 - 43 = 53.$$

But this is now too small, since we know that there are 57 students total. What's wrong here?

The problem is that we have not correctly accounted for those students who are taking all three classes. To see this, imagine that Vincent is taking all three languages. How many times do we count Vincent in our expression $29 + 34 + 33 - 15 - 16 - 12$?

We add Vincent once in each of the 29, 34, and 33 terms, since he is in all three classes. On the other hand, we subtract him once in the -15, -16, and -12 terms, since he is in all three *pairs* of classes. Therefore, in the expression $29 + 34 + 33 - 15 - 16 - 12$, we haven't counted Vincent at all!

Thus, our count $29 + 34 + 33 - 15 - 16 - 12 = 53$ doesn't count the students who are in all three classes. Since $57 - 53 = 4$, we are missing 4 students, therefore there are 4 students in all three classes.

We can use a Venn Diagram to check our answer. Using Figure 3.1, you can verify that there are $7 + 11 + 4 + 12 = 34$ students in the French class, and all of the other data given in the problem can be verified as well.

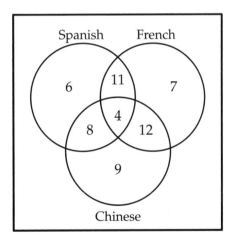

Figure 3.1: Completed Venn Diagram for Problem 3.5

☐

We can use the logic of Problem 3.5 to come up with a general formula for the number of elements in the union of 3 sets.

Problem 3.6: If A, B, and C are three sets, find an expression for the number of elements of $A \cup B \cup C$.

Solution for Problem 3.6: We start with just summing the elements in the three sets:

$$\#(A) + \#(B) + \#(C).$$

However, this sum overcounts elements that appear in more than one set. We can try to correct for this by subtracting the number of elements in each pair of sets:

$$\#(A) + \#(B) + \#(C) - \#(A \cap B) - \#(A \cap C) - \#(B \cap C).$$

If an element is in exactly two sets, it will now be counted exactly once. For example, if x is in A and B, then x will be counted once in $\#(A)$ and once in $\#(B)$, but x will also be subtracted once in $\#(A \cap B)$. So x will be added twice and subtracted once, which means that x will be counted one time overall.

Now we look at those elements in all three sets. These elements are added three times when we count the individual sets, but they are also subtracted three times when we subtract the pairs. So these elements have not yet been counted at all! We finish our count by adding them back in.

Our conclusion:

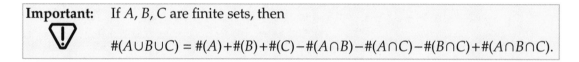

Important: If A, B, C are finite sets, then

$$\#(A \cup B \cup C) = \#(A) + \#(B) + \#(C) - \#(A \cap B) - \#(A \cap C) - \#(B \cap C) + \#(A \cap B \cap C).$$

□

Let's see an application of the 3-set PIE process.

Problem 3.7: How many positive integers less than 1000 are divisible by neither 2, 3, nor 5?

Solution for Problem 3.7: There seems to be no good direct way to count this, so we think about trying to count the opposite: how many positive integers less than 1000 are divisible by at least one of 2, 3, or 5?

The "at least" in this question should make us think about PIE. Our three sets are:

A = "Positive integers less than 1000 that are divisible by 2"
B = "Positive integers less than 1000 that are divisible by 3"
C = "Positive integers less than 1000 that are divisible by 5"

Our goal is to compute $\#(A \cup B \cup C)$. This means PIE.

What's the easiest way to compute the number of positive integers less than 1000 that are divisible by n? We simply take the largest integer less than $1000/n$.

So, using our notation above:

$$\#(A) = (\text{largest integer less than } 1000/2) = 499,$$
$$\#(B) = (\text{largest integer less than } 1000/3) = 333,$$
$$\#(C) = (\text{largest integer less than } 1000/5) = 199.$$

Now we can look at the intersections of pairs of the sets:

$$A \cap B = \text{"Positive integers less than 1000 that are divisible by 2 and 3"}$$
$$= \text{"Positive integers less than 1000 that are divisible by 6"}$$

$$A \cap C = \text{"Positive integers less than 1000 that are divisible by 2 and 5"}$$
$$= \text{"Positive integers less than 1000 that are divisible by 10"}$$

$$C \cap B = \text{"Positive integers less than 1000 that are divisible by 3 and 5"}$$
$$= \text{"Positive integers less than 1000 that are divisible by 15"}$$

We can count these using the same reasoning as above:

$$\#(A \cap B) = (\text{largest integer less than } 1000/6) = 166,$$
$$\#(A \cap C) = (\text{largest integer less than } 1000/10) = 99,$$
$$\#(B \cap C) = (\text{largest integer less than } 1000/15) = 66.$$

Finally, we know that $A \cap B \cap C$ is the set of positive integers less than 1000 that are divisible by all of 2, 3, and 5. This is the set of positive integers less than 1000 that are divisible by 30, hence

$$\#(A \cap B \cap C) = (\text{largest integer less than } 1000/30) = 33.$$

Therefore, using PIE, we can compute:

$$\#(A \cup B \cup C) = \#(A) + \#(B) + \#(C) - \#(A \cap B) - \#(A \cap C) - \#(B \cap C) + \#(A \cap B \cap C)$$
$$= 499 + 333 + 199 - 166 - 99 - 66 + 33$$
$$= 733.$$

Remember, these are the integers less than 1000 that are divisible by at least one of 2, 3, or 5. We want those that are *not* divisible by any of 2, 3, or 5, so we must subtract this answer from 999 (the total number of positive integers less than 1000) to get our final answer of $999 - 733 = 266$. $\square$

Exercises ▶

3.3.1 Dogs in the GoodDog obedience school win a blue ribbon for learning how to sit, a green ribbon for learning how to roll over, and a white ribbon for learning how to stay. There are 100 dogs in the school. Suppose:

- 73 have blue ribbons, 39 have green ribbons, and 62 have white ribbons.
- 21 have a blue ribbon and a green ribbon; 28 have a green ribbon and a white ribbon; 41 have a blue ribbon and a white ribbon.
- 14 have all three ribbons.

How many dogs have not learned any tricks?

3.3.2 How many 3-letter words (where a "word" is any string of 3 letters) have at least one A? (Solve using PIE.)

3.3.3 Vernonia High School has 85 senior boys, each of whom plays on at least one of the school's three boys varsity sports teams: football, baseball, and lacrosse. It so happens that 74 are on the football team; 26 are on the baseball team; 17 are on both the football and lacrosse teams; 18 are on both the baseball and football teams; and 13 are on both the baseball and lacrosse teams. Compute the number of senior boys playing all three sports, given that twice this number are members of the lacrosse team. *(Source: HMMT)*

3.3.4 Four coins are flipped one after the other. What's the probability of getting two consecutive tails?

3.3.5★ A new class of 180 dogs has enrolled in the GoodDog school from Problem 3.3.1. We have the following facts:

- An equal number of dogs have each of the ribbons.

- An equal number of dogs have each pair of ribbons.

- 15 dogs have all three ribbons.

- All dogs have at least one ribbon.

- The number of dogs with exactly one ribbon equals twice the number of dogs with more than one ribbon.

How many dogs have blue ribbons? **Hints:** 102, 119

3.3.6★ How many 6-digit numbers, written in decimal notation, have at least one 1, one 2, and one 3 among its digits? **Hints:** 96, 290

3.4 Counting Problems With PIE

Problem 3.8: How many positive integers less than 180 are relatively prime to 180?

Problem 3.9: The Sanders family has 3 boys and 3 girls. In how many ways can the 6 children be seated in a row of 6 chairs, so that the boys aren't all seated together and the girls aren't all seated together?

Problem 3.10: We form a 12-card deck by taking the Jacks, Queens, and Kings from a standard deck of cards (so that there are 4 cards of each rank). Arrange these 12 cards in a row at random. What is the probability that:
(a) no 4 cards of the same rank are all together?
(b) no 3 cards of the same suit are all together?

Problem 3.11: Two biologists, two chemists, and two physicists go out to dinner and sit at a round table with 6 equally spaced chairs. In how many ways can they sit so that no two scientists of the same type (for example, two biologists) are seated next to each other? (Two seatings that are merely rotations of each other are not considered distinguishably different.)

(a) First, solve the problem using casework: count the number of seatings in which the two biologists are sitting directly opposite from one another, and count the number of seatings in which the two biologists are sitting with one person between them.

(b) Now solve the problem using PIE, by counting the total number of seatings (without any restrictions), then subtracting the number of seatings with at least one pair of like scientists adjacent. (The "at least," as usual, is your signal to use PIE.)

Problem 3.12: How many 6-digit binary numbers (numbers with 0's and 1's as digits) have a string of three consecutive 1's appearing in them? (For example, 101110 and 111100 both have a string of three consecutive 1's, but 100110 doesn't).

In most cases, a counting problem won't come with the instruction "Use PIE to solve this problem." You'll have to figure out on your own that PIE is the right tool to use. In this section we'll work through a few problems that call for the use of PIE.

Problem 3.8: How many positive integers less than 180 are relatively prime to 180?

Solution for Problem 3.8: This problem should remind you of Problem 3.7.

We'll use a slight computational shortcut to make the calculations come out a little nicer than they did in Problem 3.7. Instead of determining the number of positive integers less than 180 that are relatively prime to 180, we'll instead determine the number of positive integers less than *or equal to* 180 that are relatively prime to 180. Since 180 isn't relatively prime to itself, this doesn't change our final answer.

We start by noting that $180 = 4 \times 5 \times 9 = 2^2 \times 5 \times 3^2$. Therefore, a positive integer is relatively prime to 180 if and only if it is not divisible by 2, 3, or 5. As in Problem 3.7, there's no obvious way to count the relatively prime integers directly (though see the Sidenote after this solution for more information on this), but we can use PIE to count the complement of what we want, namely the positive integers less than or equal to 180 that are divisible by at least one of 2, 3, or 5.

Hence, in words, what we are counting is:

# of integers less than or equal to 180 divisible by at least one of 2, 3, or 5 =
  # of integers less than or equal to 180 divisible by 2
  + # of integers less than or equal to 180 divisible by 3
  + # of integers less than or equal to 180 divisible by 5
  − # of integers less than or equal to 180 divisible by 2 and 3
  − # of integers less than or equal to 180 divisible by 2 and 5
  − # of integers less than or equal to 180 divisible by 3 and 5
  + # of integers less than or equal to 180 divisible by 2, 3, and 5.

Thus, our calculation is

$$\frac{180}{2} + \frac{180}{3} + \frac{180}{5} - \frac{180}{6} - \frac{180}{10} - \frac{180}{15} + \frac{180}{30} = 90 + 60 + 36 - 30 - 18 - 12 + 6 = 132.$$

Since there are 180 total positive integers less than or equal to 180, and 132 of them are not relatively prime to 180, that leaves $180 - 132 = 48$ that are relatively prime to 180. □

> **Sidenote:** Given any positive integer n, there is a fairly easy formula for computing the number of positive integers less than or equal to n that are relatively prime to n. If $p_1, p_2, \ldots, p_k$ are the distinct prime divisors of n, then
>
> $$\phi(n) = n\left(1 - \frac{1}{p_1}\right)\left(1 - \frac{1}{p_2}\right) \cdots \left(1 - \frac{1}{p_k}\right)$$
>
> is the number of positive integers less than n that are relatively prime to n. For example, as we already have computed in Problem 3.8,
>
> $$\phi(180) = 180\left(1 - \frac{1}{2}\right)\left(1 - \frac{1}{3}\right)\left(1 - \frac{1}{5}\right) = 180 \cdot \frac{1}{2} \cdot \frac{2}{3} \cdot \frac{4}{5} = 48.$$
>
> This is known as the **Euler phi function** (or **totient function**), and has many important applications in number theory, the most well-known of which is Euler's Theorem, which states that
>
> $$a^{\phi(n)} \equiv 1 \pmod{n}$$
>
> for all a that are relatively prime to n. You can prove the formula for $\phi(n)$ using PIE: try it for yourself in the Challenge Problems.

Problem 3.9: The Sanders family has 3 boys and 3 girls. In how many ways can the 6 children be seated in a row of 6 chairs, so that the boys aren't all seated together and the girls aren't all seated together?

Solution for Problem 3.9: We can do this problem in two steps. First, we determine how many ways there are to assign the 6 chairs by gender: 3 to boys and 3 to girls. Second, we count the ways to place the boys and the girls into their assigned chairs.

It seems like messy casework to try to count directly the number of legal configurations for the boys and girls, so we think instead of counting the illegal configurations: those configurations where 3 boys or 3 girls are all consecutive.

> **Concept:** Messy casework often means that there is a simpler solution using complementary counting and PIE.

With no restriction, there are $\binom{6}{3} = 20$ ways to assign the seats, since we must choose 3 of them for the boys to occupy (the girls will then occupy the other 3). Now we need to determine how many of these 20 configurations are illegal.

There are 4 configurations with the 3 boys together—we could list them as BBBGGG, GBBBGG, GGBBBG, and GGGBBB, or we could reason that there are 4 choices for where the middle boy sits: any of the seats except for the end seats.

Similarly there are 4 configurations with the 3 girls together.

But this does not mean that there are $4 + 4 = 8$ illegal configurations! We have double-counted the configurations where both the boys and the girls are together. There are two of these—BBBGGG and GGGBBB—so we must subtract 2 from our total.

Therefore there are 6 illegal configurations of boys and girls, and hence $20 - 6 = 14$ legal configurations.

Finally, in each of the 14 legal configurations, we know that there are $3! = 6$ ways to assign the boys to their designated sets, and also $3! = 6$ ways to assign the girls. Therefore, the number of possible seatings is $14 \times 6 \times 6 = 504$. □

Problem 3.10: We form a 12-card deck by taking the Jacks, Queens, and Kings from a standard deck of cards (so that there are 4 cards of each rank). Arrange these 12 cards in a row at random. What is the probability that:

(a) no 4 cards of the same rank are all together?

(b) no 3 cards of the same suit are all together?

Solution for Problem 3.10: This problem is very similar to Problem 3.9, in that we are arranging things in a row, subject to a condition that certain groups cannot be together. So for both parts of this problem, we'll use a similar strategy to the one that we used in the solution to Problem 3.9: count the number of illegal configurations, then subtract this count from the total number of configurations to get the count of our successful configurations.

(a) Note that we don't have to worry about the suits at all. The suits are totally irrelevant to this part of the problem.

First we count the number of configurations of the Jacks, Queens, and Kings in the row, without worrying about the condition. We have four J's, four Q's, and four K's to arrange in a line of 12. Without any restriction, this can be done in $\binom{12}{4}\binom{8}{4} = 34650$ ways (we choose 4 of the initial 12 spots for the J's, then 4 of the remaining 8 spots for the Q's, and the K's go in the 4 spots that are left over).

Now, we need to determine how many of these configurations are illegal. A configuration is illegal if at least one of the ranks (Jacks, Queens, or Kings) appear together as a group. As we've seen before, the "at least" is a signal to use PIE.

First, we'll count the configurations that have the 4 Jacks together. There are 9 choices for where to place the Jacks (the first Jack must be in any of positions 1 through 9), then there are $\binom{8}{4}$ ways to allocate the remaining spaces for the Queens and Kings. So there are $9 \times \binom{8}{4} = 630$ configurations with all of the Jacks together.

Similarly, by symmetry, there are 630 configurations with all the Queens together, and 630 configurations with all the Kings together.

But we have overcounted—configurations such as JJJJKKQQQQKK are counted twice, once as a configuration with the Jacks together, and once as a configuration with the Queens together. So we have to subtract off configurations in which two ranks are grouped together.

We count configurations with both the Jacks and Queens together as follows: think of arranging a block of J's, a block of Q's, and four K's into 6 slots. There are 6 choices for where to place the block of J's, then 5 choices for where to place the block of Q's; the K's will go in the remaining empty slots. Therefore, there are $6 \times 5 = 30$ configurations with the J's and Q's together. Similarly, there are 30 configurations with the J's and K's together, and 30 configurations with the Q's and K's together.

Finally, we count configurations with all three ranks together: we have a block of J's, a block of Q's, and a block of K's. All we need to choose is in what order they appear. Therefore, there are $3! = 6$ configurations with all three ranks together, which must be added back to our count.

Thus, there are a total of

$$3(630) - 3(30) + 6 = 1806$$

illegal configurations, and hence $34650 - 1806 = 32844$ legal configurations. Finally, our probability is $\frac{32844}{34650} = \frac{782}{825} \approx 94.8\%$.

(b) Similarly to part (a), we don't have to worry about the ranks of the cards here. All we have to do is keep track of the suits.

We proceed in a manner similar to part (a). We start by noting that there are $\binom{12}{3}\binom{9}{3}\binom{6}{3} = 369{,}600$ possible configurations: we choose 3 of the 12 spaces for the ♠s, then 3 of the 9 remaining spaces for the ♡s, then 3 of the remaining 6 spaces for the ◊s, then the ♣s go in the final 3 spaces.

To count the number of arrangements with all of the ♠s together, we note that there are 10 choices for where the ♠s go, then there are $\binom{9}{3}\binom{6}{3}$ ways to allocate the remaining suits. So there are $10\binom{9}{3}\binom{6}{3} = 16800$ configurations with the 3 ♠s together. By symmetry, there are also 16800 configurations for each of the other three suits being together.

Next, we look at configurations with pairs of suits together. For example, if we want all of the ♠s together and all of the ♡s together, there are $8 \times 7 \times \binom{6}{3} = 1120$ configurations (thinking of the ♠s as a block and the ♡s as a block). There are this many configurations for each of the $\binom{4}{2} = 6$ pairs of suits.

Next, we count configurations with 3 suits together. For example, if we want all of the ♠s, ♡s, and ◊s to be together, there are $6 \times 5 \times 4 = 120$ configurations. There are 120 configurations for each of the 4 choices of 3 suits to keep together.

So far, we have

$$4(16800) - 6(1120) + 4(120) = 60960$$

configurations counted. Is this count correct? Is it too high? Too low?

We know, based on our previous work with PIE, that this accurately counts all configurations with 1, 2, or 3 suits together. But it doesn't correctly count the configurations with all 4 suits together. These configurations are counted 4 times in the first term, subtracted 6 times in the second term, then added back 4 times in the third term, for a net total of being counted twice. Therefore, we need to subtract them once to get them properly counted once overall.

There are $4! = 24$ configurations with all four suits together (we merely have to choose in what order the four suits appear), so our complete count of configurations with at least one suit together is $60960 - 24 = 60936$.

This means that there are $369600 - 60936 = 308664$ configurations with no suit all together; hence, the probability of this occurring is

$$\frac{308664}{369600} = \frac{12861}{15400} \approx 83.5\%.$$

$\square$

> **Problem 3.11:** Two biologists, two chemists, and two physicists go out to dinner and sit at a round table with 6 equally spaced chairs. In how many ways can they sit so that no two scientists of the same type (for example, two biologists) are seated next to each other? (Two seatings that are merely rotations of each other are not considered distinguishably different.)

Solution for Problem 3.11: We can solve this problem using casework—let's do so, for practice.

We start by arbitrarily seating one of the biologists in any seat. Since we're dealing with a round table, this isn't really a "choice," since we can always rotate the table. Now there are two cases, depending on where we seat the other biologist.

Case 1: The other biologist sits directly opposite the first. Then to complete the seating, we must have one chemist and one physicist on either side of each biologist. We have 4 choices for where the first chemist sits (since there are 4 empty seats), then 2 choices for where the other chemist sits (since she cannot sit next to the first chemist). To finish, we have 2 choices for how to seat the two physicists. So there are a total of $4 \times 2 \times 2 = 16$ seatings in this case.

Case 2: The other biologist sits 2 seats away from the first. There are 2 choices here (2 seats to the left or 2 seats to the right). We then have 4 choices for the person to sit between the biologist. The other person of the same science must then sit directly opposite, so that the remaining two people are not next to each other. Finally, there are 2 choices for seating the remaining two people. So there are $4 \times 2 \times 2 = 16$ seatings in this case.

This gives a total of $16 + 16 = 32$ seatings.

Now let's look at the PIE solution. This is a basic PIE calculation—in words, what we want to compute is:

> # of seatings with no pair adjacent = # of seatings (with no restriction)
> − # of seatings with 1 pair adjacent
> + # of seatings with 2 pairs adjacent
> − # of seatings with all 3 pairs adjacent.

Because this is a round table, there are $6!/6 = 5! = 120$ ways to seat the six people—there are $6!$ ways to arrange 6 people, but we have to divide by 6 due to the symmetries of rotation.

Suppose we want to sit one specified pair together. We can seat them anywhere at the table (it doesn't matter where because of the rotational symmetry), and then there are $4!$ ways to seat the remaining 4 people. There are also 2 ways to seat the two people within the pair, so there are a total of $2 \cdot 4! = 48$ ways to seat all the people with one specified pair together. There are 3 different pairs, so we have to subtract $3(48)$ from our initial count of 120.

Now suppose we want to sit two specified pairs together. We can seat the first pair anywhere. We then consider the second pair as a unit, so there are $3!$ ways to seat the second pair and the other two people. There are also 2 ways to seat the people within each pair, so there are a total of $2^2 \cdot 3! = 24$ ways to seat all the people with two specified pairs together. There are 3 choices of two pairs, so we have to add 3(24) back to our running count.

Finally, if we wish to seat all three pairs together, there are 2 ways to arrange the pairs, and 2 ways to arrange the people within each pair, for a total of $2^3 \cdot 2 = 16$ ways to seat all the people with all three pairs together.

Therefore, the final answer is:

$$120 - 3(48) + 3(24) - 16 = 32.$$

□

Problem 3.12: How many 6-digit binary numbers (numbers with 0's and 1's as digits) have a string of three consecutive 1's appearing in them? (For example, 101110 and 111100 both have a string of three consecutive 1's, but 100110 doesn't).

Solution for Problem 3.12: There are four possible positions for a run of three ones: 111???, ?111??, ??111?, and ???111.

We can save ourselves a bit of work by noticing that everything of the form ?111?? must have a 1 in the first slot as well (since every 6-digit number begins with a 1), so everything of the form ?111?? is included in those numbers of the form 111???. So we really have only three types of 6-digit numbers to worry about: 111???, 1?111?, and 1??111.

There are 8 numbers of the form 111???, 4 numbers of the form 1?111?, and 4 numbers of the form 1??111 (in each case, there are 2 choices for a digit to replace each "?").

Now we count the numbers that fall into more than one of the above categories. There are 2 numbers of the form 11111?, 1 number of the form 111111, and 2 numbers of the from 1?1111.

Finally, there is one number of all three forms, namely 111111.

So the number of 6-digit numbers with at least one run of three 1's is $8 + 4 + 4 - 2 - 1 - 2 + 1 = 12$.

It is also pretty easy to list them: 100111, 101110, 101111, 110111, and all 8 numbers of the form 111???. □

Exercises ▶

3.4.1 In how many ways can we arrange the letters of the word STRATA so that the two A's are nonconsecutive and the two T's are also nonconsecutive?

3.4.2 How many positive integers less than 211 are relatively prime to 126?

3.4.3 How many positive integers less than 1000 are relatively prime to both 10 and 12?

3.4.4 3 fans each from Austin High School, Butler High School, and Central High School are seated in a row of 9 seats. In how many ways can we seat the fans if no three fans from the same school are all three seated consecutively?

3.4.5 15 students are each going to enroll in exactly one of economics, psychology, or sociology. In how many ways can they enroll, provided that no class is left empty? **Hints:** 222

3.5 PIE With Many Properties

Problem 3.13: How many positive integers less than 1000 are not divisible by a 1-digit prime?

Problem 3.14: Let A_1, A_2, ..., A_n be n sets. Find an expression for the number of elements of $A_1 \cup A_2 \cup \cdots \cup A_n$.

Problem 3.15: In how many ways can we seat 5 pairs of twins in a row of 10 chairs, such that nobody sits next to his or her twin?

In Problem 3.10 we had our first taste of a PIE problem with more than 3 sets. The general concept of PIE is the same no matter how many sets we have—we alternately overcount and undercount, and make sure that every outcome that we want gets counted exactly once.

Problem 3.13: How many positive integers less than 1000 are not divisible by a 1-digit prime?

Solution for Problem 3.13: We can more easily count the positive integers less than or equal to 1000 that are not divisible by a 1-digit prime. (This will give us the same answer as the original problem, as the number 1000 will not be in our count.)

The one-digit primes are 2, 3, 5, and 7. Let $\lfloor n \rfloor$ denote the greatest integer less than or equal to n. Then we can proceed in a similar manner as in Problem 3.8, and count the number of positive integers less than or equal to 1000 that are divisible by at least one of 2, 3, 5, or 7. The PIE calculation is:

$$
\begin{aligned}
\text{\#'s divisible by 2, 3, 5, or 7} = {}& \left\lfloor \frac{1000}{2} \right\rfloor + \left\lfloor \frac{1000}{3} \right\rfloor + \left\lfloor \frac{1000}{5} \right\rfloor + \left\lfloor \frac{1000}{7} \right\rfloor \\
& - \left(\left\lfloor \frac{1000}{6} \right\rfloor + \left\lfloor \frac{1000}{10} \right\rfloor + \left\lfloor \frac{1000}{14} \right\rfloor + \left\lfloor \frac{1000}{15} \right\rfloor + \left\lfloor \frac{1000}{21} \right\rfloor + \left\lfloor \frac{1000}{35} \right\rfloor \right) \\
& + \left(\left\lfloor \frac{1000}{30} \right\rfloor + \left\lfloor \frac{1000}{42} \right\rfloor + \left\lfloor \frac{1000}{70} \right\rfloor + \left\lfloor \frac{1000}{105} \right\rfloor \right) \\
& - \left\lfloor \frac{1000}{210} \right\rfloor.
\end{aligned}
$$

This works out to

$$500 + 333 + 200 + 142 - (166 + 100 + 71 + 66 + 47 + 28) + (33 + 23 + 14 + 9) - 4 = 772.$$

Therefore there are $1000 - 772 = 228$ positive integers less than (or less than or equal to) 1000 that are not divisible by a 1-digit prime. $\square$

Problem 3.14: Let A_1, A_2, ..., A_n be n sets. Find an expression for the number of elements of $A_1 \cup A_2 \cup \cdots \cup A_n$.

Solution for Problem 3.14: Based on what we've done before, we expect that we should successively add and subtract the sizes of the intersections of more and more sets. Specifically, we should get:

$$
\begin{aligned}
\#(A_1 \cup A_2 \cup \cdots \cup A_n) = {} & \#(A_1) + \#(A_2) + \cdots + \#(A_n) \\
& - (\#(A_1 \cap A_2) + \#(A_1 \cap A_3) + \cdots + \#(A_{n-1} \cap A_n)) \\
& + (\#(A_1 \cap A_2 \cap A_3) + \cdots + (A_{n-2} \cap A_{n-1} \cap A_n)) \\
& - (\#(A_1 \cap A_2 \cap A_3 \cap A_4) + \cdots) \\
& + \quad \vdots \\
& + (-1)^{n+1}\#(A_1 \cap A_2 \cap \cdots \cap A_n).
\end{aligned}
$$

How can we prove that this is indeed the correct formula?

As always, the goal of PIE is to make sure that every element is counted once and only once. Therefore, we can take an arbitrary element of the union of the sets, and count how many times it is counted in the above expression. We do this by considering the number of individual sets that our element is a member of. In particular, suppose element x is in exactly k of the A_i's, where $1 \le k \le n$. Let's count the number of terms in which x gets counted.

x appears in k of the sets, so it's counted $+k$ times in the first line of the formula.

x appears in $\binom{k}{2}$ of the intersections of pairs of sets, so it's counted $-\binom{k}{2}$ times in the second line of the formula.

x appears in $\binom{k}{3}$ of the intersections of triples of sets, so it's counted $+\binom{k}{3}$ times in the third line of the formula.

This pattern continues, until we get to the kth line of the formula, in which x appears in only one intersection of k sets, so it's counted $(-1)^{k+1}$ times (it's counted $+1$ if k is odd, and -1 if k is even).

Thus x gets counted a total of

$$
k - \binom{k}{2} + \binom{k}{3} - \cdots + (-1)^{k+1}
$$

times. We need to prove that this quantity equals 1, meaning that our element x gets counted exactly once.

Let's rewrite the above expression so that all the terms are binomial coefficients:

$$
\binom{k}{1} - \binom{k}{2} + \binom{k}{3} - \cdots + (-1)^{k+1}\binom{k}{k}.
$$

Only one of the binomial coefficients with top entry k is missing, namely $\binom{k}{0}$. If we add and subtract $\binom{k}{0} = 1$ to the expression, then things become clear:

$$
\left(-\binom{k}{0} + \binom{k}{1} - \binom{k}{2} + \binom{k}{3} - \cdots + (-1)^{k+1}\binom{k}{k} \right) + 1.
$$

The term in parentheses in the previous expression is simply the binomial expansion of $0^k = (1 - 1)^k$ using the Binomial Theorem. So the element x is counted exactly once by our formula. We've now shown that every element that appears in exactly k sets is counted exactly once, for all $1 \leq k \leq n$. Thus our formula is correct. $\square$

Once again, here is the formula for PIE with n sets:

> **Important:** If $A_1, A_2, \ldots, A_n$ are sets, then
>
> $$\begin{aligned}
> \#(A_1 \cup A_2 \cup \cdots \cup A_n) = {} & \#(A_1) + \#(A_2) + \cdots + \#(A_n) \\
> & - (\#(A_1 \cap A_2) + \#(A_1 \cap A_3) + \cdots + \#(A_{n-1} \cap A_n)) \\
> & + (\#(A_1 \cap A_2 \cap A_3) + \cdots + (A_{n-2} \cap A_{n-1} \cap A_n)) \\
> & - (\#(A_1 \cap A_2 \cap A_3 \cap A_4) + \cdots) \\
> & + \quad \vdots \\
> & + (-1)^{n+1} \#(A_1 \cap A_2 \cap \cdots \cap A_n).
> \end{aligned}$$

Let's try using PIE in a problem with a large number of sets:

Problem 3.15: In how many ways can we seat 5 pairs of twins in a row of 10 chairs, such that nobody sits next to his or her twin?

Solution for Problem 3.15: Trying to count this directly would involve a lot of messy casework. So instead we count the complement, which is the number of ways to seat the 10 people such that at least one set of twins sit together.

For any pair of twins, there are 9! ways to seat the twins (thought of as a block) and the other 8 people, then 2 ways to seat the twins within their block. So, given a set of twins, there are $2 \cdot 9!$ ways to seat the 10 people so that the given pair of twins sits together. Note also that there are $\binom{5}{1}$ ways to choose the pair of twins to be seated together. So the first term of our PIE calculation will be $\binom{5}{1} 2 \cdot 9!$.

For any two pairs of twins, there are 8! ways to seat the two pairs and the other 6 people, and 2 ways to seat each of the two twins within each pair, so there are $2^2 \cdot 8!$ ways to seat the 10 people with the designated pairs together. Also, there are $\binom{5}{2}$ ways to choose two pairs of twins, so the second term of our PIE calculation will be $\binom{5}{2} 2^2 \cdot 8!$.

For any three pairs of twins, there are 7! ways to seat the three pairs and the other 4 people, so there are $2^3 \cdot 7!$ to seat the 10 people with the three designated pairs together. There are $\binom{5}{3}$ choices for the three pairs of twins. Similarly, there are $2^4 \cdot 6!$ ways to seat four designated pairs together, and there are $2^5 \cdot 5!$ ways to seat all five pairs together, with $\binom{5}{4}$ and $\binom{5}{5}$ choices, respectively, for the pairs.

Therefore, there are

$$\binom{5}{1} 2 \cdot 9! - \binom{5}{2} 2^2 \cdot 8! + \binom{5}{3} 2^3 \cdot 7! - \binom{5}{4} 2^4 \cdot 6! + \binom{5}{5} 2^5 \cdot 5!$$

ways to seat the 10 people so that at least one pair of twins is together. Hence there are

$$10! - \binom{5}{1}2 \cdot 9! + \binom{5}{2}2^2 \cdot 8! - \binom{5}{3}2^3 \cdot 7! + \binom{5}{4}2^4 \cdot 6! - \binom{5}{5}2^5 \cdot 5!$$

ways to seat them so that no pair of twins is together. This can be simplified by factoring out 5!:

$$5!(30240 - 30240 + 13440 - 3360 + 480 - 32) = (120)(10528) = 1,263,360.$$

This may seem like a lot, but it's quite a bit less than $10! = 3,628,800$. If we were to seat the twins randomly, then the probability that no person is sitting next to his or her twin is

$$\frac{1,263,360}{3,628,800} = \frac{47}{135} \approx 34.8\%.$$

$\square$

Exercises

3.5.1 How many positive integers less than 529 are relatively prime to 462?

3.5.2 Yeechi has a deck of cards consisting of the 2 through 5 of hearts and the 2 through 5 of spades. She deals two cards (at random) to each of four players. What is the probability that no player receives a pair? *(Source: Mandelbrot)* **Hints:** 195

3.5.3★ Three Americans, three Canadians, three Spaniards, and three Russians are flying on a small plane that consists of 6 rows of 2 seats each. In how many ways can they be seated, so that no two people from the same country sit in the same row? **Hints:** 180

3.5.4★ Each square of a 3×3 grid of squares is painted black or white with equal probability. What is the probability that the grid does not contain a 2×2 square that is entirely white? *(Source: AIME)* **Hints:** 82, 314

3.6 Counting Items With More Than 1 of Something

Problems

Problem 3.16: My school now offers 3 new foreign languages: Arabic, Japanese, and Russian. There are 50 students enrolled in at least one of the classes. Suppose that 18 are taking both Arabic and Japanese, 15 are taking both Arabic and Russian, 13 are taking both Japanese and Russian, and 7 are taking all three languages. We wish to count how many students are taking at least two languages.

(a) Why is the answer *not* $18 + 15 + 13 - 7$?

(b) What is the answer?

Problem 3.17: If A, B, and C are three sets, how can we count the number of elements in at least two of the sets?

Problem 3.18:

(a) How many positive integers less than 2000 are divisible by at least two of 2, 3, and 5?

(b) How many positive integers less than 2000 are divisible by exactly two of 2, 3, and 5?

Problem 3.19: Five standard 6-sided dice are rolled. What is the probability that at least 3 of them show a 🎲?

Problem 3.20: Suppose $A_1, A_2, \ldots, A_7$ are sets. Determine an expression that counts the number of elements that are in at least 5 of the sets.

We've seen that PIE is a good tool for counting items that have "at least 1" of a set of properties. But what if we want to count items that have "at least 2" of a set of properties? Does PIE work in the same way?

Sort of. Let's see an example:

Problem 3.16: My school now offers 3 new foreign languages: Arabic, Japanese, and Russian. There are 50 students enrolled in at least one of the classes. If 18 are taking both Arabic and Japanese, 15 are taking both Arabic and Russian, 13 are taking both Japanese and Russian, and 7 are taking all three languages, then how many students are taking at least two languages?

Solution for Problem 3.16: We could naively use PIE as follows:

Bogus Solution: We initially count $18 + 15 + 13$ students, but that overcounts the students in all three languages, so we subtract 7, and get as our answer $18 + 15 + 13 - 7 = 39$ students in at least two languages.

This seems right, but a Venn Diagram quickly shows us the error of our ways. We can construct a partial Venn Diagram, starting with the 7 students in all three languages, and filling in the appropriate numbers for the students in two languages; the result is at right. Notice how this Venn Diagram encapsulates all the given information from the problem (except for the 50 total students, which is irrelevant). Then by summing the numbers in the diagram, we see that there are $11 + 8 + 7 + 6 = 32$ students in at least 2 languages.

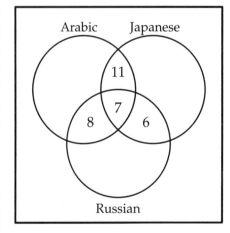

Why did our Bogus Solution not work?

Think about how many times a student in all three languages is counted in our initial count of $18 + 15 + 13$. Such a student is counted *three* times, once for each pair of classes. Since we only want to count every student once, we need to subtract this student *twice* in order to get a correct count. Therefore, the correct PIE expression is

$$18 + 15 + 13 - 2(7) = 46 - 14 = 32,$$

which matches the number that we got from the Venn Diagram. □

As we saw in Problem 3.16, using PIE properly requires that we *think* about what we're doing, and not mindlessly add and subtract numbers.

> **Concept:** Don't memorize a "formula" for PIE. Instead, think about how many times each item is counted, and make sure that each item is counted once and only once.

Let's look at the general case of the situation from Problem 3.16.

Problem 3.17: If A, B, and C are three sets, how can we count the number of elements in at least two of the sets?

Solution for Problem 3.17: We start by counting the number of elements in pairs of sets; that is,

$$\#(A \cap B) + \#(A \cap C) + \#(B \cap C).$$

However, any element that is in all three sets is in all three of these pairs, and will thus be counted 3 times. Since we only want to count it once, we must subtract twice the number of elements in all three sets. So the number of elements in at least 2 sets is

$$\#(A \cap B) + \#(A \cap C) + \#(B \cap C) - 2\#(A \cap B \cap C).$$

□

Problem 3.18:
(a) How many positive integers less than 2000 are divisible by at least two of 2, 3, and 5?
(b) How many positive integers less than 2000 are divisible by exactly two of 2, 3, and 5?

Solution for Problem 3.18:

(a) We can use the expression that we just found in Problem 3.17. The answer is

$$\left\lfloor \frac{1999}{6} \right\rfloor + \left\lfloor \frac{1999}{10} \right\rfloor + \left\lfloor \frac{1999}{15} \right\rfloor - 2\left\lfloor \frac{1999}{30} \right\rfloor = 333 + 199 + 133 - 2(66) = 533.$$

(b) We simply need to subtract, from our answer to part (a), the number of elements that are divisible by all three of 2, 3, and 5. There are $\left\lfloor \frac{1999}{30} \right\rfloor = 66$ of these, so the answer is $533 - 66 = 467$.

□

We've done "at least 2." Let's see if we can do "at least 3."

Problem 3.19: Five standard 6-sided dice are rolled. What is the probability that at least 3 of them show a 🎲?

Solution for Problem 3.19: As hopefully you've come to expect by now, the phrase "at least" in the problem statement is our signal that we want to think about using PIE.

First, of course, there are $6^5 = 7776$ equally likely possible outcomes for rolling the 5 dice. Now we count the successful outcomes, those in which at least three dice show ⚁.

We start by counting outcomes in which a specific set of three dice show ⚁. There are $\binom{5}{3} = 10$ choices for 3 dice out of 5 to come up ⚁. For any chosen 3 dice, there are 6^2 ways for those three dice to be ⚁ and the other two dice to be anything, since there are 6 choices for each of the "anything" dice. This gives an initial count of $10 \cdot 6^2 = 360$ successful outcomes.

However, this overcounts outcomes like ⚁⚁⚁⚁⚀, where four dice come up ⚁. These outcomes are counted 4 times in our above count, once for each subset of 3 ⚁'s in the set of 4 ⚁'s. Since we only want to count these outcomes once, we must subtract these outcomes 3 times from our original count. There are $\binom{5}{4} = 5$ choices for 4 dice out of 5 to show ⚁, and once we have chosen the 4 dice, the 5th die is arbitrary. Thus, there are $5 \cdot 6 = 30$ outcomes with four dice showing ⚁. As we discussed above, we need to subtract this 3 times from our original count of 360, so at this point we have $360 - 3(30) = 270$ successful outcomes.

We now know that this count of 270 accurately counts the outcomes in which we have exactly 3 ⚁'s or exactly 4 ⚁'s. But how many times is ⚁⚁⚁⚁⚁ counted?

This outcome (all ⚁'s) is counted $+10$ times in our 360 term, since it appears once for each of the $\binom{5}{3} = 10$ subsets of three dice out of the five. It is then counted $-3(5) = -15$ times in our -90 term, since it appears three times for each of the $\binom{5}{4} = 5$ subsets of four dice out of the five. Therefore, it is counted a total of $10 - 15 = -5$ times. We need it counted (positively) once, so we need to add it back 6 times, which means that we need to add 6 to our count.

Thus there are $360 - 90 + 6 = 276$ successful outcomes, and the probability of success is

$$\frac{276}{7776} = \frac{23}{648} \approx 3.55\%.$$

□

Now that we've done "at least 2" and "at least 3," you may wonder if we can count "at least k" for any k. The answer is yes, we can, but to write a general formula is fairly difficult, and we're going to defer it to later in the book. However, it's not too hard to figure out the formula for a specific example. It does require us to think a little bit and not blindly add and subtract. Let's try one:

Problem 3.20: Suppose $A_1, A_2, \ldots, A_7$ are sets. Determine an expression that counts the number of elements that are in at least 5 of the sets.

Solution for Problem 3.20: We begin by summing all of the intersections of 5 sets. There are $\binom{7}{5} = 21$ of these intersections:

$$\underbrace{\#(A_1 \cap A_2 \cap A_3 \cap A_4 \cap A_5) + \#(A_1 \cap A_2 \cap A_3 \cap A_4 \cap A_6) + \cdots + \#(A_3 \cap A_4 \cap A_5 \cap A_6 \cap A_7)}_{\binom{7}{5}=21 \text{ terms}} \qquad (*)$$

This sum counts once each element that appears in exactly 5 sets, since each such element only appears in one term of $(*)$. We will abbreviate this sum using the following somewhat sloppy notation:

$$\sum_{21 \text{ terms}} \#(\text{intersections of 5 sets}).$$

If an element appears in exactly 6 sets, it is counted 6 times in (∗), since there are $\binom{6}{5} = 6$ ways to choose 5 of the 6 sets that the element appears in. Since we only want to count these elements once, we need to subtract the sum of the 6-fold intersections 5 times:

$$-5\,(\underbrace{\#(A_1 \cap A_2 \cap \cdots \cap A_6) + \#(A_1 \cap A_2 \cap \cdots \cap A_7) + \cdots + \#(A_2 \cap A_3 \cap \cdots \cap A_7))}_{\binom{7}{6}=7 \text{ terms}} \qquad (\ast\ast)$$

Once again, we will abbreviate this term with the notation

$$-5 \cdot \underset{7 \text{ terms}}{\sum} \#(\text{intersections of 6 sets}).$$

Finally, we look at the elements that appear in all 7 sets. These elements appear in every term of both (∗) and (∗∗), so they are added 21 times in (∗) and subtracted $5 \cdot 7 = 35$ times in (∗∗); hence they have been counted a net $21 - 35 = -14$ times. Since we want to count them exactly once, we need to add them back 15 times.

So our "formula" is

$$\#(\text{elements in at least 5 sets}) = \underset{21 \text{ terms}}{\sum} \#(\text{intersections of 5 sets})$$

$$- 5 \cdot \underset{7 \text{ terms}}{\sum} \#(\text{intersections of 6 sets})$$

$$+ 15 \cdot \#(\text{intersection of 7 sets}).$$

□

Exercises

3.6.1 How many positive integers less than or equal to 3150 have at least three different prime factors in common with 3150?

3.6.2 The four sets $A, B, C,$ and D satisfy

$$\#(A \cap B) = 166, \quad \#(A \cap C) = 100, \quad \#(A \cap D) = 71, \quad \#(B \cap C) = 66, \quad \#(B \cap D) = 47, \quad \#(C \cap D) = 28,$$
$$\#(A \cap B \cap C) = 33, \quad \#(A \cap B \cap D) = 23, \quad \#(A \cap C \cap D) = 14, \quad \#(B \cap C \cap D) = 9,$$
$$\#(A \cap B \cap C \cap D) = 4.$$

(a) How many elements belong to at least two of the sets $A, B, C,$ and D?

(b) How many elements belong to exactly two of the sets $A, B, C,$ and D?

3.6.3 Four standard 6-sided dice are rolled.

(a) What is the probability that at least 3 of them are ⚃ or greater?

(b) What is the probability that exactly 3 of them are ⚃ or greater?

Hints: 65

3.6.4 Let A, B, C, D, E be five sets, and let S_2 be the set of elements that appear in least two of the five sets. Find constants p, q, r, and s such that

$$\#(S_2) = p[\#(A \cap B) + \#(A \cap C) + \#(A \cap D) + \#(A \cap E) + \#(B \cap C)$$
$$+ \#(B \cap D) + \#(B \cap E) + \#(C \cap D) + \#(C \cap E) + \#(D \cap E)]$$
$$+ q[\#(A \cap B \cap C) + \#(A \cap B \cap D) + \#(A \cap B \cap E) + \#(A \cap C \cap D) + \#(A \cap C \cap E)$$
$$+ \#(A \cap D \cap E) + \#(B \cap C \cap D) + \#(B \cap C \cap E) + \#(B \cap D \cap E) + \#(C \cap D \cap E)]$$
$$+ r[\#(A \cap B \cap C \cap D) + \#(A \cap B \cap C \cap E) + \#(A \cap B \cap D \cap E) + \#(A \cap C \cap D \cap E) + \#(B \cap C \cap D \cap E)]$$
$$+ s\,\#(A \cap B \cap C \cap D \cap E).$$

Hints: 117

3.7   Some Harder PIE Problems

Problem 3.21: 7 people are having a water balloon fight. At the same time, each of the 7 people throws a water balloon at one of the other 6 people, chosen at random. What is the probability that there are 2 people who throw balloons at each other?

Problem 3.22: In how many ways can we arrange 5 A's, 7 B's, and 4 C's into a 16-letter word, such that there are at least three 'CA' pairs occurring in the word (in order words, there are at least 3 occurrences of a 'C' immediately followed by an 'A')?

Problem 3.23: In how many ways can we choose 4 vertices of a convex n-gon (where $n > 4$) to form a convex quadrilateral, such that at least 1 side of the quadrilateral is a side of the n-gon?

Problem 3.24: I have a coat with area 5. The coat has 5 patches on it. Each patch has area at least 2.5. Prove that 2 patches exist with common area of at least 1. *(Source: PSS)*

| **WARNING!!** ☢ | This is a very hard problem. It requires both a solid understanding of PIE and a good insight as to how to properly use it. Think about the problem for a little while, but don't be discouraged if you don't make much progress. |

In this section we'll be looking at problems that call for a nontrivial use of PIE.

Problem 3.21: 7 people are having a water balloon fight. At the same time, each of the 7 people throws a water balloon at one of the other 6 people, chosen at random. What is the probability that there are 2 people who throw balloons at each other?

Solution for Problem 3.21: First, we determine the total number of outcomes. Since each person has 6 equally likely choices (each can be throwing at any of the other 6 people), and there are 7 people, we have 6^7 total possibilities.

Now we'd like to count the "successful" outcomes: those outcomes in which at least 2 people are throwing balloons at each other. As we've seen many times before, the "at least" is your signal to consider either PIE or complementary counting. However, a quick look at the complement—trying to count those cases in which no pair is throwing at each other—seems to lead to a calculation that degenerates into some really messy casework (try it and see for yourself if you don't believe me). So we'll try counting this directly, and for that we'll use PIE.

First, we count the number of outcomes in which any particular pair is throwing at each other. There are $\binom{7}{2} = 21$ pairs of people. Besides the pair throwing at each other, the other 5 people each have 6 choices each. So there are $21 \cdot 6^5$ total possibilities.

However, this overcounts those outcomes in which two or more pairs are throwing at each other. So we have to correct for this overcount. This is the point in the solution where it is most likely that you would make a mistake:

> **Bogus Solution:** There are $\binom{7}{2}\binom{5}{2} = 21 \cdot 10 = 210$ ways to choose two pairs of people, since there are $\binom{7}{2} = 21$ ways to choose one pair from 7 people and then $\binom{5}{2} = 10$ ways to choose a second pair from the remaining 5 people.

Seems right, but there's a subtle error. The error is that this counts the number of ways to choose a "first" pair and then choose a "second" pair. What we really want to count is the number of ways to choose two pairs, without regard to order. So we must divide $\binom{7}{2}\binom{5}{2}$ by 2 to correct for this overcount. Thus there are $210/2 = 105$ ways to choose two pairs of people.

Once we have chosen our two pairs, there are three people left over who each have 6 choices of person to throw at. Thus there are $105 \cdot 6^3$ total possibilities with two pairs. Since each two-pair case is counted twice in our original $21 \cdot 6^5$ count, we have to subtract our new count once, so that our running total at this point is

$$(21 \cdot 6^5) - (105 \cdot 6^3).$$

How many times do we count possibilities in which there are 3 pairs of people throwing at each other? These possibilities are counted 3 times in the first term (once for each of the 3 pairs), and subtracted 3 times in the second term (once for each 2-pair subset of the 3 pairs). Thus they currently aren't being counted at all, and we have to add them back in.

There are $\binom{7}{2} = 21$ ways to choose the first pair from the 7 people, $\binom{5}{2} = 10$ ways to choose the second pair from the 5 remaining people, and $\binom{3}{2} = 3$ ways to choose the third pair from the 3 remaining people, so there are $21 \cdot 10 \cdot 3$ ways to choose 3 pairs in order. However, we don't care about the order of the pairs, so we must divide by 3! to correct for this. Therefore there are $21 \cdot 10 \cdot 3/3! = 105$ ways to choose 3 pairs. The 7th, unpaired, person has 6 choices for whom to throw at, so this gives a total of $105 \cdot 6$ possibilities with 3 pairs.

Therefore, there are

$$(21 \cdot 6^5) - (105 \cdot 6^3) + (105 \cdot 6) = 141{,}246$$

possible outcomes in which there is at least one pair throwing at each other.

Finally, the probability that this occurs is

$$\frac{(21 \cdot 6^5) - (105 \cdot 6^3) + (105 \cdot 6)}{6^7} = \frac{23541}{6^6} = \frac{7847}{15552} \approx 50.5\%.$$

□

> **Problem 3.22:** In how many ways can we arrange 5 A's, 7 B's, and 4 C's into a 16-letter word, such that there are at least three 'CA' pairs occurring in the word (in order words, there are at least 3 occurrences of a 'C' immediately followed by an 'A')?

Solution for Problem 3.22: It is possible to solve this problem using casework, where the cases are:

Case 1: Words where all four C's are followed by an A. This is the number of arrangements of 4 (CA)'s, 1 A, and 7 B's, so there are $\binom{12}{4} \times 8 = 3960$ arrangements in this case.

Case 2: Words where 3 C's are followed by an A and the 4th C is followed by a B. This is the number of arrangements of 3 (CA)'s, 1 (CB), 2 A's, and 6 B's, so there are $\binom{12}{3} \times 9 \times \binom{8}{2} = 55440$ arrangements in this case.

Case 3: Words where 3 C's are followed by an A and the 4th C is followed by a C. This is the number of arrangements of 1 (CCA), 2 (CA)'s, 2 A's and 7 B's, so there are $12 \times \binom{11}{2} \times \binom{9}{2} = 23760$ arrangements in this case.

Case 4: Words where 3 C's are followed by an A and the 4th C appears at the end. This is the number of arrangements of 3 (CA)'s, 2 A's, and 7 B's, so there are $\binom{12}{3} \times \binom{9}{2} = 7920$ arrangements in this case.

So there are a total of $3960 + 55440 + 23760 + 7920 = 91080$ arrangements.

But we can more simply solve this problem using PIE. If we count the number of arrangements of 3 (CA)'s, 1 C, 2 A's, and 7 B's, we see that there are $\binom{13}{3} \times 10 \times \binom{9}{2} = 102960$ arrangements, but this counts every arrangement with 4 CA's 4 times. So we must subtract 3 times the number of arrangements with 4 CA's, which as we saw above is 3960. Hence there are $102960 - 3(3960) = 91080$ arrangements. □

Counting problems involving PIE sometimes appear in geometric settings:

> **Problem 3.23:** In how many ways can we choose 4 vertices of a convex n-gon (where $n > 4$) to form a convex quadrilateral, such that at least 1 side of the quadrilateral is a side of the n-gon?

Solution for Problem 3.23: We first count how many quadrilaterals have a particular side of the n-gon as one of its sides. Once we have a side of the n-gon, we need to choose 2 of the remaining $n - 2$ vertices to complete the quadrilateral (we can't choose the two vertices that are the endpoints of our already existing side). So there are $\binom{n-2}{2}$ such quadrilaterals. Since there are n sides of the n-gon that we could have started with, this gives us an initial count of $n\binom{n-2}{2}$.

This initial count, however, overcounts those quadrilaterals that have 2 or more of its sides as sides of the original n-gon, so we have to correct for this. There are essentially two cases to consider for quadrilaterals with 2 sides on the original n-gon: its 2 sides (on the original n-gon) are either consecutive or nonconsecutive.

We can pick two consecutive sides in n ways. This gives 3 vertices of our quadrilateral. Then, we have to choose a fourth vertex, which we can do in $n - 3$ ways. Thus there are a total of $n(n - 3)$ quadrilaterals of this type.

Alternatively, we can pick two nonconsecutive sides: we have n choices for the first side, then $n - 3$ choices for the second non-adjacent side. We then divide by 2 since this counts every pair of non-adjacent sides twice, once in each order. Therefore we have $n(n - 3)/2$ pairs of non-adjacent sides. This gives us all 4 vertices of the quadrilateral, so there are no more choices to be made.

Thus, adding these two cases, we see that there are $n(n-3)+n(n-3)/2 = \frac{3}{2}n(n-3)$ such quadrilaterals. These quadrilaterals are counted twice in our original count, so we need to subtract them once.

Finally, any quadrilateral with 3 sides on the original n-gon must have all 3 of those sides adjacent (otherwise we use up too many vertices). There are n of these quadrilaterals, since there are n choices for the middle of the three adjacent sides on the n-gon. These quadrilaterals are counted 3 times in our original count (once for each side on the original n-gon), but are subtracted 3 times in our count of quadrilaterals with 2 sides on the n-gon (once for each pair of sides on the original n-gon). So we need to add these back once.

Thus the number of quadrilaterals is

$$n\binom{n - 2}{2} - \frac{3}{2}n(n - 3) + n = \frac{n(n - 2)(n - 3) - 3n(n - 3) + 2n}{2} = \frac{n^3 - 8n^2 + 17n}{2}.$$

$\square$

As a quick check, we can verify that this formula works for $n = 5, 6, 7,$ and 8. For $n \leq 7$, the answer is just $\binom{n}{4}$ (since any choice of 4 vertices will give a quadrilateral with at least one side on the original n-gon). The answer for $n = 8$ is $\binom{8}{4} - 2 = 68$, since the only way that we can choose 4 vertices on an octagon that will give a quadrilateral with no sides on the octagon is if we choose every other vertex, and there's only 2 ways we can do that. Checking our formula, we see that it gives 5 when $n = 5$, 15 when $n = 6$, 35 when $n = 7$, and 68 when $n = 8$, so we probably didn't make any obvious mistake.

> **Concept:** If it's possible to do a quick check of your answer to a complicated problem, it's usually a good idea to do so.

We'll conclude our study of PIE with a very difficult problem. Solving this problem requires a solid understanding of the principle behind PIE, and not just a "memorize a formula" knowledge of PIE.

The statement of the problem is deceptively simple.

> **Problem 3.24:** I have a coat with area 5. The coat has 5 patches on it. Each patch has area at least 2.5. Prove that 2 patches exist with common area of at least 1. *(Source: PSS)*

Solution for Problem 3.24: It may not be immediately clear how to proceed. We can start by introducing some notation and listing facts that we know. Let's call the patches P_1, P_2, P_3, P_4, P_5, and let's use brackets to denote area (for example, $[P_1]$ is the area of patch P_1).

What do we know? What are we trying to prove?

> **Concept:** In proof problems, one way to start is by listing what you know and what you're trying to prove.

We know that $[P_1] \geq 2.5$, $[P_2] \geq 2.5$, etc. We also know that the coat has area 5. We're trying to prove that one of $[P_1 \cap P_2]$, $[P_1 \cap P_3]$, ..., $[P_4 \cap P_5]$ is at least 1.

How do we relate all of these items?

We know that the area of the union of the patches is no bigger than the entire coat. Therefore $5 \geq [P_1 \cup P_2 \cup P_3 \cup P_4 \cup P_5]$.

Now it's starting to look like PIE! We can write the area of the union of the patches using PIE. Although we've only used PIE up to now to count discrete elements of sets, the concept works just as well for areas. (Just think of a region as the set consisting of all of the points in the region.)

To shorten what we have to write in what follows, let's introduce some notation for the terms in our PIE expression.

$$
\begin{aligned}
A &= \text{sum of areas of patches} &&= [P_1] + [P_2] + \cdots + [P_5] \\
B &= \text{sum of areas of overlaps of 2 patches} &&= [P_1 \cap P_2] + [P_1 \cap P_3] + \cdots + [P_4 \cap P_5] \\
C &= \text{sum of areas of overlaps of 3 patches} &&= [P_1 \cap P_2 \cap P_3] + \cdots + [P_3 \cap P_4 \cap P_5] \\
D &= \text{sum of areas of overlaps of 4 patches} &&= [P_1 \cap P_2 \cap P_3 \cap P_4] + \cdots + [P_2 \cap P_3 \cap P_4 \cap P_5] \\
E &= \text{sum of areas of overlaps of all patches} &&= [P_1 \cap P_2 \cap P_3 \cap P_4 \cap P_5].
\end{aligned}
$$

Then, using PIE, we get that

$$[P_1 \cup \cdots \cup P_5] = A - B + C - D + E,$$

so because the total area of the coat is 5, we have the inequality

$$5 \geq A - B + C - D + E. \tag{3.7.1}$$

We're trying to show that at least one pair of patches has common area at least 1. If all 10 pairs of patches were to have area less than 1, then when we sum them, we would get $B < 10$. We would like to show that this cannot happen, so we'd like to prove that $B \geq 10$. (This is an example of the **Pigeonhole Principle**, which we'll cover in great detail in Chapter 5.)

So it makes sense to rewrite our inequality (3.7.1) in terms of B:

$$B \geq A + C - D + E - 5. \tag{3.7.2}$$

We also know that $A = [P_1] + \cdots + [P_5] \geq 5(2.5) = 12.5$, since each patch has area at least 2.5. Therefore we can replace A in (3.7.2) to get

$$B \geq 12.5 + C - D + E - 5 = 7.5 + C - D + E. \tag{3.7.3}$$

This is good, but not good enough: we want to show that $B \geq 10$. But we seem to have used all the information that was provided to us, so what are we going to do?

PIE got us this far, so let's use it again! Let's work on B directly by counting the area of the coat covered by at least two patches. We denote this area by X. Don't fall into this trap:

Bogus Solution:

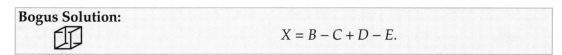

$$X = B - C + D - E.$$

No! As we saw in Section 3.6, PIE is not so simple when we're counting "at least 2" of something. We have to think about how many times each region gets counted. For example, B counts the triple-overlap regions 3 times, so we have to subtract C *twice* in order to get them counted once. Similarly, B counts the quadruple-overlap regions $\binom{4}{2} = 6$ times, but $-2C$ subtracts them $2(4) = 8$ times, so we need to add them back *three times* to get them counted once. Finally, the five-way overlaps are counted $\binom{5}{2} - 2\binom{5}{3} + 3\binom{5}{4} = 10 - 20 + 15 = 5$ times, so we need to subtract $4E$.

Therefore, we know by PIE that

$$X = B - 2C + 3D - 4E. \tag{3.7.4}$$

So what? We don't know anything about X, right?

Wrong! We know that $X \le 5$ (since X can't be bigger than the whole coat). This fact, although seemingly trivial, is very important! Thus we have

$$5 \ge B - 2C + 3D - 4E. \tag{3.7.5}$$

We can use this to eliminate one of the variables in (3.7.3). Let's solve (3.7.5) for C:

$$C \ge \frac{1}{2}(B + 3D - 4E - 5),$$

and plug this into (3.7.3):

$$B \ge 7.5 + C - D + E$$
$$\ge 7.5 + \frac{1}{2}(B + 3D - 4E - 5) - D + E$$
$$= 5 + \frac{1}{2}B + \frac{1}{2}D - E,$$

which simplifies to give

$$\frac{1}{2}B \ge 5 + \frac{1}{2}D - E,$$

or

$$B \ge 10 + D - 2E. \tag{3.7.6}$$

Remember, we wanted to prove that $B \ge 10$, so we're very close. All we need to show to finish the problem is that $D - 2E \ge 0$.

But this is yet another use of PIE! Note that D counts the area in the overlap of all 5 patches *five* times, once for each subset of 4 patches. Therefore $D \ge 5E$, which certainly means that $D \ge 2E$, and hence $D - 2E \ge 0$.

Thus we've proven that $B \ge 10$, which is what we needed to show, so we're done! $\square$

Exercises

3.7.1 Six children are playing dodgeball. Each child has a ball. At the sound of the whistle, each child chooses another child at random to chase. What is the probability that there is at least one pair of children who are chasing each other?

3.7.2 In the grid at right, we wish to go from corner A to corner B, moving only up and to the right one unit at a time. How many such paths include an edge of the shaded square?

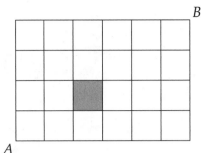

3.7.3 Let S be a set with six elements. Let $\mathcal{P}$ be the set of all subsets of S. Subsets A and B of S, not necessarily distinct, are chosen independently and at random from $\mathcal{P}$. Find the probability that B is contained in at least one of A or $S \setminus A$. (Recall that the set $S \setminus A$ is the set of all elements of S that are not in A.) *(Source: AIME)* **Hints:** 129, 282

3.7.4★ In how many ways can we choose 5 vertices of a convex n-gon (where $n > 5$) to form a convex pentagon, such that at least 2 sides of the pentagon are sides of the n-gon? **Hints:** 225, 238

3.7.5★ Prove that, in Problem 3.24, for all such coats on which no pair of patches overlaps an area of more than 1, no point on the coat is covered by more than 3 patches. **Hints:** 337, 219

3.8 Summary

➤ The principle of inclusion and exclusion (or PIE) is used to count the number of items that are in the union of two or more sets. This can be thought of as counting items that have "at least one" of a number of properties.

➤ If A and B are two sets, then

$$\#(A \cup B) = \#(A) + \#(B) - \#(A \cap B).$$

➤ If A, B, and C are three sets, then

$$\#(A \cup B \cup C) = \#(A) + \#(B) + \#(C) - \#(A \cap B) - \#(A \cap C) - \#(B \cap C) + \#(A \cap B \cap C).$$

➤ More generally, if $A_1, A_2, \ldots, A_n$ are sets, then

$$\begin{aligned}
\#(A_1 \cup A_2 \cup \cdots \cup A_n) = {} & \#(A_1) + \#(A_2) + \cdots + \#(A_n) \\
& - (\#(A_1 \cap A_2) + \#(A_1 \cap A_3) + \cdots + \#(A_{n-1} \cap A_n)) \\
& + (\#(A_1 \cap A_2 \cap A_3) + \cdots + (A_{n-2} \cap A_{n-1} \cap A_n)) \\
& - (\#(A_1 \cap A_2 \cap A_3 \cap A_4) + \cdots) \\
& + \quad \vdots \\
& + (-1)^{n+1} \#(A_1 \cap A_2 \cap \cdots \cap A_n).
\end{aligned}$$

➤ PIE can also be used whenever we have to count items that have "at least k" of a set of properties. Anytime we need to count the items that are in least k sets from a collection of n sets, there will be a PIE expression to calculate it.

➤ Don't blindly apply PIE. Think carefully about how many times each element gets counted. Make sure that elements in exactly 1 set, exactly 2 sets, etc., are each counted once and only once.

Here are a couple of problem-solving concepts regarding PIE:

> **Concept:** Messy casework often means that it's simpler to use complementary counting and PIE.

> **Concept:** If a problem asks you to count how many items have "at least" one property, that's a good sign that you may want to use PIE. Similarly, if a problem asks you to count how many items have "none" of several properties, that may be a sign to use complementary counting with PIE.

> **Concept:** Don't memorize a "formula" for PIE. Instead, think about how many times each item is counted, and make sure that each item is counted once and only once.

We also saw one general good piece of advice when trying to do proofs:

> **Concept:** In proof problems, one way to start is by listing what you know and what you're trying to prove.

REVIEW PROBLEMS

3.25 How many 4-letter words (consisting of any sequence of 4 letters, possibly repeated) start or end with a vowel? (For the purposes of this problem, consider A, E, I, O, and U to be vowels, and consider Y to be a consonant.)

3.26 How many 3-digit numbers have two consecutive digits the same?

3.27 Is it possible that among a group of 20 ninth-graders, 15 of them play lacrosse, 12 of them play soccer, and 6 of them play both? Why or why not?

3.28 When I go to work, there's a 20% probability that I'll forget my office keys, and a 30% probability that I'll forget my wallet. If there's a 5% probability that I forget both, then what's the probability that I arrive at work with both my keys and my wallet?

3.29 How many 4-letter "words" (any combination of 4 letters) have no two consecutive letters identical?

(a) Solve the problem using PIE.

(b) Solve the problem using constructive counting.

(c) Can you algebraically explain why your two answers from (a) and (b) are the same? (Of course we know that they must be the same, since they're just two different ways of counting the same thing, but can you explain it in terms of algebra?)

3.30 Sam can only remember 10-digit numbers if the first four digits are either exactly the same as the next four digits of the number or the last four digits of the number. For example, Sam can remember 1234123456 and 3444533444, but not 3344443334. How many 10-digit numbers can Sam remember?

3.31 My state uses a sequence of three letters followed by a sequence of three numbers as its standard license plate pattern (for example, GMQ829). Given that each three-letter three-digit arrangement is equally likely, find the probability that such a license plate will contain at least one palindrome (a three-letter arrangement or a three-digit arrangement that reads the same left-to-right as it does right-to-left). *(Source: AIME)*

3.32 Let $\pi = (x_1, x_2, x_3, x_4, x_5, x_6, x_7)$ be a permutation of the numbers $(1, 2, 3, 4, 5, 6, 7)$ (recall that a permutation is a rearrangement of the numbers, in which each number appears exactly once). Find the number of such permutations π in which $x_n = n$ for some odd integer n.

3.33 Twenty five of King Arthur's knights are seated at their customary round table. Three of them are chosen—all choices being equally likely—and are sent off to slay a troublesome dragon. Find the probability that at least two of the three had been sitting next to each other. *(Source: AIME)*

3.34 What is the probability that a 13-card bridge hand (dealt at random from a standard 52-card deck) has a *void* (meaning it has no cards of some suit)?

3.35 How many 5-digit sequences have a digit that appears at least 3 times? (For instance, 03005 and 22922 are examples of such sequences.)

Challenge Problems

3.36 Consider two events A and B. Find $P(A \text{ or } B)$ in terms of $P(A)$, $P(B)$, and $P(A \text{ and } B)$. **Hints:** 275

3.37 There are N students at Grant High School. Let $S(F)$ be the number of students at Grant who speak French, $S(J)$ be the number of students who speak Japanese, and $S(A)$ be the number who speak Arabic. Let $S(AF)$ be the number who speak both French and Arabic; define $S(AJ)$ and $S(FJ)$ similarly. Prove that $3N + S(AF) + S(AJ) + S(FJ) \geq 2S(A) + 2S(F) + 2S(J)$. **Hints:** 167

3.38 Given any set S, let $s(S)$ be the number of subsets of S (including S and the empty set). If X, Y, and Z are sets such that $s(X) + s(Y) + s(Z) = s(X \cup Y \cup Z)$ and $\#(X) = \#(Y) = 100$, what is the minimum possible value of $\#(X \cap Y \cap Z)$? *(Source: AMC)* **Hints:** 193, 98

3.39

(a) Prove that for any positive integer k less than 9,

$$9^k - \binom{9}{1}8^k + \binom{9}{2}7^k - \cdots - \binom{9}{7}2^k + \binom{9}{8} = 0.$$

(b) What happens if $k = 9$?

Hints: 19, 209, 233

3.40   Find the number of positive integers that are divisors of at least one of 10^{10}, 15^7, and 18^{11}. *(Source: AIME)* **Hints:** 138

3.41   There are n chairs at a table, each with a name card with the name of one of n people (with one name card for each person). The n people sit at the table. Let D_n be the number of ways the n people can sit at the table such that not a single person is sitting in the correct seat. (D_n is called the number of **derangements** of an n-member set.)

(a)   Use the Principle of Inclusion-Exclusion to find an expression for D_n.

(b)   Find a counting argument to show that $\sum_{k=0}^{n} \binom{n}{k} D_{n-k} = n!$.

3.42   In a five-team tournament, each team plays one game with every other team. Each team has a 50% chance of winning any game it plays. (There are no ties.) Find the probability that the tournament will produce neither an undefeated team nor a winless team. *(Source: AIME)* **Hints:** 320

3.43★   The goal of this problem is to derive the function for the number of positive integers less than or equal to n which do not have a factor besides 1 in common with n. Recall that this function is called the **Euler phi function**, denoted $\phi(n)$, and was discussed in the box on page 60.

(a)   Let the prime factorization of n be $n = p_1^{e_1} p_2^{e_2} \cdots p_k^{e_k}$. How many positive integers less than or equal to n are divisible by p_1? By p_2? By p_i for $1 \leq i \leq k$?

(b)   What's wrong with just subtracting all the numbers we find in (a) from n to get $\phi(n)$ (the number of positive integers less than or equal to n that have no factor besides 1 in common with n)?

(c)   How can we use PIE to correct for our error in part (b)?

(d)   After writing an expression for $\phi(n)$ using PIE, compare your expression to the expansion of the product

$$n\left(1 - \frac{1}{p_1}\right)\left(1 - \frac{1}{p_2}\right)\cdots\left(1 - \frac{1}{p_k}\right).$$

What can you conclude?

3.44★   Define a regular n-pointed **star** to be the union of n line segments $P_1P_2, P_2P_3, \ldots, P_nP_1$ such that:

- the points $P_1, P_2, \ldots, P_n$ are coplanar and no three of them are collinear,

- each of the n line segments intersects at least one of the other line segments at a point other than an endpoint,

- all of the angles at $P_1, P_2, \ldots, P_n$ are congruent,

- all of the n line segments $P_1P_2, P_2P_3, \ldots, P_nP_1$ are congruent, and

- the path $P_1P_2, P_2P_3, \ldots, P_nP_1$ turns counterclockwise at an angle of less than 180 degrees at each vertex.

There are no regular 3-pointed, 4-pointed, or 6-pointed stars. All regular 5-pointed stars are similar, but there are two non-similar regular 7-pointed stars. How many non-similar regular 1000-pointed stars are there? *(Source: AIME)* **Hints:** 95, 169

I've learned that something constructive comes from every defeat. – Tom Landry

CHAPTER 4

Constructive Counting and 1-1 Correspondences

4.1 Introduction

You may recall the concept of **constructive counting** from the book *Introduction to Counting & Probability*. The basic idea is that in order to count the number of a items in a certain set, we think about how we would construct an item belonging to that set. During the construction, we keep track of the number of choices that we have at each step. Although this sounds simple, it is a very powerful way to count! In fact, many of the counting problems that we've solved up to now have essentially used this idea.

This chapter consists of constructive counting problems: we're trying to count the number of elements of some set, and we do so by thinking about how we can construct the elements of the set. We've seen lots of problems of this type before, especially in the *Introduction to Counting & Probability* book, but the problems in this section will generally be harder than the ones you've seen before.

We'll also introduce a new tool that's closely related to constructive counting, called a **1-1 correspondence** (read as "one-to-one correspondence"). Here's the basic idea: we want to count the items in some set A. For whatever reason, set A is hard to count. But suppose that we can find another set B such that:

- B has the same number of elements as A, and

- B is easy to count.

Then to count A, all we have to do is count B, and we have our answer! Simple as that!

A "1-1 correspondence" is the tool that we use to do the first part of the above: finding a set B that has the same number of elements as A. We do this by showing that every element in A somehow "matches" a *corresponding* element in B, and vice versa. Since the elements of A and B come in matching pairs, we know that the sets have the same number of elements.

We can think of using a 1-1 correspondence as a more general version of one of our basic problem-solving strategies: if we don't know how to solve a problem, try to find a simpler, related problem that we *do* know how to solve. That's really what a 1-1 correspondence allows us to do: if we have a set that's hard to count, try to find a related set that's easy to count, and count the easy set instead! As you might expect, the tricky part is finding the "easy" set and showing that it's the same size as the "hard" set.

4.2 Some Basic Problems

Problem 4.1: How many license plates consist of 1 number followed by 3 letters followed by 3 numbers?

Problem 4.2: How many 5-digit palindromes are there? (A **palindrome** is a number that reads the same way forwards and backwards. For example, 27872 and 48484 are palindromes, but 28389 and 12541 are not.)

Problem 4.3: How many 7-digit numbers have no two adjacent digits equal?

Problem 4.4: Four points are chosen at random from the grid at right. What is the probability that the four points are the vertices of a rectangle whose sides are parallel to the sides of the grid?

The general idea with constructive counting is that we build (or "construct") the items that we're trying to count, and while doing so, keep track of the number of choices that we have at each step in the construction.

The problems in this section are all relatively basic examples of constructive counting, and should be review for you if you've mastered the *Introduction to Counting & Probability* textbook.

Problem 4.1: How many license plates consist of 1 number followed by 3 letters followed by 3 numbers?

Solution for Problem 4.1: We have 10 choices for the first number, then 26 choices for each of the 3 letters, then 10 choices for each of the last three numbers, for a total of $10 \cdot 26^3 \cdot 10^3 = 10^4 \cdot 26^3 = 175,760,000$ possible license plates. $\square$

Problem 4.2: How many 5-digit palindromes are there? (A **palindrome** is a number that reads the same way forwards and backwards. For example, 27872 and 48484 are palindromes, but 28389 and 12541 are not.)

Solution for Problem 4.2: We construct the number from left-to-right. We have 9 choices for the first digit (since it can't be 0), then 10 choices for the second digit, then 10 choices for the third digit. But now we're out of choices—the fourth digit must match the second, and the last digit must match the first. Therefore, there are $9 \cdot 10 \cdot 10 = 900$ such numbers. $\square$

Problem 4.3: How many 7-digit numbers have no two adjacent digits equal?

Solution for Problem 4.3: As in Problem 4.2, we construct these numbers from left-to-right. We have 9 choices for the first digit (since it cannot be 0). Then, no matter what we choose for the first digit, we have 9 choices for the second digit—any digit except the one that we chose for the first digit. Similarly, for each subsequent digit, we have 9 choices, since each digit can be any of 0 through 9 as long as it does not match the previous digit. Therefore, there are 9 choices for every digit, and hence there are $9^7 = 4,782,969$ such numbers. $\square$

Problem 4.4: Four points are chosen at random from the grid at right. What is the probability that the four points are the vertices of a rectangle whose sides are parallel to the sides of the grid?

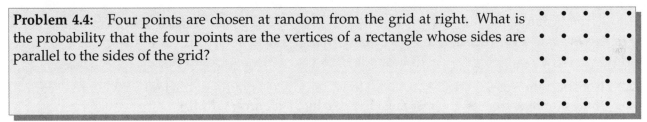

Solution for Problem 4.4: There are $\binom{25}{4}$ ways in which we can choose 4 vertices (without regard to order). We need to determine how many of these sets of 4 vertices produce a valid rectangle. How can we construct such a rectangle? There are a couple of different ways we could proceed.

Solution 1: A rectangle whose sides are parallel to the sides of the grid is determined by its two opposite corners. There are $\binom{25}{2} = 300$ pairs of points in the grid, but there are two things we have to worry about. First, if we pick two points in the same row or same column, then we won't get a rectangle. There are 5 rows and 5 columns, and each has $\binom{5}{2} = 10$ pairs of points, so we have to subtract $(10)(10) = 100$ from our count, leaving $300 - 100 = 200$ pairs of points left. Second, each rectangle has two pairs of opposite corners, so we've overcounted by a factor of 2, and hence our final count is $200/2 = 100$.

Solution 2: A rectangle whose sides are parallel to the sides of the grid can be determined by choosing two rows to form the horizontal sides and choosing two columns to form the vertical sides. We can choose two rows in $\binom{5}{2} = 10$ ways and choose two columns in $\binom{5}{2} = 10$ ways, so the number of rectangles is $(10)(10) = 100$.

Therefore, 100 of the $\binom{25}{4}$ sets of 4 vertices form valid rectangles, and hence our probability is $\frac{100}{\binom{25}{4}} = \frac{2}{253}$.

We could also constructively count the probability directly, without counting the number of rectangles. For this, we think about selecting the vertices in order, one at a time. The first vertex is arbitrarily chosen. For the second vertex, there are two exclusive possibilities:

Case 1: The second vertex is in the same row or column as the first vertex. This occurs with probability $\frac{8}{24} = \frac{1}{3}$. Then, in order to produce a valid rectangle, the third vertex must be on one of the lines containing one of the first two vertices and perpendicular to the line containing the first two vertices. There are 8 such points on these lines, and 23 points remaining to choose from, so the probability of choosing one is $\frac{8}{23}$. Finally, only one point of the 22 remaining will complete the rectangle, and the probability of choosing it is $\frac{1}{22}$.

Case 2: The second vertex is not in the same row or column as the first vertex. This occurs with probability $\frac{16}{24} = \frac{2}{3}$. These two points must be opposite corners of our rectangle, so the final two points that we choose must be the other two corners. They are chosen with probability $\frac{2}{23} \cdot \frac{1}{22}$.

Therefore, the probability of choosing the four corners of a valid rectangle is:

$$\frac{1}{3} \cdot \frac{8}{23} \cdot \frac{1}{22} + \frac{2}{3} \cdot \frac{2}{23} \cdot \frac{1}{22} = \frac{12}{3 \cdot 23 \cdot 22} = \frac{2}{253}.$$

$\square$

 **Exercises**

4.2.1 A dot is marked at each vertex of a triangle ABC. Then 2, 3, and 7 more dots are marked on the sides $\overline{AB}$, $\overline{BC}$, and $\overline{CA}$, respectively. How many triangles have their vertices at these dots? *(Source: HMMT)*

4.2.2 Consider the set $S = \{1, 2, 3, \ldots, 34\}$. How many ways are there to choose (without regard to order) three numbers from S whose sum is divisible by 3? *(Source: ARML)*

4.2.3 How many orderings of the letters in MISSISSIPPI read the same forwards as backwards?

4.2.4 Nine tiles are numbered $1, 2, 3, \ldots, 9$. Each of three players randomly selects and keeps three of the tiles, and sums those three values. Find the probability that all three players obtain an odd sum. *(Source: AIME)* **Hints:** 131

4.2.5★ Robert has 4 indistinguishable gold coins and 4 indistinguishable silver coins. Each coin has an engraving of a face on one side, but not on the other. He wants to stack the eight coins on a table into a single stack so that no two adjacent coins are face to face. Find the number of possible distinguishable arrangements of the 8 coins. *(Source: AIME)* **Hints:** 13

4.3 Harder Constructive Counting Problems

 **Problems**

Problem 4.5: We wish to compute the sum of all of the 5-digit palindromes.
(a) How many 5-digit palindromes are there?
(b) How many have '1' as the last digit? How many have '2' as the last digit? And so on?
(c) What is the units digit of the sum of all of the 5-digit palindromes?
(d) Can you extend your reasoning from parts (a)-(c) above to find the sum of all of the 5-digit palindromes?

Problem 4.6: In a special sort of lottery called *reverse keno*, a player may buy a ticket on which he selects 10 numbers from 1–100 (inclusive). Then, 10 of the numbers are drawn at random. The player wins if his ticket contains *none* of the numbers which are drawn.

(a) What is the probability that a ticket wins?

(b) Is it possible to carefully select numbers on 10 tickets so that I am guaranteed that one of them will win, regardless of what numbers are drawn?

(c) What if I have 12 tickets? Can I now guarantee that one of them will win?

Problem 4.7:

(a) How many 2-digit numbers have distinct digits and are multiples of 9?

(b) How many 10-digit numbers have all digits distinct and are multiples of 11111?

   (i) What is the sum of the digits of any 10-digit number that has all its digits distinct? What can you conclude?

   (ii) Use an argument similar to part (a) and your conclusion from (i) to count the numbers. Don't forget that a number cannot start with 0.

(Source: Russia)

Problem 4.8: In the course of a day, star tennis player Martina Combinova receives 10 different tennis rackets from fans who want her to sign and return the rackets. At various points during the day, Martina takes a break from whatever she is doing to sign some of the rackets. Whenever she decides to sign a racket, she grabs the most recently arrived racket, takes a few strokes with it, then signs it and sends it back. During lunch, Martina's coach tells her that she has signed the 9th racket that arrived. Given that the order of the rackets' arrival is fixed, how many possible post-lunch racket signing sequences are there? *(Source: AIME)*

Problem 4.5: What is the sum of all of the 5-digit palindromes?

Solution for Problem 4.5: In Problem 4.2, we found that there are 900 of these numbers. But how does that help us sum them up?

One idea is to sum each digit separately. Since we constructed the digits one at a time, it makes sense to think that we might be able to sum them one at a time.

> **Concept:** If doing the entire problem at once seems too complicated to handle, try breaking it up into manageable parts.

Start with the first digit. We know that there are 9 choices, so each choice appears in $900/9 = 100$ numbers. Therefore, the first digits sum to $100(1 + 2 + 3 + \cdots + 9) = 4500$. Since the last digits match the first digits, they also sum to 4500.

Now on to the second digit. We know that there are 10 choices, so each choice appears in $900/10 = 90$ numbers. Therefore, the second digits sum to $90(0 + 1 + 2 + \cdots + 9) = 4050$. Similarly, since the third and fourth digits are constructed in the same way, they also sum to 4050.

Therefore, all of the numbers sum to

$$4500(10^4) + 4050(10^3) + 4050(10^2) + 4050(10^1) + 4500(1) = 4500(10001) + 4050(1110) = 49{,}500{,}000.$$

Another way we could approach this is to calculate the *average* 5-digit palindrome. Based on our construction, we know that the average first and last digit is 5 (the average of 1 through 9), and the average of each of the middle three digits is 4.5 (the average of 0 through 9). So the average number is

$$5(10^4) + 4.5(10^3) + 4.5(10^2) + 4.5(10^1) + 5(1) = 55000,$$

and hence, since there are 900 such numbers, the sum of all of them is $(900)(55000) = 49{,}500{,}000.$ □

> **Concept:** In order to determine the sum of a set of numbers, we can often use a
> two-step process:
> 1. Count the number of elements in the set.
> 2. Determine the average element of the set.
> Then, we multiply the answers from our two steps to get the sum of the
> elements in the set. Note that this is essentially what we do when we sum
> an arithmetic series, such as $1 + 2 + \cdots + n$. There are n elements in the
> series, and the average is $\frac{n+1}{2}$, so the sum is $\frac{n(n+1)}{2}$.

Sometimes just figuring out if a construction exists or not is a big part of the problem.

> **Problem 4.6:** In a special sort of lottery called *reverse keno*, a player may buy a ticket on which he
> selects 10 numbers from 1–100 (inclusive). Then, 10 of the numbers are drawn at random. The player
> wins if his ticket contains *none* of the numbers which are drawn.
>
> (a) What is the probability that a ticket wins?
>
> (b) Is it possible to carefully select numbers on 10 tickets so that I am guaranteed that one of them
> will win, regardless of what numbers are drawn?
>
> (c) What if I have 12 tickets? Can I now guarantee that one of them will win?

Solution for Problem 4.6:

(a) There are $\binom{100}{10}$ equally likely possible drawings. To count the number of winning drawings, we
think about how we would construct a drawing for which our ticket wins. Our ticket wins if the
10 numbers drawn are all among the 90 numbers that are not on the ticket. Thus, the probability of
winning is

$$\frac{\binom{90}{10}}{\binom{100}{10}} = \frac{90 \times 89 \times \cdots \times 81}{100 \times 99 \times \cdots \times 91} \approx 33.05\%.$$

(b) We need to think about whether we can construct 10 tickets that guarantee a win for any possible
drawing. We may find the opposite problem easier: whether, given any 10 tickets, we can construct
a drawing that loses on all 10 tickets. In fact, since every drawing has 10 numbers, we might get
very unlucky and have the first number drawn on ticket #1, the second number drawn on ticket #2,
and so on, so that each ticket contains one of the 10 numbers drawn, and thus we would lose on all
tickets. So it is not possible to guarantee a win using 10 tickets.

(c) Given our answer to part (b), there is some hope that we might be able to construct a way to pick numbers for 12 tickets so that every drawing must miss at least one of the tickets (since there are only 10 numbers in a drawing). We fail if there is some drawing that matches a number on all 12 tickets.

Our first observation is that if a drawing loses on all tickets, then there must be a drawn number that appears on more than one ticket, since there are only 10 numbers in the drawing but there are 12 tickets. So our first idea might be to try to make sure that no numbers appear on more than one ticket—if we could guarantee this, then there's no way that any drawing could lose. However, we can't arrange this: we must choose a total of 120 numbers for our 12 tickets (10 numbers per ticket), but we only have 100 numbers to choose from. Therefore, some numbers must appear on more than 1 ticket.

What happens if any number appears on more than two tickets? Then we're in big trouble. For example, suppose that 72 appears on tickets #1, #2, and #3. If the drawing consists of 72 plus one number from each of the other nine tickets #4 through #12, then we lose on all 12 tickets. So if we hope to guarantee a win, we cannot put the same number on more than two tickets. But as we discussed above, we can't avoid duplicating numbers on different tickets. In fact, there must be at least 20 numbers that appear on two tickets.

This means that there must be some number (call it x) that appears on two different tickets. There are 9 other numbers on the first of these tickets, and also 9 other numbers on the second of these tickets. Note that these sets of 9 numbers might overlap, but even if these give us 18 different numbers, and each of them is a 2-ticket number, there's still at least one 2-ticket number left over (since there are at least 20 of them total and we've only accounted for at most $1 + 9 + 9 = 19$ of them). Therefore, there must be another number (call it y) that appears on two *other* tickets, different from the tickets containing x. But now we're doomed: if the drawing comes up x, y, and one number from each of the remaining 8 tickets, we lose.

So we cannot guarantee a win with 12 tickets.

□

As a hard challenge, see if you can figure out some system, using as few tickets as possible, that does guarantee at least one winning ticket regardless of the numbers drawn.

> **Problem 4.7:**
> (a) How many 2-digit numbers have distinct digits and are multiples of 9?
>
> (b) How many 10-digit numbers have all digits distinct and are multiples of 11111?
> (*Source: Russia*)

Solution for Problem 4.7: In both of these parts, we'll use the notation that an overbar indicates that we should treat each letter as a digit, and not as a product. For example, if $a = 6$ and $b = 3$, then $\overline{ab}$ is 63, not 18 (the product of a and b).

(a) Since all of the numbers are small, we could solve this part simply by listing them:

$$18, 27, 36, 45, 54, 63, 72, 81, 90,$$

so there are 9 such numbers. (Note that we didn't list 99, since it doesn't have distinct digits.)

We notice that our list has one number for each possible choice of first digit. That might make us wonder why, and if there's a more clever solution.

Suppose that $\overline{ab}$ is a 2-digit multiple of 9 and $a \neq b$. We know that $a + b$ must be a multiple of 9, so the only possibility is $a + b = 9$. (We can't have $a + b = 18$ since then we'd have to have $a = b = 9$, and that's not allowed.) Thus $b = 9 - a$. We conclude that for every choice of a between 1 and 9 (inclusive), we get one valid number, by setting $b = 9 - a$. Therefore there are 9 numbers.

> **Sidenote:** If you aren't familiar with why $a + b$ must be a multiple of 9 in order for $\overline{ab}$
> ♪ to be a multiple of 9, write
>
> $$\overline{ab} = 10a + b = 9a + (a + b).$$
>
> Clearly $9a$ is a multiple of 9, so for the whole expression to be a multiple of 9, $(a + b)$ must be a multiple of 9 as well.

(b) Since we are looking for 10-digit numbers with all digits different, we know that each digit 0 through 9 must be used exactly once. Surely that must help somehow.

We can use the same observation that we used in part (a), which is that a number is a multiple of 9 if and only if the sum of its digits is a multiple of 9. If our digits are 0 through 9, their sum is $(0 + 1 + \cdots + 9) = 45$, which is a multiple of 9. So, every 10-digit number with all different digits is a multiple of 9, and since 9 and 11111 are relatively prime, we may conclude that every 10-digit multiple of 11111 with all different digits is a multiple of 99999. How can we use this observation?

Note that $99999 = 100000 - 1$. If we write a 10-digit number n as $n = \overline{uv}$, where u and v are each 5-digit numbers, then

$$n = \overline{uv} = 100000u + v = 99999u + (u + v).$$

So n is a multiple of 99999 if and only if $u + v$ is. This means that $u + v = 99999$, since that's the only multiple of 99999 that could possibly be the sum of $u + v$.

Therefore, once we choose u, we must have $v = 99999 - u$ in order for $\overline{uv}$ to be a multiple of 99999. How can we construct such a number?

Note that the digits come in pairs that sum to 9: the units digit of u must sum with the units digit of v to make 9, the tens digits of u and v must sum to 9, and so on. Also, there are 5 pairs of digits that sum to 9: 0 and 9, 1 and 8, 2 and 7, 3 and 6, and 4 and 5.

We have to allocate the five pairs of digits that sum to 9 to the five positions in u and v. This can be done in 5! ways. We also have to choose which digit in each pair goes to u and which goes to v, so this is 2 choices for each pair.

This might lead you to finish the problem by concluding:

> **Bogus Solution:**   Hence, the answer is $5!(2^5) = (120)(32) = 3840$.

The problem is that a 10-digit number can't begin with 0. Since in our count above, any digit is equally likely to end up in any position, 1/10 of the numbers that we constructed above start with 0. That's not allowed. So only 9/10 of the numbers we constructed are actually allowed, hence our final answer is $(9/10)(5!)(2^5) = (9/10)(3840) = 3456$. $\square$

The last problem in this section is tough. Remember that your first rule of problem solving should be to read the problem carefully!

Problem 4.8: In the course of a day, star tennis player Martina Combinova receives 10 different tennis rackets from fans who want her to sign and return the rackets. At various points during the day, Martina takes a break from whatever she is doing to sign some of the rackets. Whenever she decides to sign a racket, she grabs the most recently arrived racket, takes a few strokes with it, then signs it and sends it back. During lunch, Martina's coach tells her that she has signed the 9th racket that arrived. Given that the order of the rackets' arrival is fixed, how many possible post-lunch racket signing sequences are there? *(Source: AIME)*

Solution for Problem 4.8: Before diving into the solution, it's important to understand what's going on here. Rackets #1 through #10 arrive at various points of the day. Whenever she signs one, Martina signs the most recently-arrived racket.

Here's an example:

- Rackets #1, #2, and #3 arrive

- Martina signs #3

- Rackets #4, #5, #6, #7 arrive

- Martina signs #7 and #6

- Racket #8 arrives

- Martina signs #8

- Rackets #9 and #10 arrive

- Martina signs #10, #9, and #5

- Lunch!

- Martina signs #4, #2, and #1.

The post-lunch signing order in this example is: #4, #2, #1. The question is: how many different post-lunch orderings are possible?

Concept: Before diving into a problem, make sure that you:

- Read the problem carefully.

- Understand what the problem is asking.

- If necessary, work through an example to get a feel for the problem.

We start by listing what we know and what we don't know. We know that #9 has already arrived and been signed. This means that #1 through #8 have already arrived, but we don't know which of them (if any) have already been signed. We also don't know if #10 has arrived yet, or if it has been signed.

So there are 9 rackets left that could potentially be signed after lunch. Rackets #1 through #8, if remaining to be signed, must be signed in descending order (the opposite order in which they arrived). On the other hand, racket #10 might already have been signed, or might arrive any time, and thus could be signed at any time (or not at all).

Therefore, we are trying to count the number of lists containing none, some, or all of the numbers 1–8 and 10, where the numbers 1–8 must appear in descending order, but 10 can appear anywhere (or nowhere).

Constructing the ways that 1 through 8 can appear is relatively easy—we only need to choose which of the numbers are in our list, since once we choose which numbers are in the list, they must be placed in descending order. For each of these possible lists of rackets 1–8, the 10 can appear at any point in the list, or not at all. However, the number of choices for where the 10 can appear is dependent on the number of elements already in the list. If there are k numbers in the list before adding 10, then the 10 can appear in any of the $k + 1$ "slots" between numbers (including the first and last slot), or may not appear at all, for $k + 2$ possibilities.

So we need to sum over the different possible lengths of the lists of numbers from 1 through 8 (*before* possibly adding the 10). There are $\binom{8}{k}$ ways to list k numbers out of 1–8, and then $k + 2$ ways to include 10 in the list (or leave it off).

Adding over the possible values of k, we see that the total number of possible orders is

$$2\binom{8}{0} + 3\binom{8}{1} + 4\binom{8}{2} + \cdots + 10\binom{8}{8} = \sum_{k=0}^{8}(k + 2)\binom{8}{k}.$$

We could work this sum out by hand, but is there a more clever way?

In fact, there are two clever methods!

Method 1: We can pull $2\left(\binom{8}{0} + \binom{8}{1} + \cdots + \binom{8}{8}\right)$ out, so that our sum is

$$2\left(\binom{8}{0} + \cdots + \binom{8}{8}\right) + \left(0\binom{8}{0} + 1\binom{8}{1} + \cdots + 8\binom{8}{8}\right) = 512 + \left(0\binom{8}{0} + 1\binom{8}{1} + \cdots + 8\binom{8}{8}\right),$$

using the fact that $\binom{8}{0} + \cdots + \binom{8}{8} = 2^8 = 256$. Now we apply a bit of algebra:

$$k\binom{8}{k} = k\frac{8!}{k!(8 - k)!} = \frac{8!}{(k - 1)!(8 - k)!} = 8\binom{7}{k - 1}.$$

Therefore, our sum is

$$512 + 8\left(\binom{7}{0} + \binom{7}{1} + \cdots + \binom{7}{7}\right) = 512 + 8(128) = 1536.$$

Method 2: Let

$$S = 2\binom{8}{0} + 3\binom{8}{1} + 4\binom{8}{2} + \cdots + 10\binom{8}{8}.$$

Use the identity $\binom{8}{k} = \binom{8}{8-k}$ to rewrite S as

$$S = 2\binom{8}{8} + 3\binom{8}{7} + 4\binom{8}{6} + \cdots + 10\binom{8}{0}.$$

Add these two expressions for S together to get

$$2S = 12\left(\binom{8}{0} + \binom{8}{1} + \cdots + \binom{8}{8}\right) = 12 \cdot 2^8 = 12(256),$$

and divide by 2 to get $S = 6(256) = 1536.$ $\square$

> **Sidenote:** The rackets in Problem 4.8 behave like a structure called a **stack**. A stack is characterized by the property that the last item that arrives is the first item that is processed (in this case, the "processing" is the signing of the racket). This is also sometimes called a **FILO** structure (FILO stands for "First In, Last Out") or a **LIFO** structure ("Last In, First Out"), and is used extensively in computer programming.

Exercises

4.3.1 Suppose that palindromes with n digits are formed using only the digits 1 and 2 and that each palindrome contains at least one of each digit. Compute the least value of n such that the number of palindromes formed exceeds 2002. *(Source: ARML)* **Hints:** 218

4.3.2 We wish to color the integers $1, 2, 3, \ldots, 10$ in red, green, and blue, so that no two numbers a and b, with $a - b$ odd, have the same color. (We do not require that all three colors be used.) In how many ways can this be done? *(Source: HMMT)* **Hints:** 75

4.3.3 Let $S = \{1, 2, 3, \ldots, 24, 25\}$. Compute the number of elements in the largest subset of S such that no two elements in the subset differ by the square of an integer. *(Source: ARML)*

4.3.4 How many pairs of positive integers (m, n) are there such that the least common multiple of m and n is 21,600?

4.3.5 An n-digit positive integer is *cute* if its n digits are an arrangement of the set $\{1, 2, \ldots, n\}$ and its first k digits form an integer that is divisible by k, for $k = 1, 2, \ldots, n$. For example, 321 is a cute 3-digit integer because 1 divides 3, 2 divides 32, and 3 divides 321. How many cute 6-digit numbers are there? *(Source: AMC)*

4.3.6★ A classroom consists of a 5×5 array of desks, to be filled by anywhere from 0 to 25 students, inclusive. No student will sit at a desk unless either all other desks in its row or all others in its column are filled (or both). Considering only the set of desks that are occupied (and not which student sits at each desk), how many possible arrangements are there? *(Source: HMMT)* **Hints:** 190

Extra! *The greatest real thrill that life offers is to create, to construct, to develop something useful.*
— Alfred P. Sloan

4.4   1-1 Correspondence Basics

Informally, we say that two finite sets A and B are in **1-1 correspondence** if they have the same number of elements; that is, if $\#(A) = \#(B)$. But this begs the question—how do we tell if A and B have the same number of elements?

We do so by matching up the elements of A and B, such that for every element of A there is a **corresponding** element of B, and for every element of B there is a **corresponding** element of A. In other words, the elements of A and B come in matching pairs, one element of A matching with one element of B.

Here's a more formal definition:

Definition: We say that the sets A and B are in **1-1 correspondence** if there exists a function f, mapping elements of A to elements of B, that satisfies both of the following properties:

(a) f is **one-to-one** or **1-1**: if $f(x) = f(y)$, then $x = y$. (This also goes by the fancier name **injective**.)

(b) f is **onto**: for every $b \in B$, there exists $a \in A$ such that $f(a) = b$. (This also goes by the fancier name **surjective**.)

The function f is also sometimes called a **bijection**.

Think about what the two parts of the above definition mean. Part (a) means that no two distinct elements of A can correspond to the same element of B. Part (b) means that every element of B has to match to some element of A. When we put these together, we see that every element of A matches to *exactly* one element of B, and no elements of B get "skipped" in this matching.

Concept: Don't worry about the terminology. The main thing to keep in mind is that a 1-1 correspondence between A and B has to pair up elements of A with elements of B, such that every element of A matches up with exactly one element of B, and every element of B matches up with exactly one element of A. Hence the name "1-1 correspondence."

Sometimes we write $a \leftrightarrow f(a)$ to describe the 1-1 correspondence, meaning that element a in set A matches with element $f(a)$ in set B.

Sidenote: **1-1 correspondences between infinite sets**

Our formal definition above works just as well for infinite sets as for finite sets. However, we don't "count" infinite sets, so if sets A and B are in 1-1 correspondence, it doesn't necessarily mean that they have the "same number" of elements. For one thing, $\#(A)$ is undefined if A is an infinite set. But even weirder things can happen. For example, can we show that the sets $\{1, 2, 3, \ldots\}$ and $\{0, 1, 2, 3, \ldots\}$ are in 1-1 correspondence? Certainly: take an element in the first set and subtract 1 to get an element in the second set. This satisfies our definition. Yet, somehow the first set seems "smaller" than the second one. More about this on page 98.

Problems

Problem 4.9: Which of the following pairs of sets are in 1-1 correspondence?

(a) Integers from 1–10 (inclusive) and integers from 23–32 (inclusive).

(b) Integers from 1–10 (inclusive) and integers from 1–100 (inclusive) that are perfect squares.

(c) Integers from 1–10 (inclusive) and unordered pairs of integers from 1–5 (inclusive).

(d) Integers from 1–10 (inclusive) and even integers from (−10)–10 (inclusive).

(e) 2-digit positive integers in base 10 and 2-digit positive integers in base 8.

Problem 4.10: Show that the 7-step paths from A to B in the grid below are in 1-1 correspondence with arrangements of the letters RRRRUUU.

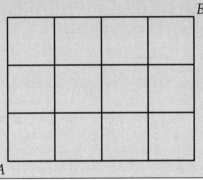

Problem 4.11: How many 4-letter words can we form such that each word has 4 different letters in increasing alphabetical order (such as FHMR)?

In the most basic examples, showing that two sets are in 1-1 correspondence is just a matter of finding a way to match every element in the first set to exactly one element in the second set, and vice versa.

Problem 4.9: Which of the following pairs of sets are in 1-1 correspondence?

(a) Integers from 1–10 (inclusive) and integers from 23–32 (inclusive).

(b) Integers from 1–10 (inclusive) and integers from 1–100 (inclusive) that are perfect squares.

(c) Integers from 1–10 (inclusive) and unordered pairs of integers from 1–5 (inclusive).

(d) Integers from 1–10 (inclusive) and even integers from (−10)–10 (inclusive).

(e) 2-digit positive integers in base 10 and 2-digit positive integers in base 8.

Solution for Problem 4.9:

(a) The elements match up $1 \leftrightarrow 23$, $2 \leftrightarrow 24$, etc., up through $10 \leftrightarrow 32$; in general, $x \leftrightarrow 22 + x$. This matches up each element of 1–10 (inclusive) with exactly one element of 23–32 (inclusive), and vice versa. So the sets are in 1-1 correspondence.

(b) The 1-1 correspondence is $x \leftrightarrow x^2$, which can also be written as $\sqrt{y} \leftrightarrow y$.

(c) We can list the unordered pairs from 1–5, and match them with integers 1–10, as follows:

$$1 \leftrightarrow \{1,2\} \quad 2 \leftrightarrow \{1,3\} \quad 3 \leftrightarrow \{1,4\} \quad 4 \leftrightarrow \{1,5\} \quad 5 \leftrightarrow \{2,3\}$$
$$6 \leftrightarrow \{2,4\} \quad 7 \leftrightarrow \{2,5\} \quad 8 \leftrightarrow \{3,4\} \quad 9 \leftrightarrow \{3,5\} \quad 10 \leftrightarrow \{4,5\}$$

There are, of course, many other ways that you could have matched up the sets. It doesn't matter that there's no "formula" for the matching: as long as we can match each element of one set with exactly one element of the other set, and vice versa, we have a 1-1 correspondence.

(d) These sets are not in 1-1 correspondence. There are 10 integers from 1–10 (inclusive), but 11 even integers from (−10)–10 (inclusive), so no matter how we try to match them, there will always be one integer in the second set "left over."

(e) This is also not a 1-1 correspondence. There are $9 \cdot 10 = 90$ integers of the form $\overline{ab}_{10}$ but only $7 \cdot 8 = 56$ integers of the form $\overline{ab}_8$. So if we try to match up the 90 integers in base 10 with the 56 integers in base 8, we'll quickly run out of base 8 integers to use.

□

Notice in Problem 4.9 that all of our examples that were in 1-1 correspondence with the integers 1–10 (inclusive) had 10 elements. In part (a), there are 10 integers between 23 and 32 (inclusive); in part (b), there are 10 perfect squares between 1 and 100 (inclusive); in part (c), there are $\binom{5}{2} = 10$ unordered pairs of integers between 1 and 5 (inclusive). This is the basic feature of 1-1 correspondence.

> **Concepts:** Two finite sets can be placed in 1-1 correspondence if and only if they have the same number of elements.

1-1 correspondences are sometimes used to relate two different ways of representing the same information, as in the next example.

Problem 4.10: Show that the 7-step paths from A to B in the grid below are in 1-1 correspondence with arrangements of the letters RRRRUUU.

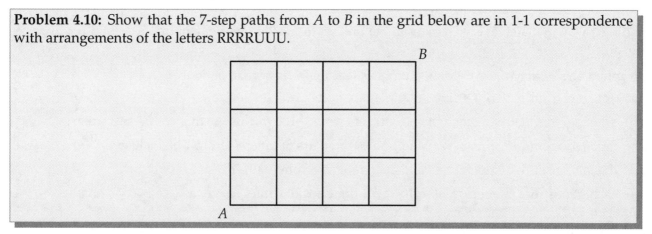

Solution for Problem 4.10: In order to travel in 7 steps from A to B, we must take 4 steps to the right and 3 steps up. Each different arrangement of these 7 steps will give a different path. So, given a path, we can write it as a 7-step sequence of R's and U's, where the letter R corresponds to a step to the right, and the letter U corresponds to a step up. For example, see Figure 4.1 below:

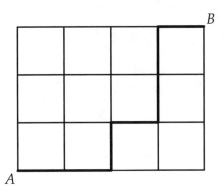

Figure 4.1: Path corresponding to RRURUUR

Conversely, every sequence of 4 R's and 3 U's corresponds to a unique path with 4 right steps and 3 up steps. Therefore the two sets are in 1-1 correspondence. □

The point of Problem 4.10 is that while we may not necessarily know how to count paths, we certainly know how to count sequences of 4 R's and 3 U's: there are $\binom{7}{3} = 35$ of them. Therefore, there are 35 paths from A to B.

Generally, in counting problems, we use a 1-1 correspondence to replace a difficult-to-count set with a set that's easier to count. The next problem shows a basic example of this.

Problem 4.11: How many 4-letter words can we form such that each word has 4 different letters in increasing alphabetical order (such as FHMR)?

Solution for Problem 4.11: Directly counting 4-letter words whose letters are in alphabetical order (for example, using constructive counting) leads to extremely nasty casework. For example, if the word starts with A, then there are 25 choices for the second letter; if this second letter is B, then there are 24 choices for the third letter, however if this second letter is C, then there are only 23 choices for the third letter. Furthermore, if the word starts with B, then there are only 24 choices for the second letter... the casework would go on like this seemingly forever!

Fortunately, we can create a nice 1-1 correspondence between the set

$$A = \{\text{4-letter words whose letters are all different and in alphabetical order}\}$$

and the set

$$B = \{\text{(unordered) sets of 4 distinct letters}\}.$$

The correspondence is simple: given a 4-letter word in A, we simply use the letters in the word as our element of B. To go the other way, given any four distinct letters in B, we just list them in alphabetical order to get a word in A. Every word in A corresponds uniquely to a set of 4 letters in B, and every set of 4 letters in B gives us a unique word in A.

So A and B are in 1-1 correspondence, and therefore #(A) = #(B). But B is easy to count—it's just $\binom{26}{4} = 14{,}950$. Hence there are 14,950 elements of A. □

Obviously, the examples in this section were pretty easy, and included nothing that you didn't already know how to count. In the next couple of sections, we'll do some more complicated examples.

Exercises

4.4.1 Which of the following pairs of sets are in 1-1 correspondence? If the pair is in 1-1 correspondence, show the correspondence.

(a) $\{1, 2, 3, \ldots, 12\}$ and $\{66, 68, 70, \ldots, 88\}$

(b) $\{3, 6, 9, \ldots, 60\}$ and $\{4, 9, 14, \ldots, 104\}$

(c) $\{0, 1, \ldots, 15\}$ and $\mathcal{P}(\{1, 2, 3, 4\})$

(d) {ways to choose 2 items from a group of 5, where order does not matter} and {ways to choose 1 item from a group of 10}

(e) {3-digit numbers with no 2} and {3-digit numbers with no 4}

(f) {2-digit numbers with no 5} and {2-digit numbers with no 0}

4.4.2 How many 3-letter "words" have their middle letter later in the alphabet than both of the other two letters? (For instance, CYJ and FKF are examples of such words, but RPB and WWL are not.)

4.4.3 Compute the number of distinct paths not passing through point $(2, 2, 2)$ that travel from point $(0, 0, 0)$ to point $(4, 4, 4)$ in 12 steps, changing a coordinate by 1 at each step. *(Source: ARML)*

4.4.4 If A is in 1-1 correspondence with B, and B is in 1-1 correspondence with C, show that A is in 1-1 correspondence with C.

4.4.5 Can a finite set A be in 1-1 correspondence with a proper subset $B \subset A$? (Recall that a subset $B \subset A$ is **proper** if $B \neq A$.)

4.4.6★

(a) Show that the odd divisors of 42 are in 1-1 correspondence with the even divisors of 42.

(b) Show that the odd divisors of 28 are *not* in 1-1 correspondence with the even divisors of 28.

(c) For what positive integers n are the odd divisors of n in 1-1 correspondence with the even divisors of n? **Hints:** 224, 183

> **Sidenote:** **More about 1-1 correspondences and infinite sets**
> As we saw on page 94, some strange things can happen when we look at 1-1 correspondences between infinite sets. In particular, Exercise 4.4.5 above is not necessarily true if the set A is infinite; for example, the sets $\{0, 1, 2, 3, \ldots\}$ and $\{1, 2, 3, \ldots\}$ are in 1-1 correspondence even though the latter is a proper subset of the former. Any infinite set that is in 1-1 correspondence with the set of positive integers is called **countable**. However, not every infinite set is countable. It turns out that $\mathbb{Q}$ (the set of rational numbers) is countable but that $\mathbb{R}$ (the set of real numbers) is uncountable.
>
> More about this on page 104.

4.5 More Complicated 1-1 Correspondences

Problem 4.12: What is the maximum number of intersection points of diagonals inside a convex n-gon? For example, the picture at right shows a pentagon with 5 such intersection points circled.

(a) Compute this by hand for $n = 3, 4, 5, 6, 7$ by drawing a picture and carefully counting the diagonal points.

(b) Do you see a pattern in the numbers that you found in part (a)?

(c) Use your answer to (b) to find an appropriate 1-1 correspondence, and finish the problem.

Problem 4.13: Let $n = 2^{12}3^85^6$. How many divisors of n^2 are less than n?

(a) Try the question for small values of n, such as $n = 4$, $n = 5$, and $n = 6$. Do you notice anything interesting?

(b) What other set is in 1-1 correspondence with divisors of n^2 less than n?

(c) Finish the problem.

Problem 4.14: Let k be an odd number. Show that there are fewer odd divisors of $4k$ than even divisors of $4k$.

Problem 4.15: The tennis club has 100 members, and Stephanie is the president. Which is larger: the number of ways to choose a 10-person advisory committee (which does not contain Stephanie) to advise Stephanie on what new rackets to buy, or the number of 11-person committees (which may include Stephanie) to plan an upcoming tournament? *Note: you should be able to do this problem without a calculator!*

In the last section, the problems were pretty easy. In this section, things get a bit more difficult.

Problem 4.12: What is the maximum number of intersection points of diagonals inside a convex n-gon? For example, the picture at right shows a pentagon with 5 such intersection points circled.

Solution for Problem 4.12: For a problem like this, where we're trying to find a formula in terms of a positive integer n, it often helps to try some small values of n and see if we discover a pattern.

> **Concept:** When stuck on a "find a general formula"-type problem, experiment with some small values, and look for a pattern.

We can draw some pictures for some small polygons. You should be able to draw this for triangles, quadrilaterals, and pentagons yourself; Figure 4.2 below shows a hexagon with 15 intersection points and a heptagon with 35 intersection points:

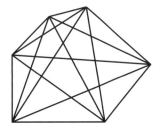 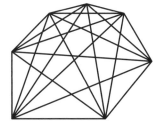

Figure 4.2: Convex hexagon and heptagon with diagonals

From our examples, we have the following chart:

n	3	4	5	6	7
Points	0	1	5	15	35

It's probably going to be pretty messy to draw an octagon! And maybe it's hard to see a pattern in the above numbers. But where do we often look for counting numbers? Pascal's Triangle, of course!

Concept: Look inside Pascal's Triangle to find counting numbers.

So let's go ahead and draw Pascal's Triangle, and we'll highlight the numbers that we found:

$$
\begin{array}{ccccccccccccccc}
&&&&&&&1&&&&&&&\\
&&&&&&1&&1&&&&&&\\
&&&&&1&&2&&1&&&&&\\
&&&&1&&3&&3&&1&&&&\\
&&&\boxed{1}&&4&&6&&4&&\boxed{1}&&&\\
&&1&&\boxed{5}&&10&&10&&\boxed{5}&&1&&\\
&1&&6&&\boxed{15}&&20&&\boxed{15}&&6&&1&\\
1&&7&&21&&\boxed{35}&&\boxed{35}&&21&&7&&1\\
\end{array}
$$

Aha, now a clear pattern emerges! The boxed numbers on the left sides of each row are $\binom{4}{0}$, $\binom{5}{1}$, $\binom{6}{2}$, $\binom{7}{3}$, etc. But the (same) boxed numbers on the right side are even easier to write as binomial coefficients: $\binom{4}{4}$, $\binom{5}{4}$, $\binom{6}{4}$, $\binom{7}{4}$, etc. Now we can see pretty clearly that the answer looks like $\binom{n}{4}$. So now that we think that we know what the answer is, how can we prove it?

We know one "obvious" set with $\binom{n}{4}$ elements: the possible choices of sets of 4 vertices out of the n total vertices. Perhaps we can show that this set is in 1-1 correspondence with intersection points of diagonals inside the n-gon.

Concept: When you suspect that a set that you're trying to count has the same number of elements as a set that's easy to count, look for a 1-1 correspondence between the set you have and the "easy" set.

In fact, we can find the 1-1 correspondence. Given any intersection point of diagonals, we can get 4 vertices by taking the two pairs of endpoints of the diagonals intersecting at that point. Conversely, any 4 vertices uniquely determine two intersecting diagonals, by looking at the (unique) convex quadrilateral with those 4 vertices. For example, Figure 4.3 below shows an intersection point (circled) and its corresponding 4 vertices (in bold):

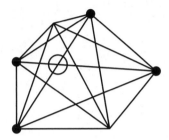

Figure 4.3: Hexagon with intersection point and 4 vertices in correspondence

So we've established a 1-1 correspondence between

$$\{\text{intersection points of diagonals}\} \leftrightarrow \{\text{sets of 4 vertices}\},$$

and since it is easy to see that the latter set has $\binom{n}{4}$ elements, we can conclude that the former set does as well.

Note that we've glossed over one significant detail: what happens if more than 2 diagonals intersect at the same point? Then we don't have our 1-1 correspondence, but this means that the set on the left in our above correspondence is strictly smaller than the set on the right, so that the number of intersection points is strictly less than $\binom{n}{4}$.

Hence $\binom{n}{4}$ is the maximum number of intersection points of diagonals inside a convex n-gon. $\square$

Problem 4.13: Let $n = 2^{12}3^85^6$. How many divisors of n^2 are less than n?

Solution for Problem 4.13: A logical place to start is to count the number of divisors of n^2. Notice that $n^2 = 2^{24}3^{16}5^{12}$. A divisor of n^2 is a number of the form $2^a3^b5^c$, where $a \leq 24$, $b \leq 16$, and $c \leq 12$. Therefore, since a can be anything from 0 to 24 (inclusive), and we have similar choices for b and c, there are $25 \cdot 17 \cdot 13 = 5525$ factors of n^2 (including n^2 itself).

So what's wrong with the following?

> **Bogus Solution:** A factor is less than n if and only if $a \leq 12$, $b \leq 8$, and $c \leq 6$, and there are $13 \cdot 9 \cdot 7 = 819$ of these. This includes n itself, so we must subtract 1, and we get a final answer of 818.

The problem is that there are a lot of divisors of n^2 that are less than n but that are not themselves divisors of n. For example, 2^{13} is not a divisor of n, but it is a divisor of n^2, and it is a lot smaller than n. Deciding what choices of a, b, c make $2^a3^b5^c < n$ is a quite difficult number theory problem.

There doesn't seem to be any obvious way to directly count what we want. So instead, we might look for a 1-1 correspondence, from our set of divisors of n^2 less than n to some other set that we know how to count. Is there an "obvious" set that is in 1-1 correspondence to the set of divisors of n^2 less than n? Remember: a 1-1 correspondence is essentially a pairing between elements of two sets, so is there anything that we can pair a divisor of n^2 less than n to?

> **Concept:** 1-1 correspondences are essentially pairings. So look for a set of things to pair elements of your set to.

Indeed, divisors of n^2 come in pairs: if $n^2 = xy$ for some integers x and y, then either $x = y = n$, or $x < n$ and $y > n$, or $x > n$ and $y < n$. In particular, a divisor $x < n$ of n^2 is naturally paired with the divisor $\frac{n^2}{x} > n$.

Therefore, we have a 1-1 correspondence between the sets

$$\{\text{divisors of } n^2 \text{ less than } n\} \leftrightarrow \{\text{divisors of } n^2 \text{ greater than } n\}.$$

Since these sets are of equal size, and together they total 5524 elements (since they contains all of the divisors of n^2 except for n), we conclude that they each have $5524/2 = 2762$ elements. Hence the answer to our problem is 2762. $\square$

The last step of the previous problem indicates another important way in which we can use 1-1 correspondences.

> **Important:** If you can partition a finite set A into two non-overlapping subsets B and C (meaning that $B \cup C = A$ and $B \cap C = \emptyset$), and you can find a 1-1 correspondence between B and C, then B and C are each exactly one-half the size of A.

A slight variation of this is in the next problem.

Problem 4.14: Let k be an odd number. Show that there are fewer odd divisors of $4k$ than even divisors of $4k$.

Solution for Problem 4.14: Let's try a simple example to get a feel for things. Suppose $k = 5$, so $4k = 20$. The odd divisors of 20 are 1 and 5. The even divisors of 20 are 2, 4, 10, and 20. Indeed, we see that there are only two odd divisors whereas there are four even divisors.

Let's look at another example. Suppose $k = 15$, so $4k = 60$. The odd divisors of 60 are 1, 3, 5, and 15. The even divisors of 60 are 2, 4, 6, 10, 12, 20, 30, and 60. We see that there are four odd divisors and eight even divisors.

A pattern seems to have emerged. In both examples, there are exactly twice as many even divisors as odd divisors. This might suggest trying to split the divisors of $4k$ into *three* groups, one of which is the odd divisors, and showing a "1-to-1-to-1" correspondence between the three groups. Then each group will be equal size, and the odd divisors will be exactly one-third of all the divisors, whereas the even divisors, which make up the other two groups, will be the other two-thirds of the divisors.

A little experimentation will indicate how we should form the three groups. We notice that every divisor of $4k$ contains either 0, 1, or 2 powers of 2 as a factor (since $4k = 2^2 \cdot k$, and k is odd). So those are our three groups: odd divisors (which contain no power of 2), divisors that are multiples of 2 but not of 4 (which contain one power of 2), and divisors that are multiples of 4 (which contain two powers of 2).

Let's draw a chart for our initial two examples $k = 5$ and $k = 15$:

k	Odd	Divisors of $4k$	
		Multiples of 2, not 4	Multiples of 4
5	$1, 5$	$2, 10$	$4, 20$
15	$1, 3, 5, 15$	$2, 6, 10, 30$	$4, 12, 20, 60$

Now it's clear what the 1-to-1-to-1 correspondence is. If m is an odd divisor of $4k$, then $2m$ is the corresponding divisor of $4k$ that's a multiple of 2 but not of 4, and $4m$ is the corresponding divisor of $4k$ that's a multiple of 4. And it's also clear that all of these correspondences are reversible: for example, if we have a divisor of $4k$ that's a multiple of 4, then we can divide it by 4 to get an odd divisor of $4k$.

Therefore, not only have we proven the original problem statement, we've in fact established something stronger: the odd divisors of $4k$ make up exactly one-third of the total number of divisors of $4k$. (If you know some number theory, try to prove this using number-theory techniques as well.) □

In the next problem, we're trying to determine not that two sets are of equal size, but which of two sets is the larger. Although the next problem also has algebraic solutions, think about how we might use a correspondence to solve it.

Problem 4.15: The tennis club has 100 members, and Stephanie is the president. Which is larger: the number of ways to choose a 10-person advisory committee (which does not contain Stephanie) to advise Stephanie on what new rackets to buy, or the number of 11-person committees (which may include Stephanie) to plan an upcoming tournament? *Note: you should be able to do this problem without a calculator!*

Solution for Problem 4.15: Both of these sets are relatively easy to count. The racket-advisory committee needs 10 members from a set of 99, so there are $\binom{99}{10}$ ways to form the committee. The tournament-planning committee needs 11 members from a set of 100, so there are $\binom{100}{11}$ ways to form the committee. So which is larger, $\binom{99}{10}$ or $\binom{100}{11}$?

Since we know that Stephanie is never on the 10-person committee, we can always add her and get an 11-person committee. Similarly, given an 11-person committee with Stephanie on it, we can always remove her and get a 10-person committee without her. Therefore, we have a 1-1 correspondence

$$\{\text{10-person committees without Stephanie}\} \leftrightarrow \{\text{11-person committees with Stephanie}\}$$

However, the number of 11-person committees with Stephanie is certainly less than the total number of 11-person committees (with no restriction on the membership). Therefore, the number of 11-person committees is larger than the number of 10-person committees without Stephanie.

As we mentioned earlier, there are algebraic solutions to this as well. We can use Pascal's identity in reverse:

$$\binom{100}{11} - \binom{99}{10} = \binom{99}{11} > 0,$$

or we can write out the definitions of the binomial coefficients:

$$\binom{100}{11} = \frac{100!}{11!89!} = \frac{100}{11} \cdot \frac{99!}{10!89!} = \frac{100}{11}\binom{99}{10} > \binom{99}{10}.$$

However, the 1-1 correspondence solution has the slight advantage that we perhaps see more clearly *why* the statement is true. □

 **Exercises**

4.5.1 Annie the Ant starts at the lattice point $(0,0)$ and each minute moves 1 space up or 1 space to the right (with equal probability). Benny the Beetle starts at $(5,7)$ and each minute moves 1 space down or 1 space to the left (with equal probability). What is the probability that they meet? *(Source: AMC)*

4.5.2 Compute $\displaystyle\sum_{n_{60}=0}^{2} \sum_{n_{59}=0}^{n_{60}} \cdots \sum_{n_2=0}^{n_3} \sum_{n_1=0}^{n_2} \sum_{n_0=0}^{n_1} 1$. *(Source: HMMT)*

4.5.3 I roll 101 fair 6-sided dice.

(a) What is the probability that an even number of them come up even?

(b) Find the smallest integer s such that the probability that the sum is greater than s is less than $\frac{1}{2}$.

4.5.4★ Suppose n points are selected on the circumference of a circle and all $\binom{n}{2}$ chords connecting a pair of these points are drawn. Given that no three chords pass through the same point inside the circle, find the number of triangles that are formed by portions of the chords inside the circle and do not have any of the n points as vertices. **Hints:** 73

Sidenote: **More about 1-1 correspondences between infinite sets**

Given any set S, finite or infinite, we can show that S is not in 1-1 correspondence with its power set $\mathcal{P}(S)$. The proof uses a very clever argument called the **Cantor diagonalization argument**, due to the German mathematician Georg Cantor.

Suppose on the contrary that there is a map $f : S \to \mathcal{P}(S)$ establishing the 1-1 correspondence. Define the set

$$A = \{s \in S \mid s \notin f(s)\}.$$

In words, A contains each element s that is not contained in its corresponding subset $f(s)$ in $\mathcal{P}(S)$. The question then is: what element $a \in S$ corresponds to $A \in \mathcal{P}(S)$? In other words, what element $a \in S$ has $f(a) = A$? There must be such an element if f gives a 1-1 correspondence. But this leads to a paradox: if $a \in A$, then by definition $a \notin f(a) = A$, and if $a \notin A = f(a)$, then again by definition $a \in A$.

In other words, the set $\mathcal{P}(S)$ is always "bigger" than the set S, even if S is infinite. We'll continue this discussion in Challenge Problem 4.47 on page 117.

4.6 Clever 1-1 Correspondences

Problems

Problem 4.16: The game **Chomp** is played as follows: We start with a 5×7 array of cookies, as in the picture below. The players alternate turns, and on each turn, the player chooses any cookie remaining on the board and removes (or "chomps") that cookie along with all the cookies above and/or to the right of the selected cookie. For example, a possible first move is:

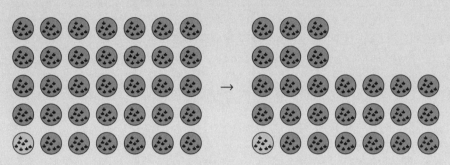

The cookie in the lower-left corner of the board is poison: the player who is forced to chomp it loses. We wish to determine how many positions of the board are possible in the game.

(a) In English, how can we describe a possible position of the game?

(b) How does your English description from (a) lead to a mathematical description that we can count?

(c) Count the number of possible positions.

Definition: A **partition** of a positive integer n is a decomposition of n into a sum of positive integers (not necessarily distinct), where we don't care about the order of the integers in the sum.

For example, the partitions of 3 are 3, $1 + 2$, and $1 + 1 + 1$. Note that $1 + 2$ and $2 + 1$ are considered the same partition, since we don't care about the order of the integers in the sum.

Problem 4.17: List all of the partitions of 4, 5, and 6.

Problem 4.18: Prove that the number of partitions of an integer n into 3 parts is equal to the number of partitions of the integer $2n$ into 3 parts, where each part is less than n.

(a) Try to prove it for $n = 3$, $n = 4$, $n = 5$, $n = 6$.

(b) Try to prove it for general n.

Problem 4.19:

(a) List the partitions of 8 into exactly 3 parts.

(b) List the partitions of 8 in which the largest term is 3.

(c) How many partitions are there in your lists from (a) and (b)?

(d) Prove the general result: the number of partitions of n into exactly r parts is equal to the number of partitions of n in which the largest term is r.

Problem 4.20:

(a) Steve flips 1 coin and Marissa flips 2 coins. What is the probability that Marissa flips more heads than Steve does?

(b) Steve flips 2 coins and Marissa flips 3 coins. What is the probability that Marissa flips more heads than Steve does?

(c) Steve flips 499 coins and Marissa flips 500 coins. What is the probability that Marissa flips more heads than Steve does?

Problem 4.16: The game **Chomp** is played as follows: We start with a 5×7 array of cookies, as in the picture below. The players alternate turns, and on each turn, the player chooses any cookie remaining on the board and removes (or "chomps") that cookie along with all the cookies above and/or to the right of the selected cookie. For example, a possible first move is:

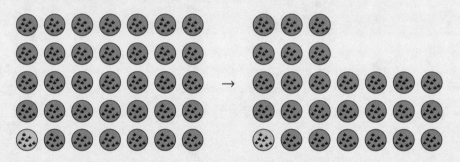

The cookie in the lower-left corner of the board is poison: the player who is forced to chomp it loses. How many possible positions of the board are possible in the game?

Solution for Problem 4.16: Constructive counting might get complicated, because the legal positions are a bit hard to describe. In particular, if we start to focus on possible *moves*, as opposed to possible *positions*, we get into really nasty casework, and we also have the problem that the same position can often be reached by a variety of different sequences of moves.

So not really knowing what else to do, let's focus on a particular legal position to try to get a handle on the problem.

> **Concept:** When unsure how to proceed, look at some examples and see if you can find a pattern.

For example, the following might be the position after each player has moved twice:

Extra! *C is for cookie, that's good enough for me.* – Cookie Monster

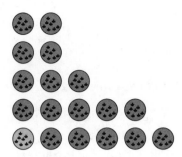

How can we describe such a position? What's the most "obvious" thing that we notice?

We definitely notice that the number of cookies in each row is nondecreasing as we go from top to bottom. More specifically, each row has at least as many cookies in it as the row immediately above.

This gives us our first 1-1 correspondence:

$$\{\text{Legal Chomp positions}\} \leftrightarrow \left\{ \begin{array}{l} \text{Arrangements in which the number of cookies in} \\ \text{each row is nondecreasing from top to bottom} \end{array} \right\}$$

Note that we haven't shown all the steps necessary to prove that this is a 1-1 correspondence, but by now you should be able to manage that on your own.

So fine, we have a 1-1 correspondence, but how do we count those arrangements?

Maybe it will help if we put the cookies in a grid, and include spaces for the cookies already chomped.

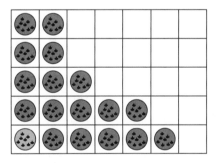

Hmmm...let's draw in the "border" of the cookies:

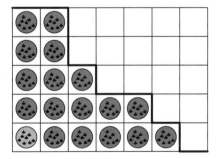

Aha! We see that the border of the cookies is a path from the top-left corner to the bottom-right corner, where the steps of the path are down or to the right. This is true for every arrangement of rows of cookies in nondecreasing numbers: as we move along a row, we're moving right, and as we go down from one row to the next, we're moving down. We know that we never have to move left or up since the rows are nondecreasing.

Therefore, we have the following 1-1 correspondences:

$$\{\text{Legal Chomp positions}\} \quad \leftrightarrow \quad \left\{\begin{array}{l}\text{Arrangements in which the number of cookies in each}\\ \text{row is nondecreasing from top to bottom}\end{array}\right\}$$

$$\leftrightarrow \quad \left\{\begin{array}{l}\text{Paths from the top-left corner to the bottom-right corner}\\ \text{of a } 5 \times 7 \text{ grid, where all the steps are down or to the}\\ \text{right}\end{array}\right\}$$

Fortunately, this last set is easy to count! We need to take 12 steps total, and must choose 5 of them to be down, so the number of paths is $\binom{12}{5} = 792$.

Thus there are 792 legal Chomp positions, including the start of the game (where all 35 cookies are present) and the end of the game (where all the cookies are gone and one player is poisoned). □

Sidenote: **So how do I win?**

We've just counted the number of legal positions in a Chomp game on a 5×7 grid. But we didn't really discuss the game itself—in particular, if both players play intelligently, which player has a winning strategy? Chomp is a very interesting game, in that it turns out that the first player always has a winning strategy (except on the trivial 1×1 board), but except for certain board sizes, nobody knows what that strategy is! We'll discuss this more in Chapter 13.

For the next few problems, we'll introduce the notion of a **partition**. This comes up quite often in counting and number theory problems.

Definition: A **partition** of a positive integer n is a decomposition of n into a sum of positive integers (not necessarily distinct), where we don't care about the order of the integers in the sum.

For example, the partitions of 3 are 3, $2 + 1$, and $1 + 1 + 1$. Note that $1 + 2$ and $2 + 1$ are considered the same partition, since we don't care about the order of the integers in the sum.

Problem 4.17: List all of the partitions of 4, 5, and 6.

Solution for Problem 4.17: The partitions of 4 are 4, $3 + 1$, $2 + 2$, $2 + 1 + 1$, and $1 + 1 + 1 + 1$. There are 5 different partitions of 4.

The partitions of 5 are 5, $4 + 1$, $3 + 2$, $3 + 1 + 1$, $2 + 2 + 1$, $2 + 1 + 1 + 1$, and $1 + 1 + 1 + 1 + 1$. There are 7 different partitions of 5.

The partitions of 6 are 6, $5 + 1$, $4 + 2$, $4 + 1 + 1$, $3 + 3$, $3 + 2 + 1$, $3 + 1 + 1 + 1$, $2 + 2 + 2$, $2 + 2 + 1 + 1$, $2 + 1 + 1 + 1 + 1$, and $1 + 1 + 1 + 1 + 1 + 1$. There are 11 different partitions of 6. □

Determining the number of partitions of an arbitrary positive integer n is difficult, and there is no nice formula for this. However, there are many nice relations among partitions; the next two problems will give two examples.

Problem 4.18: Prove that the number of partitions of an integer n into 3 parts is equal to the number of partitions of the integer $2n$ into 3 parts, where each part is less than n.

Solution for Problem 4.18: As we often do, to get a handle on this problem, let's look at a few small values of n and see if we notice anything interesting.

Concept: When trying to prove something for all positive integers n, first try a few small values of n and see if you notice a pattern.

n	Partitions of n into 3 parts	Partitions of $2n$ into 3 parts less than n
2	None	None
3	$1 + 1 + 1$	$2 + 2 + 2$
4	$2 + 1 + 1$	$3 + 3 + 2$
5	$3 + 1 + 1$	$4 + 4 + 2$
	$2 + 2 + 1$	$4 + 3 + 3$
6	$4 + 1 + 1$	$5 + 5 + 2$
	$3 + 2 + 1$	$5 + 4 + 3$
	$2 + 2 + 2$	$4 + 4 + 4$

Besides confirming that the result seems to be true—there are the same number of partitions in each column—do we notice anything?

Since we want to show that the two sets are equal, it makes sense to look for a 1-1 correspondence between them. So the question becomes: how does a partition of n with 3 parts correspond to a partition of $2n$ with 3 parts all less than n?

Let's experiment with the $n = 6$ case. We'd like to exhibit a 1-1 correspondence

$$\{4 + 1 + 1, 3 + 2 + 1, 2 + 2 + 2\} \leftrightarrow \{5 + 5 + 2, 5 + 4 + 3, 4 + 4 + 4\}$$

and do it in such a way that it generalizes for arbitrary n.

The observation that should jump out is that each set has one partition with three equal terms ($2 + 2 + 2$ and $4 + 4 + 4$), one partition with three distinct terms ($3 + 2 + 1$ and $5 + 4 + 3$), and one partition with two equal terms and a third different term ($4 + 1 + 1$ and $5 + 5 + 2$). It makes sense that these are going to end up as the pairs in our correspondence.

Further observation leads us to the insight that the corresponding terms in our above pairs all add to 6. This leads to our correspondence.

$$4 + 1 + 1 \quad \leftrightarrow \quad (6 - 4) + (6 - 1) + (6 - 1) = 2 + 5 + 5$$
$$3 + 2 + 1 \quad \leftrightarrow \quad (6 - 3) + (6 - 2) + (6 - 1) = 3 + 4 + 5$$
$$2 + 2 + 2 \quad \leftrightarrow \quad (6 - 2) + (6 - 2) + (6 - 2) = 4 + 4 + 4$$

So our general conjecture is that given a partition of n with 3 parts, we can subtract each part from n to get a partition of $2n$ with 3 parts all less than n, and vice versa. Now we have to prove it.

Suppose that $n = a + b + c$ is a partition of n. Then

$$(n - a) + (n - b) + (n - c) = 3n - (a + b + c) = 3n - n = 2n,$$

so $(n-a)+(n-b)+(n-c)$ is a partition of $2n$. Furthermore, since a, b, and c are all positive integers strictly between 0 and n, so are $n - a$, $n - b$, and $n - c$. Therefore, each partition of n with 3 parts corresponds to a partition of $2n$ with 3 parts, each less than n.

But we're not done! We need to show that the correspondence works in the other direction as well.

> **Important:** When proving a 1-1 correspondence, you need to demonstrate that the correspondence is indeed 1-1, meaning that it must be reversible: you need to be able to go back and forth between the two sets. If you only show one direction, you're not done yet!

Conversely, suppose that $2n = d + e + f$ is a partition of $2n$, with each of d, e, and f less than n. Then

$$(n - d) + (n - e) + (n - f) = 3n - (d + e + f) = 3n - 2n = n,$$

so $(n - d) + (n - e) + (n - f)$ is a partition of n. Furthermore, since d, e, and f are all positive integers strictly between 0 and n, so are $n - d$, $n - e$, and $n - f$. Therefore each partition of $2n$ with 3 parts all less than n corresponds to a partition of n with 3 parts.

Thus we have established a 1-1 correspondence between the sets

$$\{\text{Partitions of } n \text{ with 3 parts}\} \leftrightarrow \{\text{Partitions of } 2n \text{ with 3 parts all less than } n\},$$

and hence these sets have an equal number of elements. □

> **Problem 4.19:** Prove that the number of partitions of n into exactly r parts is equal to the number of partitions of n in which the largest term is r, for all positive integers $1 \le r \le n$.

Solution for Problem 4.19: In this problem, it's a bit trickier to list examples, since we have to pick values for both n and r. So let's just pick one pair of values for n and r; our goal is to pick values small enough so that they're easy to work with, but big enough so that any pattern (if it exists) will hopefully emerge.

Let's pick $n = 8$ and $r = 3$, and list the partitions in each set.

Partitions of 8 into 3 parts	Partitions of 8 with largest term 3
$6 + 1 + 1$	$3 + 3 + 2$
$5 + 2 + 1$	$3 + 3 + 1 + 1$
$4 + 3 + 1$	$3 + 2 + 2 + 1$
$4 + 2 + 2$	$3 + 2 + 1 + 1 + 1$
$3 + 3 + 2$	$3 + 1 + 1 + 1 + 1 + 1$

It's a little more difficult to see a pattern here than it was in the last problem.

One pattern that you might notice is that list of the largest terms $(6, 5, 4, 4, 3)$ in each of the partitions in the left column matches the list of the number of terms $(3, 4, 4, 5, 6)$ in each of the partitions in the right column. This is sufficiently interesting that we might pick different values of n and/or r and see if this still holds. Let's try another example, with $n = 9$ and $r = 3$.

Partitions of 9 into 3 parts	Partitions of 9 with largest term 3
$7 + 1 + 1$	$3 + 3 + 3$
$6 + 2 + 1$	$3 + 3 + 2 + 1$
$5 + 3 + 1$	$3 + 3 + 1 + 1 + 1$
$5 + 2 + 2$	$3 + 2 + 2 + 2$
$4 + 4 + 1$	$3 + 2 + 2 + 1 + 1$
$4 + 3 + 2$	$3 + 2 + 1 + 1 + 1 + 1$
$3 + 3 + 3$	$3 + 1 + 1 + 1 + 1 + 1 + 1$

It still works: the largest terms $(7,6,5,5,4,4,3)$ of the partitions in the left column match the number of terms in the partitions in the right column.

This gives us an important clue towards how to build our 1-1 correspondence. In fact, we can build an algorithm that describes the correspondence. It's easiest to describe with an example, so let's show the correspondence between

$$5 + 3 + 1 \leftrightarrow 3 + 2 + 2 + 1 + 1.$$

Start with the 5 in the left partition, and write it as $1 + 1 + 1 + 1 + 1$. Now take the 3 and add 1 to each of the first 3 terms, giving $2 + 2 + 2 + 1 + 1$. Finally, take the 1 and add 1 to the first term, giving $3 + 2 + 2 + 1 + 1$.

To go the other way, start with $3 + 2 + 2 + 1 + 1$, and take the number of terms (5) as the first term in our new partition. Subtract 1 from each term: $2 + 1 + 1$, and take the number of terms remaining (3) as the next term in our new partition. Subtract 1 again from each term: 1, and take the number of terms remaining (1) as the next term in our new partition. Subtract 1 again, and the original partition is gone. The numbers we've extracted along the way gives us our new partition: $5 + 3 + 1$.

Let's run through this algorithm again, with a different choice of partition:

Old	New	Old	New
Given: partition with 3 parts		Given: partition with largest part 3	
$4 + 3 + 2$		$3 + 3 + 2 + 1$	
Start with 4 1's		4 terms; subtract 1 from each term	
$3 + 2$	$1 + 1 + 1 + 1$	$2 + 2 + 1$	4
Add 1 to first 3 terms		3 terms; subtract 1 from each term	
2	$2 + 2 + 2 + 1$	$1 + 1$	$4 + 3$
Add 1 to first 2 terms		2 terms; subtract 1 from each term	
	$3 + 3 + 2 + 1$		$4 + 3 + 2$

This is a bit wordy. Fortunately, we can more easily describe this correspondence using a tool called a **Ferrers diagram**.

> **Concept:** Often, we can better understand complicated concepts by using a suitable
> diagram.

We can represent a partition with r parts as r rows of dots, in which each row contains a number of dots equal to the number in the partition. Similarly, we can represent a partition whose largest element is r as r rows of dots, in which each *column* has a number of dots equal to a number in the partition.

$$
\begin{array}{ccccccc}
4 & \to & \bullet & \bullet & \bullet & \bullet \\
3 & \to & \bullet & \bullet & \bullet & \uparrow \\
2 & \to & \bullet & \bullet & \uparrow & 1 \\
  &   & \uparrow & \uparrow & 2 & \\
  &   & 3 & 3 & &
\end{array}
$$

Thus, we have 1-1 correspondences

$$
\{\text{Partitions of } n \text{ with } r \text{ parts}\} \overset{\text{Read as rows}}{\leftrightarrow} \{\text{Ferrers diagrams with } r \text{ rows}\} \overset{\text{Read as columns}}{\leftrightarrow} \left\{\begin{array}{l}\text{Partitions of } n \\ \text{whose largest} \\ \text{element is } r\end{array}\right\}
$$

Hence the numbers of partitions in each set are equal. □

We will explore partitions further in Chapter 14.

> **Problem 4.20:** Steve flips 499 coins and Marissa flips 500 coins. What is the probability that Marissa flips more heads than Steve does?

Solution for Problem 4.20:   As usual, we can look at some small examples to try to get a feel for the problem.

If Steve has 1 coin and Marissa has 2 coins, then since there are 3 flips total, there are $2^3 = 8$ possible outcomes. We can easily list the outcomes in which Marissa "wins" by flipping more heads than Steve:

$$(T,HT), (T,TH), (T,HH), (H,HH).$$

In each pair, the first entry is Steve's flip and the second entry is Marissa's two flips. Since Marissa wins 4 out of 8 times, the probability of her winning is $\frac{4}{8} = \frac{1}{2}$.

If Steve has 2 coins and Marissa has 3 coins, there are now $2^5 = 32$ possible outcomes. We could try to list them all out, but instead we can do a bit of casework.

If Steve flips 0 heads (which he can do in 1 way), then Marissa wins as long as she avoids TTT. This gives her 7 ways to win.

If Steve flips 1 head (which he can do in 2 ways), then Marissa wins if she flips 2 or 3 heads, which she can do in 4 ways. This gives her 8 ways to win.

If Steve flips 2 heads (which he can do in 1 way), then Marissa must flip HHH to win. This gives her 1 way to win.

So Marissa has $7 + 8 + 1 = 16$ successful outcomes, and the probability of her winning is $\frac{16}{32} = \frac{1}{2}$.

This is probably not a coincidence, so we now conjecture that as long as Marissa has one more coin than Steve, her probability of winning is $\frac{1}{2}$. This suggests looking for a 1-1 correspondence between the winning outcomes and the non-winning outcomes. (This is similar to the method we used in Problem 4.13.)

The natural way to get a 1-1 correspondence between coin-flipping outcomes is to reverse the outcome of every coin flip (in other words, change every H to T and every T to H). This does indeed give a 1-1 correspondence between winning and non-winning outcomes, as follows: Suppose we have a winning outcome where Steve has a heads and Marissa has b heads, with $a < b$. When we reverse heads and tails, Steve now has $499 - a$ heads and Marissa has $500 - b$ heads. We would like to show that this is a non-winning outcome for Marissa, meaning that Steve has as least as many heads as Marissa does. We see this as follows:

$$
\begin{aligned}
499 - a \geq 500 - b \quad &\Leftrightarrow \quad -a \geq 1 - b, \\
&\Leftrightarrow \quad a \leq b - 1, \\
&\Leftrightarrow \quad a < b,
\end{aligned}
$$

where the last equivalence follows since a and b are integers. This works the other way as well: if we start with an outcome in which Steve has as least as many heads as Marissa, and reverse all the heads and tails, then we have an outcome in which Marissa has more heads than Steve.

So winning and non-winning outcomes are in 1-1 correspondence, hence each make up half of the total outcomes, and Marissa has a $\frac{1}{2}$ chance of winning. $\square$

Exercises

4.6.1 Use an argument similar to that in Problem 4.16 to find the number of legal positions in a game of Chomp played on an $m \times n$ board, where m and n are positive integers.

4.6.2 Show that the number of partitions of n into parts of even size equals the number of partitions of n into parts that occur an even number of times. **Hints:** 236

4.6.3 A class of 10 students took a math test. Each problem was solved by exactly 7 of the students. If the first nine students each solved 4 problems, how many problems did the tenth student solve? *(Source: HMMT)* **Hints:** 76

4.6.4★ In the diagram at right, the side length of the large equilateral triangle is 3. There are 15 parallelograms of various sizes that are formed by segments in the grid (one example is shown in bold). Find a formula for the number of parallelograms that are formed by an analogous triangular grid with side length n. *(Source: Canada)* **Hints:** 266, 338, 342

4.6.5 Let $\lfloor x \rfloor$ denote the greatest integer less than or equal to x. (For example, $\lfloor 3 \rfloor = 3$ and $\lfloor 6.73 \rfloor = 6$.) Prove that for any positive integer n,

$$
\left\lfloor \frac{n+1}{2} \right\rfloor + \left\lfloor \frac{n+2}{4} \right\rfloor + \left\lfloor \frac{n+4}{8} \right\rfloor + \left\lfloor \frac{n+8}{16} \right\rfloor + \cdots = n.
$$

Hints: 283, 14

4.6.6 Are there more partitions of 2006 into only even parts or into only odd parts?

4.6.7★ Let S be the set $\{1, 2, \ldots, n\}$. Let k be the number of subsets T of S such that the elements of T have an integer average. Prove that $n + k$ is even. *(Source: Putnam)* **Hints:** 112, 340

4.7 Summary

➤ The basic idea of constructive counting is to think about how you would construct the items that you're trying to count, while keeping track of the number of possibilities at each stage of the construction.

➤ Often, a big part of the problem is simply figuring out if a construction exists or not.

➤ Whenever we can match every element in set A to exactly one element in set B and vice versa, we have a 1-1 correspondence.

➤ Two finite sets can be placed into 1-1 correspondence if and only if they have the same number of elements.

➤ We generally use 1-1 correspondences to count a hard-to-count set, by placing the set in 1-1 correspondence with an easy-to-count set, which then necessarily has the same number of elements as our original set.

➤ To prove that two sets A and B are in 1-1 correspondence, we have to show that every element of A corresponds to exactly one element of B, and that every element of B corresponds to exactly one element of A. Showing the correspondence in only one direction is not sufficient.

➤ When you suspect that a set that you're trying to count has the same number of elements as a set that's easy to count, look for a 1-1 correspondence between the set you have and the "easy" set.

➤ A **partition** of a positive integer n is a decomposition of n into a sum of positive integers (not necessarily distinct), where we don't care about the order of the integers in the sum. Many problems involving partitions can be solved using Ferrers diagrams.

Here are some general problem-solving strategies that are applicable to constructive counting:

> **Concept:** If doing the entire problem at once seems too complicated to handle, try breaking it up into manageable parts.

> **Concept:** Sometimes, it's easier to think about how you would construct the items that you *don't* want to count, and subtract the count of these items from the total.

> **Concept:** When stuck on a "find a general formula"-type problem, experiment with some small values, and look for a pattern.

> **Concept:** Often, we can better understand complicated concepts by using a suitable diagram.

> **Concept:** Before diving into a problem, make sure that you:
>
> - Read the problem carefully.
>
> - Understand what the problem is asking.
>
> - If necessary, work through an example to get a feel for the problem.

REVIEW PROBLEMS

4.21 For how many ordered triples (a, b, c) of positive integers less than 10 is the product abc divisible by 20? *(Source: HMMT)*

4.22 How many nonempty subsets of $\{1, 2, 3, \ldots, 12\}$ have the property that the sum of the largest element and the smallest element is 13? *(Source: HMMT)*

4.23 Each face of a cube is painted either red or blue, each with probability $\frac{1}{2}$. The color of each face is determined independently. What is the probability that the painted cube can be placed on a horizontal surface so that the four vertical faces are all the same color? *(Source: AMC)*

4.24 What is the probability that a randomly chosen divisor of 30^{39} is a multiple of 30^{29}?

4.25 Forty slips are placed into a hat, each bearing a number 1, 2, 3, 4, 5, 6, 7, 8, 9, or 10, with each number entered on four slips. Four slips are drawn from the hat at random without replacement. What is the probability that two of the slips bear a number a and the other two bear a number $b \neq a$? *(Source: AMC)*

4.26 A hotel packed a breakfast for each of three guests. Each breakfast should have consisted of three types of rolls, one each of nut, cheese, and fruit rolls. The preparer wrapped each of the nine rolls, and, once they were wrapped, the rolls were indistinguishable from one another. She then randomly put three rolls in a bag for each of the guests. Find the probability that each guest got one roll of each type. *(Source: AIME)*

4.27 Suppose we have a 10×10 grid of points, such as the set of lattice points with both coordinates between 0 and 9 (inclusive). How many squares (with sides parallel to the sides of the grid) can be formed by connecting 4 of these points?

4.28 Consider the sequence $1, 4, 5, 16, 17, 20, 21, 64, 65, \ldots$, which is formed by including only positive integers that can be expressed as the sum of distinct powers of 4. What is the 50th term in this sequence?

4.29 Show that, for any nonempty finite set S, the number of subsets of S with an even number of elements is equal to the number of subsets of S with an odd number of elements.

4.30 Show that the number of partitions of an integer n into at most r parts is equal to the number of partitions of n into parts of at most r.

4.31 Let $f(n,k)$ denote the number of partitions of n into k parts. Prove that

$$f(n,k) = f(n-1, k-1) + f(n-k, k).$$

4.32 On the Cartesian plane, Johnny wants to travel from $(0,0)$ to $(5,1)$, and he wants to pass through all twelve lattice points (x, y) such that $0 \leq x \leq 5$ and $0 \leq y \leq 1$. On each step, Johnny can go from one point to any other point via the straight line segment connecting the two points. How many paths are there from $(0,0)$ to $(5,1)$, passing through all 12 points, such that the path never crosses itself? One such path is shown below. *(Source: HMMT)*

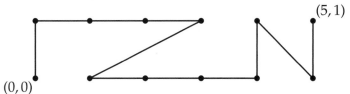

Challenge Problems

4.33 Eight knights are randomly placed on a chessboard (not necessarily on distinct squares). A knight on a given square attacks all the squares that can be reached by moving either (1) two squares up or down followed by one square left or right, or (2) two squares left or right followed by one square up or down. Find the probability that every square, occupied or not, is attacked by some knight. *(Source: HMMT)* **Hints:** 331

4.34 A true-false test has 10 questions. Suppose that if you answer any five questions "true" and the remaining five questions "false," then your score is guaranteed to be at least four. How many answer keys are there for which this is true? *(Source: HMMT)* **Hints:** 34, 176

4.35 In how many ways can 4 purple balls and 4 green balls be placed into a 4×4 grid such that every row and column contains one purple ball and one green ball? Only one ball may be placed in each box, and rotations and reflections of a single configuration are considered different. *(Source: HMMT)* **Hints:** 109, 198

4.36 How many 4×4 matrices whose entries are each 1 or -1 are such that the sum of the entries in each row is 0 and the sum of the entries in each column is 0? *(Source: AIME)* **Hints:** 94, 24

4.37 A *lattice point* is a point (x, y) such that x and y are both integers. Suppose we color each of the lattice points within the square $0 \leq x, y \leq 10$ either black or white. Find the number of colorings in which each of the 100 unit squares (bounded by the colored lattice points) has exactly two white vertices. **Hints:** 107

4.38 10 points in the plane are given, with no 3 collinear. 4 distinct segments joining pairs of these points are chosen at random, all such segments being equally likely. Find the probability that some 3 of the segments form a triangle whose vertices are among the 10 given points. *(Source: AIME)*

4.39 Consider the set $S = \{1, 2, 3, 4, 5, \ldots, 100\}$. How many subsets of this set with 2 or more elements satisfy:

(i) the terms of the subset form an arithmetic sequence, and

(ii) we cannot include another element from S with this subset to form an even longer arithmetic sequence?

Hints: 311, 47

4.40 Define the *alternating sum* of a set of positive integers as follows: List the elements of the set in decreasing order. Take the first number in the list, subtract the second, add the third, subtract the fourth, and so on. For example, the alternating sum of the set $\{3, 5, 11, 7\}$ is $11 - 7 + 5 - 3 = 6$. Find the sum of the alternating sums of all of the subsets of $\{1, 2, 3, 4, 5, 6, 7, 8, 9, 10\}$. *(Source: AIME)* **Hints:** 133, 270

4.41 In a classroom, 34 students are seated in 5 rows of 7 chairs. The place at the center of the room is unoccupied. A teacher decides to reassign the seats such that each student will occupy a chair adjacent to his/her present one (i.e. move one desk forward, back, left, or right). In how many ways can this reassignment be made? *(Source: HMMT)* **Hints:** 187, 164

4.42 In how many ways can we place a positive number of rooks on an 8×8 chessboard, such that no two lie in the same row or the same column, and so that none of the rooks lies to the left of and below another rook? *(Source: HMMT)* **Hints:** 323

4.43 Eight congruent equilateral triangles, each of a different color, are used to construct a regular octahedron. How many distinguishable ways are there to construct the octahedron? (Two colored octahedrons are distinguishable if neither can be rotated to look just like the other.) *(Source: AMC)* **Hints:** 307, 181

4.44★ Is there a 2000-element subset of the set $\{1, 2, 3, \ldots, 3000\}$ such that no element in the subset is exactly double another element of the subset? **Hints:** 113, 155

4.45★ Find the number of ways to color each square of a 2007×2007 square grid either black or white such that each row and each column has an even number of black squares. *(Source: Mandelbrot)* **Hints:** 89, 151

4.46★ Three numbers, a_1, a_2, a_3, are drawn randomly, without replacement, from the set $\{1, 2, 3, \ldots, 1000\}$. Three other numbers, b_1, b_2, b_3, are then drawn randomly, without replacement, from the remaining set of 997 numbers. What is the probability that, after suitable rotation, a brick of dimensions $a_1 \times a_2 \times a_3$ can be enclosed in a box of dimension $b_1 \times b_2 \times b_3$, with the sides of the brick parallel to the sides of the box? *(Source: AIME)* **Hints:** 297, 300

4.47★ This problem continues our discussion (from page 104) of 1-1 correspondences between infinite sets. Recall that any infinite set that can be placed into 1-1 correspondence with the set of all positive integers is called **countable**, and any infinite set that cannot is called **uncountable**.

(a) Prove that $\mathbb{Z}$ (the set of all integers) is countable. **Hints:** 26

(b)★ Prove that $\mathbb{Q}$ (the set of rational numbers) is countable. **Hints:** 223

(c)★ Prove that $\mathbb{R}$ (the set of real numbers) is uncountable. (You may use the fact, discussed on page 104, that no set S is can be placed in 1-1 correspondence with its power set $\mathcal{P}(S)$.) **Hints:** 101, 144

Discourse is fleeting, but junk mail is forever. – Joe Bob Briggs

CHAPTER **5**

The Pigeonhole Principle

5.1 Introduction

Once again, we have a simple concept with a fancy name: the **Pigeonhole Principle**. It actually has a fancier name, the **Dirichlet Principle**, after the German mathematician Johann Peter Gustav Lejeune Dirichlet, and an even fancier name, the **Dirichlet Box Principle**, but it is usually referred to (in the United States) as the Pigeonhole Principle.

Imagine a mailroom with a bunch of small mailboxes (sometimes called *pigeonholes*). If we have, say, 100 mailboxes in the room, and 101 letters arrived today, then we know that at least 2 of the letters must go in the same pigeonhole. More generally, if there are more letters than pigeonholes, then some pigeonhole must get more than one letter. That's really all there is to it!

Although the Pigeonhole Principle may seem "obvious," it is extremely useful in solving a wide variety of counting problems.

5.2 It's Just Common Sense!

Problem 5.1: Suppose that I have 5 balls and 4 boxes. Prove that, no matter how I place the balls into boxes, at least one box must contain more than 1 ball.

The Pigeonhole Principle couldn't be more simple. It merely states that if we have more objects than slots to place them in, then at least one slot must contain more than one object.

Here's a basic example:

Problem 5.1: Suppose that I have 5 balls and 4 boxes. Prove that, no matter how I place the balls into boxes, at least one box must contain more than 1 ball.

Solution for Problem 5.1: We can prove this very easily by contradiction. Suppose that the 4 boxes have $a, b, c,$ and d balls in them, respectively. If each box has no more than 1 ball, then $a \le 1, b \le 1, c \le 1,$ and $d \le 1$. Therefore, $a + b + c + d \le 1 + 1 + 1 + 1 = 4$. But we know that $a + b + c + d = 5$, so $5 \le 4$. Obviously this is a contradiction, so not all of the boxes can have at most one ball. □

We can generalize this simple example:

Important: The **Pigeonhole Principle**: If I place k balls into n boxes, where $k > n$, then at least one box must contain more than 1 ball.

The proof is essentially the same as in the specific example from Problem 5.1—we will leave it as an exercise.

Don't let all this detail obscure the fact that the Pigeonhole Principle is just common sense.

Concept: The Pigeonhole Principle is just common sense. If you have more items than boxes to place them in, then at least one of the boxes must contain more than one item.

Exercises

5.2.1 Prove the general statement of the Pigeonhole Principle: If I place n balls into k boxes, with $n > k$, then at least one box must contain more than 1 ball.

5.2.2 Prove that if I flip 3 coins, I must get at least two heads or at least two tails.

5.2.3 How many cards do we have to draw out of a standard 52-card deck in order to guarantee that we draw at least one pair (two cards of matching rank)?

5.3 Basic Pigeonhole Problems

Problems

Problem 5.2: I have a drawer with a large number of white, brown, and black socks. How many socks do I have to pull out of the drawer in order to ensure that I get a matching pair?

Extra! *I have reached an age where if someone tells me to wear socks, I don't have to.* – Albert Einstein

Problem 5.3: We wish to prove that given any 6 integers, there are 2 of them whose difference is divisible by 5.

(a) Since we want to prove that at least 2 of 6 items share a property, how many "boxes" will we need in order to apply the Pigeonhole Principle?

(b) Given that we are considering divisibility by 5, what would be a natural choice for our boxes?

(c) Finish the problem.

Problem 5.4: Given a unit square and 5 points in the square, we wish to prove that there must exist a pair of these points that are at most $\sqrt{2}/2$ distance apart.

(a) We have 5 points and we wish to show that a pair of them have some property. How many "boxes" will we need to apply the Pigeonhole Principle?

(b) How can we choose our boxes?

(c) How does our choice of boxes in (b) force two points in the same box to be at most $\sqrt{2}/2$ distance apart?

Problem 5.5: A group of 25 people are at a party. Over the course of the party, some of the attendees shake hands with each other. Prove that, at the end of the party, there exist two guests that have shaken hands with the same number of people.

Remember, the general idea of the Pigeonhole Principle is that we have more balls than boxes, so some box must contain more than one ball. Often the main difficulty is trying to decide what the "balls" and the "boxes" are.

Problem 5.2: I have a drawer with a large number of white, brown, and black socks. How many socks do I have to pull out of the drawer in order to ensure that I get a matching pair?

Solution for Problem 5.2: Since we want two socks of the same color, it makes sense to think of "colors" as our "boxes" for this problem. For example, we can imagine having 3 boxes labeled "White," "Brown," and "Black," and as we pull socks out of the drawer, we put them into the appropriate box.

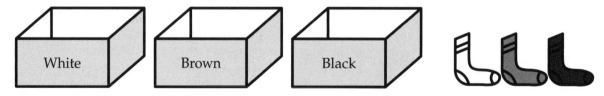

Since we have 3 colors—white, brown, and black—we need 4 "balls," or socks, to put into these boxes in order to ensure that one box has at least two. For example, if we only pull out 3 socks, we might pull out one of each color (as shown above), and thus would have 1 sock in each box. If we pull out a 4th sock, it will have to go into one of the boxes, and then that box will have 2 socks (a matching pair). Therefore, we need to pull 4 socks out of the drawer. □

Problem 5.3: Prove that given any 6 integers, there are 2 of them whose difference is divisible by 5.

Solution for Problem 5.3: Suppose x and y are integers whose difference is divisible by 5. This means that

$$\frac{x-y}{5} = \frac{x}{5} - \frac{y}{5}$$

is an integer. This only happens when x and y have the same remainder upon division by 5. More formal ways to write this are:

$$5 \mid (x-y) \quad \text{or} \quad x \equiv y \pmod 5.$$

Now our course of action is more clear. Our "boxes" for this problem are remainders when dividing by 5. Since we have 6 integers, we may conclude, by the Pigeonhole Principle, that at least two of them must have the same remainder upon division by 5 (since there are only 5 possible remainders: 0, 1, 2, 3, or 4). The difference of these two integers will be divisible by 5. □

> **Problem 5.4:** Given a unit square and 5 points in the square, show that there must exist a pair of these points that are at most $\sqrt{2}/2$ distance apart.

Solution for Problem 5.4: We're looking for a pair of points out of 5 that satisfy a certain condition, so this suggests the Pigeonhole Principle, in which the points are placed into 4 boxes, guaranteeing that at least two of the points are in the same box. We also want the condition that "two points are in the same box" to imply that "two points are at most $\sqrt{2}/2$ distance apart." Furthermore, we think of what distance $\sqrt{2}/2$ represents, and one common distance that should come to mind is the diagonal of a square of side length $\frac{1}{2}$. We have all the pieces—it remains to assemble them into a solution.

Divide the unit square into four smaller squares of side length $\frac{1}{2}$ as shown in the diagram to the right. By the Pigeonhole Principle, given any 5 points, at least two of them must be in the same small square. These two points then can be apart by at most the diagonal of a small square, which is $\sqrt{2}/2$. □

> **Concept:** Whenever we have to show that "a pair" of objects or "at least 2" objects
> share some property, that's our cue to think about the Pigeonhole Principle.

> **Problem 5.5:** A group of 25 people are at a party. Over the course of the party, some of the attendees shake hands with each other. Prove that, at the end of the party, there exist two guests that have shaken hands with the same number of people.

Solution for Problem 5.5: Since we want to prove that two guests have shaken hands with the same number of people, it makes sense to think of "# of handshakes" as our boxes. Since there are 25 people at the party, each person could shake hands with $0, 1, 2, \ldots, 23$, or 24 of the other people.

Uh-oh. There are 25 numbers in the list $0, 1, \ldots, 23, 24$, and there are 25 people. We need *more* people than numbers in order to apply the Pigeonhole Principle. Are we doomed?

No, we're not doomed, because we can't simultaneously have someone who shook no hands (and thus would be in the 0 box) and someone who shook everybody's hand (and thus would be in the 24 box). So there are really only 24 boxes, and we can apply the Pigeonhole Principle to prove that there must be two people who have shaken the same number of hands. □

> **Concept:** Sometimes, before applying the Pigeonhole Principle, we have to do a little work to reduce the number of boxes.

5.3.1 Prove that given any 11 integers, there will be at least two with the same units digit.

5.3.2

(a) What is the maximum number of rooks that can be placed on an 8×8 chessboard such that each row and column contains no more than 1 rook?

(b)★ What is the maximum number of bishops that can be placed on an 8×8 chessboard such that each diagonal contains no more than 1 bishop? **Hints:** 265

5.3.3 Prove that among any 8 positive integers that sum to 20, there must be a group of them that sums to 4.

5.3.4 What is the size of the largest subset S of $\{1, 2, 3, \ldots, 50\}$ such that no pair of distinct elements of S has a sum divisible by 7? *(Source: AMC)*

5.3.5 Prove the general version of Problem 5.3: given any positive integer n, and any set of $n + 1$ integers, there are 2 of them whose difference is divisible by n.

5.3.6 A subset B of the set of integers from 1 to 100, inclusive, has the property that no two elements of B sum to 125. What is the maximum possible number of elements in B? *(Source: AMC)* **Hints:** 115, 274

5.4 More Advanced Pigeonhole Problems

> **Problem 5.6:** Suppose that I place 25 balls into 6 boxes. Prove that one of the boxes must contain at least 5 balls.

> **Problem 5.7:** Suppose that I place n items into k boxes. What is the largest number m such that I can be guaranteed that one of the boxes contains at least m items?

> **Problem 5.8:** There are 20 children in a small mountain town. Any two of them have a common grandfather, and each child has two distinct grandfathers.
> (a) How many grandfathers can there be in the town?
> (b) Prove that there are 14 children who have a common grandfather.
> *(Source: ToT)*

> **Problem 5.9:** Given any set of ten distinct 2-digit numbers, prove that there exist two disjoint subsets (of the 10 numbers) with the same sum. *(Source: IMO)*

> **Problem 5.10:** Aimee plays at least one game of chess a day for eight weeks, but she plays no more than 11 games in any 7-day period. Show that there is some period of consecutive days in which she plays exactly 23 games.

As we've seen, the Pigeonhole Principle can be used to show that more than 1 item share some quality. But what if we want to show that more than k items share some quality, where $k > 1$ is a positive integer? Fortunately, given the right conditions, we can do that too.

Here's a basic example:

> **Problem 5.6:** Suppose that I place 25 balls into 6 boxes. Prove that one of the boxes must contain at least 5 balls.

Solution for Problem 5.6: As with the basic Pigeonhole Principle, we can think about this just using common sense. If we place 4 balls into each of the 6 boxes, that accounts for 24 balls; any other distribution of the first 24 balls will result in 5 balls in at least one box. But we still have one ball left over, so whichever box we place it in will have 5 balls.

Although this is good common-sense reasoning, it is not a formal proof, so let's formally prove the statement. Suppose that the boxes are numbered from 1 to 6, and that there are $a_1, a_2, \ldots, a_6$ balls in the respective boxes. Since there are 25 total balls, we must have

$$a_1 + a_2 + \cdots + a_6 = 25.$$

If every box contains at most 4 balls, then $a_1 \leq 4, a_2 \leq 4, \ldots, a_6 \leq 4$. Therefore

$$25 = a_1 + a_2 + a_3 + a_4 + a_5 + a_6 \leq 4 + 4 + 4 + 4 + 4 + 4 = 24,$$

giving $25 \leq 24$, a contradiction. So some box must contain at least 5 balls. $\square$

This is an example of a more general version of the Pigeonhole Principle. It basically says that some box is guaranteed to have at least a certain number of balls. Exactly how many balls are guaranteed in general? We'll work that out in the next problem.

> **Problem 5.7:** Suppose that I place n items into k boxes. What is the largest number m such that I can be guaranteed that one of the boxes contains at least m items?

Solution for Problem 5.7: We want to take our reasoning from Problem 5.6 and apply it to the general case. So what's wrong with the following argument?

> **Bogus Solution:** We can put $\dfrac{n-1}{k}$ balls into each of the k boxes. This accounts for
> $k\left(\dfrac{n-1}{k}\right) = n - 1$ of the balls. There is 1 ball left over, which must
> go in some box, and thus some box must contain at least $\dfrac{n-1}{k} + 1$
> balls.

Seems reasonable, and it works with $n = 25$ and $k = 6$ as in Problem 5.6. So why is this not quite correct?

The problem is that $\dfrac{n-1}{k}$ might not be an integer. The best we can do in our initial step is to put $\left\lfloor \dfrac{n-1}{k} \right\rfloor$ balls in each box, where $\lfloor x \rfloor$ is the "floor" function, meaning the greatest integer less than or equal to x. Then our leftover ball(s) ensure that at least one box will have $\left\lfloor \dfrac{n-1}{k} \right\rfloor + 1$ balls.

This is the best that we can do, as follows: Suppose that $n \geq k$, and write $n - 1 = qk + r$, where q is a positive integer and $0 \leq r < k$ is the remainder of $(n - 1)/k$. Then we can place $\left\lfloor \dfrac{n-1}{k} \right\rfloor + 1$ balls in the first $r + 1$ boxes and $\left\lfloor \dfrac{n-1}{k} \right\rfloor$ balls in the remaining $k - (r + 1)$ boxes. Noting that $\left\lfloor \dfrac{n-1}{k} \right\rfloor = q$, we see that this account for all n balls:

$$\begin{aligned}(r + 1)(q + 1) + (k - (r + 1))q &= (r + 1 + k - (r + 1))q + (r + 1) \\ &= kq + r + 1 \\ &= (n - 1) + 1 = n.\end{aligned}$$

Finally, if $n < k$, then there's nothing to prove: we can only guarantee at least 1 box in some box. $\square$

> **Important:** The generalized Pigeonhole Principle: If we place n balls into k boxes, then at least one box must contain at least $\left\lfloor \dfrac{n-1}{k} \right\rfloor + 1$ balls.

> **Concept:** Don't memorize this formula! It's better to use your common sense when applying the Pigeonhole Principle than to memorize a somewhat obscure formula.

> **Problem 5.8:** There are 20 children in a small mountain town. Any two of them have a common grandfather, and each child has two distinct grandfathers. Prove that there are 14 children who have a common grandfather. *(Source: ToT)*

Solution for Problem 5.8: Since we're trying to prove that a group of things all share some property in common—specifically, that 14 children have a common grandfather—this problem is a good candidate for the Pigeonhole Principle. But how can we use it? It seems clear that the children are our "balls" and the grandfathers are our "boxes," but we only have 20 children. How are we going to be able to get 14 of them in the same "box"?

The first thing to observe is that we actually have 40 "balls," since each child has two distinct grandfathers. In other words, each child is going to be placed into two boxes, one per grandfather. So we need to be able to use Pigeonhole in a way such that placing 40 balls into the boxes will result in at least 14 balls in one box.

Now we look at the numbers. We might notice that if we only had 39 balls, we could get exactly 13 balls in each of 3 boxes. We could also look at our formula from above: placing 40 balls into k boxes will guarantee that at least one box contains

$$\left\lfloor \frac{39}{k} \right\rfloor + 1$$

balls; if we want this to equal 14, we need $\left\lfloor \frac{39}{k} \right\rfloor = 13$, which suggests $k = 3$. So if there were only 3 grandfathers in the entire town, then there'd be 3 "boxes" (grandfathers) for 40 "balls" (grandchildren), and hence one of the boxes would have 14 balls, and we'd be done.

So we're left to try to prove the (perhaps surprising) claim that there are only 3 (or fewer) grandfathers. Let's see how we might do this. Suppose the 1st kid has grandfathers A and B. The 2nd kid has to have a grandfather in common with the 1st kid, so suppose it's A that they have in common (it doesn't matter; the argument works the same if we supposed it was B). So the 2nd kid has grandfathers A and C. Now there are two possibilities.

One possibility is that every kid might have grandfather A. If this is the case, then there might be more than 3 grandfathers in the town (every kid might have a unique 2nd grandfather), but that's no worry, because in this case, the problem becomes trivial: not only are there 14 kids with a common grandfather, but in fact all 20 kids have a common grandfather!

The other possibility is that there's some kid who doesn't have A as a grandfather. But since this kid must have a grandfather in common with each of the first two kids, we know that this new kid must have B and C as her grandfathers. At this point, we have kids with the following sets of grandfathers:

$$(A \text{ and } B) \qquad (A \text{ and } C) \qquad (B \text{ and } C)$$

Since every other kid must have a grandfather in common with all 3 of these children, we see that $\{A, B, C\}$ are the only possible grandfathers. If there were a 4th grandfather called D, and some child had grandfather D, then that child's other grandfather would have to be in common with all 3 of the kids listed above, but those 3 kids don't have a grandfather in common. Therefore, there are only 3 grandfathers, and we can use the Pigeonhole Principle as described earlier to complete the proof.

Here's a summary of our solution:

One possibility is that all the children have a grandfather in common, in which case the problem becomes trivial. If they don't all have a grandfather in common, then we showed that there are only 3 grandfathers in the entire town. But this means that we have 40 children "balls" to place into 3 grandfather "boxes," since each child has 2 grandfathers, and hence, by the Pigeonhole Principle, one of the grandfathers must have at least 14 grandchildren. □

We'll finish this section with a couple of harder applications of the Pigeonhole Principle.

Problem 5.9: Given any set of ten distinct 2-digit numbers, prove that there exist two disjoint subsets (of the 10 numbers) with the same sum. *(Source: IMO)*

Solution for Problem 5.9: The fact that we're trying to prove the existence of two of something with the same property (in this case, two subsets with the same sum) is our tipoff that we may want to consider using the Pigeonhole Principle. So subsets are going to be our "balls" and sums are going to be our "boxes". In order to apply the Pigeonhole Principle, we need to count how many of each of these there are.

A set with 10 elements has $2^{10} = 1024$ subsets, since each of the 10 elements can either be in or not in any particular subset. However, we can exclude the empty set (since it has sum 0) and the subset consisting of the entire set (since there cannot be another subset disjoint to this). Therefore we really have only $1024 - 2 = 1022$ subsets to consider.

Now we count the number of possible sums. The smallest possible sum is 10, coming from the subset $\{10\}$, and the largest possible sum is

$$91 + 92 + \cdots + 99 = 9(95) = 855,$$

coming from the subset $\{91, 92, \ldots, 99\}$. Therefore, there are $855 - 10 + 1 = 846$ possible sums of subsets consisting of between 1 and 9 elements of the set.

Since $1022 > 846$, we are guaranteed (by the Pigeonhole Principle) that at least 2 different subsets have the same sum. But how do we know that these subsets are disjoint?

We don't know that they're disjoint, but it's not a problem! Simply remove any elements that the two subsets have in common. Since we're removing the same elements, and the subsets originally had the same sum, they will still have the same sum after we remove the (same) elements from each. □

Note that 10 elements is the minimum value for which this argument works. If we had started with only 9 elements, then we would have $2^9 - 2 = 510$ subsets, but the sums could range from 10 to $92 + \cdots + 99 = 8(95.5) = 764$, so there would be $764 - 10 + 1 = 755$ possible sums. Hence, since $510 < 755$, we couldn't apply the Pigeonhole Principle. (This doesn't mean that the result is not true for 9 elements, it just means that our proof wouldn't work.)

> **Problem 5.10:** Aimee plays at least one game of chess a day for eight weeks, but she plays no more than 11 games in any 7-day period. Show that there is some period of consecutive days in which she plays exactly 23 games.

Solution for Problem 5.10: It's not immediately clear how to apply the Pigeonhole Principle to this problem. For one thing, we want to show that there's a period in which Aimee plays *exactly* 23 games, but the Pigeonhole Principle, as we typically use it, only guarantees *at least* some amount of something. Is there another way that we can think about this problem so that it is equivalent to showing that there are at least 2 of something?

Suppose that we keep a running total of the number of games that Aimee plays through the first i days, where $0 \le i \le 56$. Specifically, let g_i be the total number of games that she plays starting at Day 1 and going through Day i; note that $g_0 = 0$ since she plays 0 games in the first 0 days. The statement that we want to prove—that Aimee played exactly 23 games during some period—is equivalent to showing that there must exist integers $0 \le i < j \le 56$ such that $g_j - g_i = 23$. This should remind you a bit of Problem 5.3, so we'll try to use a similar approach.

Consider the remainders when each g_i is divided by 23. The possible values of these remainders are the first 23 nonnegative integers: $\{0, 1, \ldots, 22\}$. Eight weeks is 56 days, thus by the Pigeonhole Principle, there must be some remainder r which occurs on three different days. Again by the Pigeonhole Principle, two of these days must be at most 4 weeks apart: if there were more than 4 weeks between the first and second days and the second and third days, then there would be more than 8 weeks between the first and third, which is too long. So we know that there exist integers $0 \le i < j \le 56$, with $j - i \le 28$, such that $g_j - g_i$ is a multiple of 23. But we need to show that it is *exactly* 23.

There's still information in the problem statement that we haven't used!

> **Concept:** When you seem to be stuck on a problem, think if there is any information
> in the problem statement that you haven't yet used.

We also have the information that Aimee plays at most 11 games in any 7-day period. This means that in 4 weeks she will have played at most 44 games. Thus $0 < g_j - g_i \leq 44$. But $g_j - g_i$ is also a multiple of 23. The only multiple of 23 greater than 0 and less than or equal to 44 is 23. Thus $g_j - g_i = 23$, and we conclude that starting with Day $i + 1$ and ending with Day j, Aimee played exactly 23 games. $\square$

Exercises

5.4.1 Show that, in a 13-card bridge hand, there must be at least 4 cards of the same suit.

5.4.2 Prove that, for any set of 17 positive integers, there is some subset of 5 of them whose sum is a multiple of 5. **Hints:** 106

5.4.3 Every triangle with vertices among 11 given points (no 3 of which are collinear) is assigned one of four colors: amber, burgundy, chartreuse, or dark green. What is the largest number N for which we can be sure that at least N of the triangles are assigned the same color? *(Source: Mandelbrot)*

5.4.4 A pen-pal club has 12 members. In August, each member of the club sends a letter to 6 of the other members, chosen at random.

(a) Prove that some pair of members of the club will send each other letters. **Hints:** 226

(b) In September, each member of the club sends a letter to only 5 of the other members. Does the conclusion from part (a) still hold? Why or why not? **Hints:** 17

5.4.5★ Somewhere in the universe, n students are taking a 10-question math competition. Their collective performance is called *laughable* if, for some pair of questions, there exist 57 students such that either all of them answered both questions correctly or none of them answered both questions correctly. Compute the smallest n such that the performance must be laughable, no matter how the students performed on the competition. *(Source: HMMT)* **Hints:** 137, 53, 126

5.5 Summary

➤ Formally, the Pigeonhole Principle states that if we place n balls into k boxes, where $n > k$, then at least one box must contain more than 1 ball.

➤ Informally, the Pigeonhole Principle is just common sense. If you have more items than boxes to place them in, then at least one of the boxes must contain more than one item.

➤ More generally, if we have n items to be placed in k boxes, then at least one box must contain at least $\left\lfloor \dfrac{n-1}{k} \right\rfloor + 1$ balls.

➤ Don't memorize a "formula" for the Pigeonhole Formula (like the one above). Instead, use your common sense when applying it.

➤ Whenever we have to show that "a pair" of objects or "at least 2" objects share some property, that's our cue to think about the Pigeonhole Principle. More generally, it can often be used to show "at least n" objects share some property, or in problems that ask you to find the maximum number of objects satisfying some property.

➤ Sometimes, before applying the Pigeonhole Principle, we have to do a little work to reduce the number of boxes.

➤ Usually, the tricky part of applying the Pigeonhole Principle to a problem is identifying what are the "boxes" and the "balls."

REVIEW PROBLEMS

5.11 My sock drawer has lots of socks of four different colors (white, black, brown, and blue). How many socks do I have to pull out to ensure:

(a) at least one matching pair?

(b) at least two matching pairs?

5.12 The Aopslandia Grand Championship lacrosse tournament has 38 teams and lasts for several weeks. Each team will play each other team exactly once over the course of the tournament. Prove that, at any point in the tournament, there must be two teams who have completed the same number of games.

5.13 Prove that for any set of five positive integers, there are three of them whose sum is divisible by 3.

5.14 10 students took a 35-question history test. The test was very hard: each question was solved by exactly 1 student. If we know that at least one student got exactly 1 question right, at least one student got exactly 2 questions right, and at least one student got exactly 3 questions right, show that there must be a student that got at least 5 questions right.

5.15 The farmer's market down the street from my house has 7 different kinds of apples: Golden Delicious, Fuji, Granny Smith, Gala, McIntosh, Cortland, and Braeburn. I want to bake a pie, for which I need 5 apples of the same type (because a pie with more than one type of apple, well, that would be disgusting).

(a) If I randomly select apples, how many do I need to buy to guarantee that I will be able to bake my pie?

(b) If I want to bake 3 pies, now how many randomly-selected apples do I have to buy?

5.16 Prove that for any set of 11 positive integers, there is some nonempty subset of them whose sum is divisible by 11.

5.17 A group of 15 friends has $100 among them, and each person in the group has an integer number of dollars. Prove that two of them must have the same amount.

5.18 Show that if 16 people are seated in a row of 20 chairs, then some group of 4 consecutive chairs must be occupied.

5.19 What is the maximum number of kings that we can place on an 8×8 chessboard, such that no two kings are adjacent (including diagonally)?

Challenge Problems

5.20 A certain (small!) college has 20 students and offers 6 courses. Each student can enroll in any or all of the 6 courses, or none at all (which is a real waste of tuition). Prove or disprove: there must exist 5 students and 2 courses, such that either all 5 students are in both courses, or all 5 students are in neither course. *(Source: Putnam)* **Hints:** 244

5.21 Sam's band has 6 members, but only 4 of them play together in a concert. Additionally, no 3 members can play together in more than one performance. How many concerts can Sam's band give, at most? *(Source: Mandelbrot)* **Hints:** 258

5.22 Prove that for every prime p except 2 and 5, there is a power of p that ends with the digits 0001. **Hints:** 273

5.23 Let r by any real number, and $n \geq 1$ be a positive integer. Show that at least one of $r, 2r, \ldots, (n-1)r$ differs from an integer by at most $\frac{1}{n}$. **Hints:** 278

5.24 Every point in the plane is colored either red, green, or blue. Prove that there exists a rectangle in the plane such that all four of its vertices are the same color. **Hints:** 319, 289

5.25★ In a 6×6 grid, what is the largest number of squares that can be colored such that no four of the colored squares form the corners of a rectangle with vertical and horizontal sides? For example, the coloring pattern shown at right is not allowed, since the four marked squares form the corners of such a rectangle. *(Source: Mandelbrot)* **Hints:** 293, 227

5.26★ Let $X = \{x_1, x_2, \ldots, x_m\}$ be a set of m positive integers, all less than or equal to n, and let $Y = \{y_1, y_2, \ldots, y_n\}$ be a set of n positive integers, all less than or equal to m. Prove that there is a nonempty subset of X and a nonempty subset of Y with the same sum. *(Source: Putnam)* **Hints:** 130, 170, 175

5.27★ 40 teams play a tournament in which every team plays every other team exactly once. No ties occur, and each team has a 50% chance of winning any game it plays. Find the probability that no two teams win the same number of games. *(Source: AIME)* **Hints:** 260

5.28★ An international society has its members from six different countries. The list of members has 1978 names, numbered $1, 2, \ldots, 1978$. Prove that there is at least one member whose number is the sum of the numbers of two (not necessarily distinct) members from his own country. *(Source: IMO)* **Hints:** 80, 162, 108

Life is so constructed, that the event does not, cannot, will not, match the expectation. – Charlotte Brontë

CHAPTER 6

Constructive Expectation

6.1 Introduction

Recall that **expected value** is essentially a weighted average, where the values of different outcomes of an event are weighted based on their probability. If every outcome is equally likely, then the expected value is just the average of the possible outcomes.

In many situations, we can just use our basic definition of expected value:

$$E(X) = \sum_i P(X_i)V(X_i),$$

where $P(X_i)$ is the probability of event X_i occurring, and $V(X_i)$ is the value of outcome X_i.

However, often it is quite difficult to list all of the possible outcomes directly, determine their probabilities and values, and compute the expectation in the usual manner. Instead, for many problems, we can take a more constructive approach to computing expectation.

The key fact that we will use is that we can sum expected values across different events. What we mean by this is that if we have a series of events that occur in succession, then the expected value of the sum of the outcomes of the events is equal to the sum of the expected values of the outcomes of the individual events. This is true even if the events are not independent of one another. This is a powerful tool that lets us find expected values of complicated events by breaking them down into more manageable sub-events.

This is a bit hard to describe in words, but hopefully a few problems will make the concept clear.

Extra! *High achievement always takes place in the framework of high expectation. –* Charles Kettering
⫸⫸⫸⫸

6.2 Basic Examples

Problem 6.1: We simultaneously flip a penny, a nickel, a dime, and a quarter.

(a) List all 16 possible outcomes and the total value of the coins that land heads-up. What is the expected value of this total?

(b) Separately compute the expected values of each coin's contribution to the total.

(c) Sum your answers for (b) and compare this sum to your answer from (a).

Problem 6.2:

(a) We roll a pair of dice. What is the expected value of the total of the dice?

(b) We simultaneously roll 10 dice. What is the expected value of the total of the dice?

Problem 6.3: In the Showcase Showdown on the immensely popular TV game show *The Price Is Right*, if a contestant spins $1 on the Big Wheel, she wins $1,000 and gets a bonus spin in which she has a 1/20 chance of winning an additional $10,000 and a 1/10 chance of winning an additional $5,000. Suppose that a contestant has a 1/10 chance of spinning $1. What is the expected value of the contestant's winnings?

Problem 6.4: If X and Y are independent events, show that $E(X + Y) = E(X) + E(Y)$.

We can often exploit the fact that we may sum expected values across multiple events. Let's see a basic example of this.

Problem 6.1: We simultaneously flip a penny, a nickel, a dime, and a quarter. What is the expected value of the sum of the values of the coins that land heads-up?

Solution for Problem 6.1: Since each of the four coins are equally likely to come up heads or tails, there are $2^4 = 16$ equally likely outcomes. We can find the expected value by simply averaging the values of all the possible outcomes. For this, we can make a chart:

Penny	T	H	T	H	T	H	T	H	T	H	T	H	T	H	T	H
Nickel	T	T	H	H	T	T	H	H	T	T	H	H	T	T	H	H
Dime	T	T	T	T	H	H	H	H	T	T	T	T	H	H	H	H
Quarter	T	T	T	T	T	T	T	T	H	H	H	H	H	H	H	H
Total	0	1	5	6	10	11	15	16	25	26	30	31	35	36	40	41

So we see that the expected value is

$$\frac{0 + 1 + 5 + 6 + 10 + 11 + 15 + 16 + 25 + 26 + 30 + 31 + 35 + 36 + 40 + 41}{16} = \frac{328}{16} = 20.5.$$

However, we could have done this more simply by looking at each coin separately. The penny has an expected value of $\frac{1}{2}(0) + \frac{1}{2}(1) = \frac{1}{2}$. Similarly, the nickel has expected value $\frac{5}{2}$, the dime has expected value $\frac{10}{2} = 5$, and the quarter has expected value $\frac{25}{2}$. Therefore, the expected value when we flip them all is $\frac{1}{2} + \frac{5}{2} + 5 + \frac{25}{2} = \frac{41}{2} = 20.5$. $\square$

> **Concept:** The expected value of a sum of events is the sum of the expected values of the individual events.

Problem 6.2:

(a) We roll a pair of dice. What is the expected value of the total of the dice?

(b) We simultaneously roll 10 dice. What is the expected value of the total of the dice?

Solution for Problem 6.2: Rather than list all 36 possible outcomes for the pair of dice, we can use the fact that the expected value for a roll of a single die is $\frac{1}{6}(1 + 2 + 3 + 4 + 5 + 6) = 3.5$. Therefore, the expected value of the total of two dice is $2(3.5) = 7$, and the expected value of the total of 10 dice is $10(3.5) = 35$. $\square$

Problem 6.3: In the Showcase Showdown on the immensely popular TV game show *The Price Is Right*, if a contestant spins $1 on the Big Wheel, she wins $1,000 and gets a bonus spin in which she has a 1/20 chance of winning an additional $10,000 and a 1/10 chance of winning an additional $5,000. Suppose that a contestant has a 1/10 chance of spinning $1. What is the expected value of the contestant's winnings?

Solution for Problem 6.3: There are four possible outcomes:

- The contestant wins nothing with probability $\dfrac{9}{10}$.

- The contestant wins $1,000 with probability $\dfrac{1}{10} \cdot \dfrac{17}{20} = \dfrac{17}{200}$.

- The contestant wins $6,000 with probability $\dfrac{1}{10} \cdot \dfrac{1}{10} = \dfrac{1}{100}$.

- The contestant wins $11,000 with probability $\dfrac{1}{10} \cdot \dfrac{1}{20} = \dfrac{1}{200}$.

Therefore, the expected value of her win is:

$$\frac{9}{10}(\$0) + \frac{17}{200}(\$1000) + \frac{1}{100}(\$6000) + \frac{1}{200}(\$11000) = \frac{17(\$1000) + 2(\$6000) + \$11000}{200} = \frac{\$40000}{200} = \$200.$$

We can do this more simply by calculating the expected value of the regular spin and the bonus spin separately. The regular spin has expected value $\frac{1}{10}(\$1000) = \$100$. The bonus spin has expected value $\frac{1}{10}(\$5000) + \frac{1}{20}(\$10000) = \$1000$, but we have to remember that the contestant only gets the bonus spin with probability $\frac{1}{10}$. Therefore, the overall expected value is $\$100 + \dfrac{1}{10}(\$1000) = \$100 + \$100 = \$200$. $\square$

Note that in Problem 6.3, the two events that we are summing are not independent: the player only gets a bonus spin if she wins on the regular spin, so the bonus outcome is dependent on the regular outcome. Nonetheless, we saw that we could still sum the individual expected values to get the overall expected value.

Now that you've seen a few examples, you hopefully have a good feel for what we're doing. We are computing the expected value of the sum of outcomes of individual events, and we have seen (so far) that we can do this by summing the expected values of the individual events. Let's try to make this into a formal statement, and see if we can prove it.

Problem 6.4: If X and Y are independent events, show that $E(X + Y) = E(X) + E(Y)$.

WARNING!! This proof is somewhat algebraically intense. Don't worry if you don't see every detail on the first reading.

Solution for Problem 6.4: We'll need to write formulas for the expected values of X and Y, and for that, we'll need to introduce some notation. (Also, we're only going to consider the situation in which both X and Y have a finite number of possible outcomes; if X and Y have infinitely many outcomes, the result is still true, but it is somewhat difficult to prove.)

Suppose that X has m possible outcomes, which have probabilities $p_1, p_2, \ldots, p_m$ and values $x_1, x_2, \ldots, x_m$. This means that $p_1 + p_2 + \cdots + p_m = 1$ and that

$$E(X) = p_1 x_1 + p_2 x_2 + \cdots + p_m x_m = \sum_{i=1}^{m} p_i x_i.$$

Similarly, suppose that Y has n possible outcomes, which have probabilities $q_1, q_2, \ldots, q_n$ and values $y_1, y_2, \ldots, y_n$. This means that $q_1 + q_2 + \cdots + q_n = 1$ and that

$$E(Y) = q_1 y_1 + q_2 y_2 + \cdots + q_n y_n = \sum_{j=1}^{n} q_j y_j.$$

What do we know about the event "$X + Y$"? Since X and Y are independent events, we know that "$X + Y$" has mn possible outcomes: value $x_i + y_j$ occurs with probability $p_i q_j$, for all $1 \le i \le m$ and $1 \le j \le n$. So

$$\begin{aligned} E(X + Y) = {} & p_1 q_1 (x_1 + y_1) + p_1 q_2 (x_1 + y_2) + \cdots + p_1 q_n (x_1 + y_n) \\ & + p_2 q_1 (x_2 + y_1) + p_2 q_2 (x_2 + y_2) + \cdots + p_2 q_n (x_2 + y_n) \\ & + \vdots \\ & + p_m q_1 (x_m + y_1) + p_m q_2 (x_m + y_2) + \cdots + p_m q_n (x_m + y_n) \\ = {} & \sum_{i=1}^{m} \sum_{j=1}^{n} p_i q_j (x_i + y_j). \end{aligned}$$

Why is this equal to $E(X) + E(Y)$?

We can isolate the x terms in each row. For example, in the first row of the above equation for $E(X + Y)$, we have

$$p_1q_1x_1 + p_1q_2x_1 + \cdots + p_1q_nx_1 = p_1x_1(q_1 + \cdots + q_n) = p_1x_1,$$

since $q_1 + \cdots + q_n = 1$. Similarly, the x terms in the second row sum to p_2x_2, and so on. Thus all the x terms, summed over all rows, give

$$p_1x_1 + p_2x_2 + \cdots + p_mx_m = E(X).$$

We can also isolate the y terms, but this time by column. For example, in the first column of the above equation for $E(X + Y)$, we have

$$p_1q_1y_1 + p_2q_1y_1 + \cdots + p_mq_1y_1 = q_1y_1(p_1 + \cdots + p_m) = q_1y_1,$$

since $p_1 + \cdots + p_m = 1$. Similarly, the y terms in the second column sum to q_2y_2, and so on. Thus all the y terms, summed over all columns, give

$$q_1y_1 + q_2y_2 + \cdots + q_ny_n = E(Y).$$

This shows that $E(X + Y) = E(X) + E(Y)$. $\square$

> **Concept:** Don't worry if you didn't follow all of the details in the above proof—the algebra used is somewhat complicated. It's vastly more important that you understand the concept: we can sum expectations across separate events to get the expectation of the combined event.

In fact, the result $E(X + Y) = E(X) + E(Y)$ is still true even if X and Y are dependent events, as we saw as example of in Problem 6.3. It is a lot more difficult to prove, however: the algebra is a lot messier, and we'd need to introduce notation that we haven't yet developed. So we're not going to try to prove this here, but you can accept it as true from now on.

This is not limited to just 2 events: we can sum any number of events in the same fashion. To summarize:

> **Important:** If $X_1, X_2, \ldots, X_k$ are several events (independent or not), then
>
> $$E(X_1 + X_2 + \cdots + X_k) = E(X_1) + E(X_2) + \cdots + E(X_k).$$

  Exercises

6.2.1 I flip 20 coins. I then discard the coins that come up heads, and re-flip those that come up tails. What is the expected number of coins that again come up tails?

6.2.2

(a) If I roll one green die and two blue dice, what is the expected value of the sum of the values of the blue dice minus the value of the green die?

(b) If I roll 37 green dice and 38 blue dice, what is the expected value of the sum of the values of the blue dice minus the value of the sum of the green dice?

6.2.3 Kai picks a number and Jae picks a number. Kai picks w with probability p and x with probability $1 - p$. Jae picks y with probability q and z with probability $1 - q$.

(a) What is the expected value of Kai's number?

(b) What is the expected value of Jae's number?

(c) Find an algebraic expression for the expected value of the sum of Kai's and Jae's numbers by finding the probabilities of each of the 4 cases and then computing the expected value manually. Is your expression equal to the sum of the expressions you found in (a) and (b)?

6.2.4 Henry flips 10 coins and lays them on the desk. He then chooses one of the coins at random, and if it is tails, he flips it to heads. What is the expected number of heads showing? *(Source: Mandelbrot)*
Hints: 37

6.2.5★ What is the expected number of turns of a randomly-chosen 9-step path from point A to point B in the grid at right? (A *turn* is any point where the path changes direction; for example, the path shown at right has 3 turns, one at each of the marked points.) **Hints:** 150

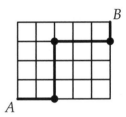

6.3 Summing Expectations Constructively

Problems

Problem 6.5: I flip a coin 10 times.

(a) What is the probability that the 1st and 2nd flips come up heads?

(b) What is the probability that the 2nd and 3rd flips come up heads?

(c) What is the expected number of pairs of consecutive tosses that come up heads? (For example, the sequence THHTHHHTHH has 4 pairs of consecutive HH's.)

Problem 6.6: I have 12 addressed letters to mail, and 12 corresponding pre-addressed envelopes. For some wacky reason, I decide to put the letters into the envelopes at random, one letter per envelope. What is the expected number of letters that get placed into their proper envelopes?

Problem 6.7: An equilateral triangle is tiled with n^2 smaller congruent equilateral triangles such that there are n smaller triangles along each of the sides of the original triangle. (The case $n = 11$ is shown at right.) For each of the small equilateral triangles, we randomly choose a vertex V of the triangle and draw an arc with that vertex as center connecting the midpoints of the two sides of the small triangle with V as an endpoint. Find the expected value of the number of full circles formed in terms of n. *(Source: USAMTS)*

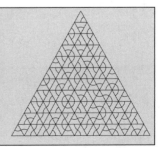

> **Problem 6.8:** There are 650 special points inside a circle of radius 16. You have a flat washer in the shape of an annulus (the region between two concentric circles), which has an inside radius of 2 and an outside radius of 3. Our goal is to show that it is always possible to place the washer so that it covers up at least 10 of the special points.
>
> (a) Suppose we can show that the expected number of points covered by a randomly-placed washer is at least 9. Explain why this would prove our result that it is possible to choose a placement of the washer that covers at least 10 points.
>
> (b) Let G be one of the special points. Determine the probability that G is covered by a randomly-placed washer.
>
> (c) Compute the expected number of points covered by a randomly-placed washer, and finish the problem.
>
> *(Source: PSS)*

Just as we can often count items by constructing them and keeping track of the number of choices along the way, we can also often compute expectations using a similar constructive method. As we saw in Problem 6.4, we have a very powerful tool: we can sum expectations, even across dependent events.

This first problem of this section shows the power of this idea.

> **Problem 6.5:** I flip a coin 10 times. What is the expected number of pairs of consecutive tosses that come up heads? (For example, the sequence THHTHHHTHH has 4 pairs of consecutive HH's.)

Solution for Problem 6.5: There are several tactics that we might think about. For instance, we might try to count the number of sequences with 1 pair of heads, with 2 pairs of heads, and so on, but this quickly devolves into a really messy PIE calculation. (If you don't believe this, try it and see.)

We might also try a straightforward constructive counting argument, but the problem with this approach is that the probabilities of the existence of pairs of heads in consecutive locations are not independent. For example, if the first two flips are HH, then there's a $\frac{1}{2}$ chance of a second pair of heads in the next slot (in other words, of the first three flips being HHH), but if the first two flips are TT, then there's 0 chance of a pair of heads in the next slot (since the first three flips will be either TTH or TTT). This makes it hard to construct the sequences without a lot of messy casework.

Fortunately, we don't have to do that!

Any particular pair of consecutive flips is HH with probability $\frac{1}{4}$. Therefore, each pair of consecutive flips contributes $\frac{1}{4}$ to the overall expected value. Since there are 9 possible pairs of consecutive flips that could be HH, the expected number of pairs of consecutive flips of heads is $9(\frac{1}{4}) = \frac{9}{4}$.

If you don't quite buy this explanation, there's another way to think about it. We know that there are $2^{10} = 1024$ possible sequences of ten coin flips (and they're all equally likely). We also know that $2^8 = 256$ of them begin with HH (since we have 2 choices for each of the remaining 8 flips). Similarly, for any particular choice of consecutive flips, we know that 256 of the sequences have HH in that position (for example, there are 256 sequences of the form ???HH?????, where ? can be anything), since we have 2 choices for each of the other 8 positions.

Therefore, when we sum the number of pairs of consecutive heads over all 1024 sequences, we find that 256 of the sequences contribute a pair of heads in the first position, 256 of the sequences contribute a pair of heads in the second position, and so on for each of the 9 possible positions. Therefore, there are 9(256) total consecutive pairs of heads among the 1024 sequences, and hence the expected value for the number of consecutive pairs of heads in any sequence is $\frac{9(256)}{1024} = 9\left(\frac{256}{1024}\right) = 9\left(\frac{1}{4}\right) = \frac{9}{4}$. □

> **Problem 6.6:** I have 12 addressed letters to mail, and 12 corresponding pre-addressed envelopes. For some wacky reason, I decide to put the letters into the envelopes at random, one letter per envelope. What is the expected number of letters that get placed into their proper envelopes?

Solution for Problem 6.6: Once again, we could try to count the number of outcomes with 0 correct letters, 1 correct letter, etc., but this would become a messy PIE calculation. We can instead sum the expected value constructively.

Each envelope individually has a $\frac{1}{12}$ chance of receiving the correct letter (since it is equally likely to receive any of the 12 letters). Therefore, each envelope contributes $\frac{1}{12}$ to the total expected number of letters in their correct envelopes. Since there are 12 envelopes, and each contributes $\frac{1}{12}$, the expected number of letters in their proper envelopes is $12(\frac{1}{12}) = 1$.

Again, if you don't feel completely comfortable with this argument, note that for each envelope, 11! of the 12! possible arrangements will have that particular envelope correctly filled. Therefore, when we sum over all 12! arrangements, we get a total of 12(11!) envelopes correctly filled, and hence the expected number of such envelopes is 12(11!)/12! = 12!/12! = 1. □

> **Sidenote:** It is natural to ask, in connection with Problem 6.6, what is the probability that none of the envelopes receives its correct letter? This is an example of a permutation called a **derangement**. We looked at derangements in Challenge Problem 3.41 when we were studying PIE. As we saw then, we can count the number of such permutations using PIE, by counting the number of arrangements in which at least 1 letter gets placed into its proper envelope. However, a nicer solution is to use **recursion**, which we will discuss in Chapter 10.

> **Problem 6.7:** An equilateral triangle is tiled with n^2 smaller congruent equilateral triangles such that there are n smaller triangles along each of the sides of the original triangle. (The case $n = 11$ is shown at right.) For each of the small equilateral triangles, we randomly choose a vertex V of the triangle and draw an arc with that vertex as center connecting the midpoints of the two sides of the small triangle with V as an endpoint. Find the expected value of the number of full circles formed in terms of n.
> *(Source: USAMTS)*

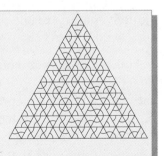

Solution for Problem 6.7: We can see that at any interior point inside the big triangle, we'll only get a circle if all 6 of the little triangles surrounding that point line up exactly correctly, as in the picture at right. Each little triangle lines up with probability $\frac{1}{3}$, so the probability of getting a full circle at any particular point is $(\frac{1}{3})^6 = \frac{1}{729}$.

Now we have to determine how many interior points there are. We can see that there is 1 interior point near the top (at the bottom of the 2nd row of triangles), then 2 interior points in the next row, and so on, down to $n - 2$ points in the bottom interior row. Therefore there are

$$1 + \cdots + (n - 2) = \frac{(n - 2)(n - 1)}{2}$$

interior points. Since each of these points contributes $\frac{1}{729}$ to the expected number of circles, we get a final answer of

$$\frac{(n - 2)(n - 1)}{1458}.$$

□

One very powerful application of expected value is in geometric optimization problems, as in the next example.

Problem 6.8: There are 650 special points inside a circle of radius 16. You have a flat washer in the shape of an annulus (the region between two concentric circles), which has an inside radius of 2 and an outside radius of 3. Show that it is always possible to place the washer so that it covers up at least 10 of the special points. *(Source: PSS)*

Solution for Problem 6.8: This is an existence problem. We wish to show that there exists some position at which we can place the washer so that it covers at least 10 points. But why is this related to expected value?

Suppose that we could compute the expected number of points covered by a randomly-placed washer, and that this value was greater than 9. What could we conclude? If the *average* washer covers more than 9 points, then there must exist *some* washer that covers 10 or more points. Because, if every washer covered 9 or fewer, how could the average possibly be greater than 9?

So our goal is to show that the expected value of the number of points covered by a washer placed at random is greater than 9. If we can show this, then we can conclude that there must be some place where we can put the washer to cover 10 or more points. Note that this is essentially a geometric version of the Pigeonhole Principle.

Calculating this expected value, however, doesn't appear to be so simple. For starters, there are infinitely many possible places to put the washer! Evaluating each possible position of the washer will literally take forever, since there are an infinite number of places we can put it.

However, there are only 650 points in the circle we have to consider. Therefore, we can take an element-by-element approach, similar to what we did in Problem 6.5. Recall in that problem, we noted that $\frac{1}{4}$ of the total sequences of flips had a pair of consecutive heads in any particular position, and we could sum this over each possible position to get a total expected value of $\frac{9}{4}$ consecutive pairs of heads. We'll do the same thing here: we'll calculate, for each possible special point, the portion of the washer placements in which that special point is covered.

Extra! *One of the common denominators I have found is that expectations rise above that which is* ⟹⟹⟹⟹ *expected.* – George W. Bush

Let P be one of our special points. P is covered by the washer if and only if the center of the washer is at least 2 units away from the point, but no more than 3 units away. This region (the region between 2 and 3 units away from P) we will call our "success" region, and it is the region between the heavy dashed circles in the diagram to the right.

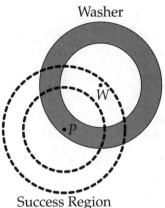

Washer

Success Region

For example, in the picture at right, a washer is shown with center W inside the "success" region, so the washer covers our point P. Any washer whose center is in the "success" region will cover point P, and conversely, any washer that covers point P will have its center in the "success" region.

Therefore, the area of the region in which the center of the washer must be in order to cover P is just the area of the "success" region. The region is the difference between a circle of radius 3 and a circle of radius 2, so its area is $3^2\pi - 2^2\pi = 5\pi$.

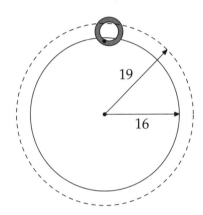

Now we consider the overall region where the center of the washer might be. We must be careful to include those cases in which the center of the washer is just outside our initial circle, but still covering special points near the circumference of the circle, as shown at left. To take into account the possibility of a washer centered outside the circle covering special points inside the circle (an example of which is shown at left), we note that the "possible" region, in which the center of our washer can be placed and still have the washer cover special points, is a circle with radius 19, not 16.

Now, we can evaluate the probability that a washer placed at random covers a given special point. Out of the $19^2\pi = 361\pi$ area in which we can place the center of our washer, there is a 5π area in which it can be placed to cover P. Therefore, the probability P is covered by a randomly placed washer is $(5\pi)/(361\pi) = 5/361$.

However, there's nothing special about the P we examined. For each of the 650 special points, the probability that a randomly placed washer covers that point is 5/361. Therefore, each special point contributes 5/361 to the expected value of the number of special points covered by a randomly placed washer. So, the total expected value of the number of special points covered by a randomly placed washer is

$$650 \cdot \frac{5}{361} \approx 9.003.$$

Since the expected value of the number of special points covered by a randomly placed washer that overlaps some portion of our circle is greater than 9, there must be some placement of the washer that covers at least 10 special points. $\square$

Exercises

6.3.1 There are 20 houses along the shore of a lake. Each is painted one of four colors at random. I leave my house (one of the 20) and walk around the lake. Each time I pass a house that is the same color as the previous house, I laugh. (This includes comparing the first house I see to mine, and comparing my house to the previous house I passed when I finish my journey.) I don't laugh at any other time during my walk. What is the expected value of the number of times that I laugh during my walk?

6.3.2 George has six ropes. He chooses two of the twelve loose ends at random (possibly from the same rope), and ties them together, leaving ten loose ends. He again chooses two loose ends at random and joins them, and so on, until there are no loose ends. Find, with proof, the expected value of the number of loops George ends up with. *(Source: USAMTS)*

6.3.3 One fair 6-sided die is rolled; let a denote the number that comes up. We then roll a dice; let the sum of the resulting a numbers be b. Finally, we roll b dice, and let c be the sum of the resulting b numbers. Find the expected value of c. *(Source: HMMT)*

6.3.4★ 51 points are inside a square of side length 1. Prove that we can cover some three of them by a circle with radius $\frac{1}{7}$. **Hints:** 67, 123

6.3.5★ For any subset $S \subseteq \{1, 2, \ldots, 15\}$, call a number n an *anchor* for S if n and $n + \#(S)$ are both members of S, where $\#(S)$ denotes the number of members of S. Find the average number of anchors over all possible subsets $S \subseteq \{1, 2, \ldots, 15\}$. *(Source: HMMT)* **Hints:** 235

6.4★ A Coat With Many Patches (Reprise)

Problem 6.9: Recall Problem 3.24:

> I have a coat with area 5. The coat has 5 patches on it. Each patch has area at least 2.5. Prove that 2 patches exist with common area of at least 1. *(Source: PSS)*

Our goal is to prove it directly using expected value, without using PIE.

(a) Define a function f such that $f(p)$ is the number of patches containing point p. If p is chosen at random, compute a lower bound for $E(f(p))$.

(b) Show that $\binom{k}{2} \geq 2k - 3$ for any integer k such that $0 \leq k \leq 5$.

(c) Use parts (a) and (b) to show that $E\left(\binom{f(p)}{2}\right) \geq 2$.

(d) Use part (c) to show that some pair of patches must have common area at least 1.

Let's revisit the coat-and-patches problem from Chapter 3. You may recall that when we considered this problem earlier, we solved it using a series of PIE computations. Now let's find a completely different solution, using expected value.

Problem 6.9: Recall Problem 3.24:

> I have a coat with area 5. The coat has 5 patches on it. Each patch has area at least 2.5. Prove that 2 patches exist with common area of at least 1. *(Source: PSS)*

Prove it directly using expected value, without using PIE.

Solution for Problem 6.9:   We will use a very clever expected value argument (due to Ravi Bopanna).

Here's the general strategy: we know that there are $\binom{5}{2} = 10$ *pairs* of patches. If the result is not true—that is, if every pair of patches has an overlap with an area of less than 1—then the 10 pairs together overlap with an area less than 10. This would mean that a randomly chosen point on the coat (of area 5) would be covered by fewer than $10/5 = 2$ pairs of patches.

So, working the other way, if we can show that a randomly-chosen point on the coat is covered (on average) by at least 2 pairs of patches, then one of the pairs of patches must have overlapping area of at least 1.

Let's try to make this argument a bit more formal. Define a function f where, for any point p on the coat, $f(p)$ is the number of patches containing point p. In particular, note that $f(p)$ is a nonnegative integer between 0 and 5 (inclusive) for all points p.

The quantity that we're interested in, though, is the number of *pairs* of patches containing p. If p is covered by $f(p)$ patches, then it is covered by $\binom{f(p)}{2}$ pairs of patches. So our goal is to show that

$$E\left(\binom{f(p)}{2}\right) \geq 2.$$

We don't know anything immediately about $\binom{f(p)}{2}$, but we do have information about $f(p)$. Because the sum of the areas of the patches is at least 12.5, and the area of the coat is 5, we know that

$$E(f(p)) \geq \frac{12.5}{5} = 2.5.$$

Since we're dealing with integer values in this problem, let's double this expression and write it as $E(2f(p)) \geq 5$.

We need to relate $\binom{f(p)}{2}$ to $f(p)$. Since we know that $f(p)$ can only take on the values $0, 1, 2, 3, 4, 5$, let's just make a chart:

$f(p)$	0	1	2	3	4	5
$2f(p)$	0	2	4	6	8	10
$\binom{f(p)}{2}$	0	0	1	3	6	10

We notice that for our middle two values of $f(p)$, we have

$$\binom{f(p)}{2} = 2f(p) - 3,$$

and furthermore, for all possible values of $f(p)$ we have

$$\binom{f(p)}{2} \geq 2f(p) - 3.$$

Now we can compute our bound for $E\left(\binom{f(p)}{2}\right)$:

$$E\left(\binom{f(p)}{2}\right) \geq E(2f(p) - 3) = 2E(f(p)) - 3 \geq 2(2.5) - 3 = 2.$$

Therefore, the expected value of the number of pairs of patches that contain a randomly chosen point is at least 2, and as described at the beginning of our solution, this means that some pair of patches must overlap with an area of at least 1. □

6.5 Summary

➤ To compute the expected value of a complicated event, think about breaking up the event into easier-to-manage components. If the value of an event is the sum of values of several intermediate components, then we can compute the expected value of each component and sum them to get the expected value of the overall event.

➤ More precisely, if $X_1, X_2, \ldots, X_n$ are events, then

$$E(X_1 + X_2 + \cdots + X_n) = E(X_1) + E(X_2) + \cdots + E(X_n).$$

➤ This works even if the events are not independent: the expected value of a sum of events is always the sum of the expected values of the individual events.

➤ Expected value can also be used in geometric optimization problems.

REVIEW PROBLEMS

6.10 I have mn identical game pieces. Each one is a square with side length 2 inches. Each piece has a quarter circle drawn on it with its center at one of the corners and with radius 1. If I put the game pieces in an $m \times n$ grid, what is the expected value of the number of full circles I form?

6.11 A 10-digit binary number with four 1's is chosen at random. What is its expected value?

6.12 Five balls numbered 1 through 5 are in a bin. You draw them out one at a time, without replacement. Every time the number on the drawn ball matches the number of the draw, you win a dollar. For example, if you draw ball #2 on the second draw, you win a dollar for that draw. What is the expected amount of your winnings?

6.13 The Happy Animals Kennel has 18 cages in a row. It will allocate 6 to dogs and 12 to cats. Let A be the number of times in the row of cages that a dog cage and a cat cage are adjacent. For example, in the arrangement *cdcdddcdcccccccdccc*, we have $A = 8$. Given that the kennel will choose an arrangement at random from among all the possible arrangements, find the expected value of A.

6.14 Select numbers a and b between 0 and 1 independently and at random, and let c be their sum. Let A, B, and C be the results when a, b, and c, respectively, are rounded to the nearest integer. (Assume that $\frac{1}{2}$ is rounded up to 1, and similarly that $n + \frac{1}{2}$ is rounded up to $n + 1$ for all nonnegative integers n.)

(a) What is the probability that $A + B = C$?

(b) Find $E(C - (A + B))$.

(c) Suppose $a_1, a_2, \ldots, a_{100}$ are chosen randomly between 0 and 1, and let $c = a_1 + a_2 + \cdots + a_{100}$. As before, let A_i be the result when a_i is rounded to the nearest integer (for all $1 \le i \le 100$), and let C be the result when c is rounded to the nearest integer. Find $E(C - (A_1 + A_2 + \cdots + A_{100}))$.

(Source: AMC)

6.15 Suppose that in the country of Aopslandia:

- 20% of families have no children

- 20% of families have exactly 1 child

- 30% of families have exactly 2 children

- 20% of families have exactly 3 children

- 10% of families have exactly 4 children

- no families have more than 4 children

A child is chosen at random. What is the expected number of siblings of the child?

Challenge Problems

6.16 Fifteen freshmen are sitting in a circle around a table, but the course assistant (who remains standing) has made only six copies of today's handout. No freshman should get more than one handout, and any freshman who does not get one should be able to read a neighbor's. If the freshmen are distinguishable but the handouts are not, how many ways are there to distribute the six handouts subject to the above conditions? *(Source: HMMT)* **Hints:** 345, 301, 87

6.17 Show that we can color the elements of the set $S = \{1, 2, \ldots, 2007\}$ with 4 colors such that any subset of S with 10 elements, whose elements form an arithmetic sequence, is not all one color. *(Source: IMO)* **Hints:** 33, 185

6.18 For any positive integer n, let $p_n(k)$ be the number of permutations of the set $\{1, 2, \ldots, n\}$ that have exactly k fixed points. (A *fixed point* of a permutation is an element i such that i is in the ith position of the permutation. For example, the permutation $(3, 2, 5, 4, 1)$ of $\{1, 2, 3, 4, 5\}$ has two fixed points: the 2 (in the 2nd spot) and the 4 (in the 4th spot).) Prove that $\sum_{k=0}^{n} k p_n(k) = n!$. *(Source: IMO)* **Hints:** 3

6.19★ There are several circles inside a square with side length 1. The sum of the circumferences of the circles is 10. Prove that there exists a line that intersects at least 4 of the circles. **Hints:** 43, 177

6.20★ A standard 52-card deck is shuffled, and cards are turned over one-at-a-time starting with the top card. What is the expected number of cards that will be turned over before we see the first Ace? (Recall that there are 4 Aces in the deck). **Hints:** 325, 52

6.21★ In a competition, there are a contestants and b judges, where $b \geq 3$ is an odd integer. Each judge rates each contestant as either "pass" or "fail." Suppose that k is a number such that, for any two judges, their ratings coincide for at most k contestants. Prove that $\dfrac{k}{a} \geq \dfrac{b-1}{2b}$. *(Source: IMO)* **Hints:** 202

I resolved to stop accumulating and begin the infinitely more serious and difficult task of wise distribution.

– Andrew Carnegie

CHAPTER 7

Distributions

7.1 Introduction

If you read *Introduction to Counting & Probability*, then you're already a bit familiar with a **distribution**. Basically, whenever we have to assign (or "distribute") indistinguishable items among distinguishable people or objects, we have a distribution problem.

In this chapter, we'll start by looking at the basic problem: distributing indistinguishable objects to distinguishable recipients, such that each recipient receives at least one object. This problem is well understood, and its solution is fairly easy, although it requires a slight bit of cleverness.

However, it turns out the solutions to many more complicated distribution problems can be put into 1-1 correspondence with the solutions of our basic distribution problem. In other problems, we can find a different 1-1 correspondence to relate a distribution to a different problem that we know how to solve. In the most difficult sort of distribution problem, we'll have to think "outside the box" and come up with a more clever solution.

An important lesson to learn about solving distribution problems is not to rely on a formula to solve them. Instead, reason out the answer each time you're presented with a distribution. This will not only minimize the possibility of incorrectly solving basic problems, but also increase your success at solving more difficult distribution problems. Also, many seemingly-difficult counting problems are really distribution problems in disguise. Once you have a better understanding of how to solve distribution problems, you will be able to use this knowledge to solve an even wider variety of counting problems.

Extra! *Of all things, good sense is the most fairly distributed: everyone thinks he is so well supplied*
⇒⇒⇒⇒ *with it that even those who are the hardest to satisfy in every other respect never desire more of it than they already have.* – René Descartes

7.2 Basic Distributions

Problem 7.1: A working crew of 4 people just finished work on your yard. You pay the crew $200, using ten $20 bills. In how many ways can the four crew members divide the 10 bills among themselves if each member must get at least one bill? Assume the crew members are distinct, but the $20 bills are not.

Problem 7.2: In how many ways can I pass out 11 identical lollipops to 6 kids, if each kid must receive at least one lollipop?

Problem 7.3: In how many ways can we distribute n indistinguishable items into k distinguishable boxes, if each box must contain at least one item?

Problem 7.4: How many quadruples (a, b, c, d) of positive integers are solutions to $a + b + c + d = 17$?

The basic idea of a distribution is simple. We have a bunch of indistinguishable objects that we want to distribute into distinguishable piles. We don't care which specific objects end up in each pile, just how many of them end up in each pile.

Here's a basic example:

Problem 7.1: A working crew of 4 people just finished work on your yard. You pay the crew $200, using ten $20 bills. In how many ways can the four crew members divide the 10 bills among themselves if each member must get at least one bill? Assume the crew members are distinct, but the $20 bills are not.

Solution for Problem 7.1: This is a distribution problem because we don't care which specific $20 bills go to which people; we only care about how many $20 bills each person gets. The bills are indistinguishable, but the people are distinguishable.

It is possible to solve this problem through a lengthy casework process. This is somewhat tedious, but leads to the interesting Hockey Stick identity. (See Chapter 13 of *Introduction to Counting & Probability* if you'd like to see this method.) Here, we will show only the faster, more clever solution.

Imagine that we arrange the ten $20 bills in a row, where each "$" denotes a bill:

$$\$ \, \$ \, \$ \, \$ \, \$ \, \$ \, \$ \, \$ \, \$ \, \$$$

We don't care which specific bills each person receives, since the bills are indistinguishable. We only care how many bills each person receives. So we can assume that the first person takes some number of bills from the left side of the row, then the next person takes some bills from those remaining at the left side of the row, and so on. For example, if person #1 gets $80, person #2 gets $40, person #3 gets

$60, and person #4 gets $20, we can think of that as:

$$\underbrace{\$\,\$\,\$}_{\#1} \ \ \underbrace{\$\,\$}_{\#2} \ \ \underbrace{\$\,\$\,\$}_{\#3} \ \ \underbrace{\$}_{\#4}$$

Better yet, we can place "dividers" between the $'s to separate them into groups:

$$\$\,\$\,\$\,\$\,|\,\$\,\$\,|\,\$\,\$\,\$\,|\,\$$$

In order to divide the 10 $'s into 4 groups, we need to insert 3 dividers. Since a divider can be placed between any pair of $'s and there are 9 consecutive pairs, we have 9 "slots" to choose from in which to insert the dividers. Note that we can place at most one divider into any slot, since we need to guarantee that each person receives a *positive* number of bills.

To use our more formal language, there is a 1-1 correspondence between

{# of ways to distribute 10 $'s among 4 people}   ↔   {# of ways to insert 3 dividers into 9 slots}.

The right side of the above correspondence is easy to count: we can place 3 dividers into 9 slots in $\binom{9}{3} = 84$ ways, since we simply have to choose 3 of the 9 slots in which to place dividers. So there are 84 ways to distribute the money. □

Let's see another example before going to the general case.

Problem 7.2: In how many ways can I pass out 11 identical lollipops to 6 kids, if each kid must receive at least one lollipop?

Solution for Problem 7.2: We can arrange the 11 lollipops in a row (we will use ✖ to represent a lollipop):

$$\text{✖✖✖✖✖✖✖✖✖✖✖}$$

To divide this into 6 groups, we need to insert 5 dividers. The first kid then gets the leftmost group, the next kid gets the next group, and so on. For example, to give 2 lollipops to each of the first two kids, 1 to each of the next three, and 4 to the last kid, we insert dividers as:

$$\text{✖✖|✖✖|✖|✖|✖|✖✖✖✖}$$

There are 10 slots into which we can insert the dividers, and we must choose 5 of them. Therefore there are $\binom{10}{5} = 252$ ways to give out the lollipops. □

Having seen a couple of examples, we should have an idea how to handle the general problem.

Problem 7.3: In how many ways can we distribute n indistinguishable items into k distinguishable boxes, if each box must contain at least one item?

Solution for Problem 7.3: We use the same general idea that we used in the two previous examples. Imagine arranging the n items in a line. In order to divide them into k groups, we need to insert $k-1$ dividers. There are $n-1$ slots between items into which the dividers can be inserted, and we need to choose $k-1$ of these slots.

Formally, we have a 1-1 correspondence:

$$\left\{ \begin{array}{l} \text{Ways to distribute } n \text{ indistinguishable items} \\ \text{into } k \text{ distinguishable boxes, with each box} \\ \text{getting at least 1 item} \end{array} \right\} \leftrightarrow \left\{ \begin{array}{l} \text{Ways to insert } k-1 \text{ dividers into } n-1 \\ \text{slots} \end{array} \right\}.$$

Since there are $\binom{n-1}{k-1}$ ways to choose $k-1$ of the $n-1$ slots to receive dividers, we conclude using the 1-1 correspondence described above that there are $\binom{n-1}{k-1}$ ways to distribute the items. $\square$

We can summarize our results as follows:

> **Important:** The number of ways to distribute n indistinguishable items into k distinguishable boxes, where each box must receive at least one item, is
>
> $$\binom{n-1}{k-1}.$$

> **Concept:** As we often say, don't memorize the above formula! Instead, understand the concept that leads to the formula. That is, when presented with a distribution problem, your thought process should be to think of dividers and slots, not to try to recall the above formula from memory.

Distributions often turn up in counting the number of solutions to certain Diophantine equations. (A **Diophantine equation** is an equation in which we are looking for integer solutions.) Let's see an example.

> **Problem 7.4:** How many quadruples (a, b, c, d) of positive integers are solutions to $a + b + c + d = 17$?

Solution for Problem 7.4: This is exactly the same as distributing 17 indistinguishable items into 4 distinguishable boxes, with the condition that each box must get at least one item. The "items" are copies of the number 1, and the "boxes" are a, b, c, and d. For example, setting $a = 6$ means that we are placing six of the "1" "items" into the "box" labeled "a." The fact that we are looking for positive integer solutions means that each "box" must receive at least one 1.

More formally, we have the correspondences

$$\{\text{Positive integer solutions to } a + b + c + d = 17\} \leftrightarrow \left\{ \begin{array}{l} \text{Ways to distribute 17 1's into four variables,} \\ \text{such that each variable receives at least one 1} \end{array} \right\}$$

$$\leftrightarrow \{\text{Ways to insert 3 dividers into 16 slots}\}$$

Therefore there are $\binom{16}{3} = 560$ solutions. $\square$

Exercises

7.2.1 In how many ways can 10 kindergarten children eat 30 cookies, if each child must eat at least one cookie?

7.2.2 Suppose the head of my 4-person work crew insists that she get at least 3 of the 10 $20 bills that will be distributed to the crew. Each of the other members will get at least one. In how many ways can the bills be distributed?

7.2.3   Find the number of positive integer solutions to the equation $a + b + c = 100$.

7.2.4   In how many ways can 8 licorice sticks and 10 chocolate bars be distributed to 5 kids, if:

(a)   each kid must receive at least one piece of candy of each type?

(b)★   each kid must receive at least one piece of candy, but no kid can receive both types of candy?
Hints: 171

7.3   Distributions With Extra Conditions

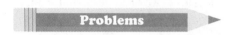

Problem 7.5: How many solutions does the equation $v + w + x + y + z = 21$ have, where v, w, x, y, z are all nonnegative integers?

(a)   Why is the answer not $\binom{20}{4}$, using the formula that we learned in the previous section?

(b)   How can we change this problem to look like a problem that we considered in the previous section?

(c)   What is the answer to this problem?

Problem 7.6: In how many ways can we distribute n indistinguishable balls into k distinguishable boxes, if some box(es) may remain empty?

Problem 7.7: In how many ways can I distribute 20 candy canes to 7 children, if each child must receive at least one, and two of the children are twins who insist on receiving the same amount?

(a)   What are the possibilities for the amounts of candy that I can give to the twins?

(b)   For each possible amount in (a), in how many ways can I distribute the rest of the candy to the other 5 children?

(c)   Use your answers to (b) to finish the problem.

Problem 7.8: In how many ways can I give out 15 pieces of candy to 4 kids, with each kid getting at least one piece, if the oldest kid insists on receiving more pieces than any other kid?

(a)   Explain why this solution does not work: The number of ways to distribute 15 candies to 4 kids is $\binom{14}{3} = 364$. But only $\frac{1}{4}$ of these have the oldest kid with the most candy (by symmetry), so the number of distributions satisfying the condition of the problem is $364/4 = 91$.

(b)   What are the possible amounts the candy that we could give to the oldest kid?

(c)   For each possible amount in (b), in how many ways can we distribute the rest of the candy to the other kids?

(d)   Finish the problem.

Problem 7.9: How many solutions does $a + b + c + d = 27$ have in nonnegative integers, if a must be even and d must be a power of 3?

In the last section, we discussed the basic distribution problem of counting the number of ways to distribute n indistinguishable items into k distinguishable boxes, such that each box has at least one item. However, many distribution problems will have different conditions on the types of allowable distributions. The most common of these is eliminating the condition that every box must contain at least one item. Let's see an example of this:

Problem 7.5: How many solutions does the equation $v + w + x + y + z = 21$ have, where v, w, x, y, z are all nonnegative integers?

Solution for Problem 7.5: We can't use our straightforward "dividers and slots" approach here, because we have to allow for the fact that some of the variables might be 0. But can we convert it to one of our earlier distributions?

Concept: When faced with a problem that you don't initially know how to do, try to convert it into a problem that you do know how to do.

What if we force the variables to be positive?

Let $v' = v + 1$, $w' = w + 1$, and so on. If $v, w, \ldots$ are nonnegative, then $v', w', \ldots$ are positive, and vice versa. Then

$$v' + w' + x' + y' + z' = (v + 1) + (w + 1) + (x + 1) + (y + 1) + (z + 1) = (v + w + x + y + z) + 5 = 26.$$

Aha—now we want to count the *positive* solutions to $v' + w' + x' + y' + z' = 26$. We know how to do that!

As usual, we could write this as a correspondence:

{Nonnegative solutions to $v + w + x + y + z = 21$} $\leftrightarrow$ {Positive solutions to $v' + w' + x' + y' + z' = 26$}

$\leftrightarrow$ {Ways to insert 4 dividers into 25 slots}

So there are $\binom{25}{4} = 12650$ solutions. $\square$

We can use our insight from Problem 7.5 to solve the general form of this distribution problem.

Problem 7.6: In how many ways can we distribute n indistinguishable balls into k distinguishable boxes, if some box(es) may remain empty?

Solution for Problem 7.6: Here's the general idea:

1. Add k balls to our group of balls, so that we have $n + k$ balls total.

2. Distribute the $n + k$ balls into the k boxes, leaving no box empty.

3. Remove 1 ball from each box.

Clearly this operation is reversible, so this establishes 1-1 correspondences:

$$\left\{ \begin{array}{l} \text{Distributions of } n \text{ balls to } k \text{ boxes,} \\ \text{with some box(es) possibly empty} \end{array} \right\} \quad \leftrightarrow \quad \left\{ \begin{array}{l} \text{Distributions of } n + k \text{ balls to } k \text{ boxes,} \\ \text{with no boxes empty} \end{array} \right\}$$

$$\leftrightarrow \quad \{\text{Insertion of } k - 1 \text{ dividers into } n + k - 1 \text{ slots}\} .$$

Therefore we can do this distribution in $\binom{n + k - 1}{k - 1}$ ways. $\square$

> **Important:** The number of ways to distribute n indistinguishable balls in k distinguishable boxes, where some box(es) may remain empty, is
>
> $$\binom{n + k - 1}{k - 1}.$$

> **Concept:** You know what I'm going to say (I hope): don't memorize this formula. Understand where it comes from.

We've now handled the two most common distributions: those with no restrictions at all, and those in which no box can be left empty. But many distribution problems tack on more exotic conditions. Let's do a couple of examples.

Problem 7.7: In how many ways can I distribute 20 candy canes to 7 children, if each child must receive at least one, and two of the children are twins who insist on receiving the same amount?

Solution for Problem 7.7: Unfortunately, there's no "slick" way of solving this problem. We're going to have to get our hands dirty with some casework.

The twins can each receive between 1 and 7 pieces of candy (if they receive more than 7 each, then there's not enough candy left for the other 5 kids). This leaves the rest of the candy to be distributed to the other 5 kids, which we know how to count.

Here's a chart showing the cases:

# of candies given to each of the twins	# of candies remaining to be given to the other 5 kids	# of ways to distribute remaining candy
1	18	$\binom{17}{4}$
2	16	$\binom{15}{4}$
3	14	$\binom{13}{4}$
4	12	$\binom{11}{4}$
5	10	$\binom{9}{4}$
6	8	$\binom{7}{4}$
7	6	$\binom{5}{4}$

So the number of possible distributions is

$$\binom{17}{4} + \binom{15}{4} + \binom{13}{4} + \binom{11}{4} + \binom{9}{4} + \binom{7}{4} + \binom{5}{4} = 2380 + 1365 + 715 + 330 + 126 + 35 + 5 = 4956.$$

There doesn't seem to be any good way to simplify this answer to a nicer expression, so there probably wasn't a simpler method that we could have used. □

> **Concept:** If you get a "nice" answer, then there's often a relatively simple combinatorial explanation for the simpler answer, and a relatively simple method of obtaining it. However, if you get an "ugly" answer, then there's probably not a simple method for getting it.

> **Problem 7.8:** In how many ways can I give out 15 pieces of candy to 4 kids, with each kid getting at least one piece, if the oldest kid insists on receiving more pieces than any other kid?

Solution for Problem 7.8: It might be tempting to use this quick "shortcut":

> **Bogus Solution:** The number of ways to distribute 15 candies to 4 kids is $\binom{14}{3} = 364$. But only $\frac{1}{4}$ of these have the oldest kid with the most candy (by symmetry), so the number of distributions satisfying the condition of the problem is $364/4 = 91$.

This doesn't work since not every distribution will have a unique kid with the most candy—there might be a tie. For example, if I gave 5 to each of the first two kids, 3 to the third kid, and 2 to the last kid, then there's no kid with more pieces than any other kid. This isn't allowed: the oldest kid must have *more* pieces than any other kid. If we could somehow compute the number of distributions that have a unique kid with the most candy, then it would be valid to take $\frac{1}{4}$ of that number to get our answer. Unfortunately, there's no easy way to compute that.

So instead, once again, we appeal to casework. This time, our cases will be determined by the number of candies that we give to the oldest (greedy) kid.

We can give the oldest kid any amount of candies between 5 and 12 inclusive. If we give him more than 12, then we don't have enough candy left so that each of the other 3 kids gets at least one piece. On the other hand, if we give him 4 or fewer, then there are at least 11 pieces left, so by the Pigeonhole Principle at least one of the other 3 kids must receive at least 4 pieces, which is not allowed.

If we give the oldest kid between 7 and 12 pieces (inclusive), then we can distribute the rest of the candy to the other 3 kids however we want, since there's no way to give one of these kids as many as the oldest. (Remember that each kid must get at least 1 piece.) So these cases give

$$\binom{2}{2} + \binom{3}{2} + \cdots + \binom{7}{2} = 1 + 3 + 6 + 10 + 15 + 21 = 56$$

possible distributions.

> **Sidenote:** Note that, in the line above, we could have used the Hockey Stick identity:
> ♪
> $$\binom{2}{2} + \binom{3}{2} + \cdots + \binom{7}{2} = \binom{8}{3} = 56.$$
>
> (If you're not familiar with the Hockey Stick identity, we'll see it in greater detail in Chapter 12.) As we just mentioned at the end of the last problem, a simple answer like $\binom{8}{3}$ often means a simple explanation. Can you find a simple combinatorial explanation to why, if we give the oldest kid between 7 and 12 pieces (inclusive), the number of ways to distribute the candy is just $\binom{8}{3}$?

If we give the oldest kid 6 pieces, then there are 9 pieces left. Normally, this would result in $\binom{9}{2} = 28$ distributions, but we cannot distribute the candies in groups of 6,2,1 or 7,1,1 in any order, since that would give some kid at least as many candies as the oldest kid has. So we must exclude these $3! + \binom{3}{1} = 9$ distributions (since they can be in any order), and hence this case contributes $28 - 9 = 19$ distributions.

If we give the oldest kid 5 pieces, then there are 10 pieces left. If none of the three remaining kids can have more than 4 pieces, then each kid must have at least 2 pieces (check this for yourself). The only way to do this is to distribute the candies in groups of $4, 4, 2$ on in groups of $4, 3, 3$. Each of these groups can be arranged in 3 ways (for each group, we merely have to choose which kid gets the different number of pieces), so there are $3 \times 2 = 6$ valid distributions in this case. As a check, we can list them (since there are only 6 of them):

$$5, 4, 4, 2, \quad 5, 4, 2, 4, \quad 5, 2, 4, 4, \quad 5, 4, 3, 3, \quad 5, 3, 4, 3, \quad 5, 3, 3, 4.$$

Therefore, the number of distributions is $56 + 19 + 6 = 81$. □

> **Problem 7.9:** How many solutions does $a + b + c + d = 27$ have in nonnegative integers, if a must be even and d must be a power of 3?

Solution for Problem 7.9: Once again, there's no nice way to do this except by getting into casework. We generally want to start by looking at the most restrictive condition first, which in this case is the condition that d must be a power of 3.

> **Concept:** When dealing with restrictions, it usually is best to deal with the most restrictive condition first.
> ⊙━▭

This means that d must be 27, 9, 3, or 1. We can now make a chart.

d	Possible values for a	# of choices for b,c	Number of solutions
27	0	1	1
9	$0, 2, \ldots, 18$	$19, 17, \ldots, 1$	100
3	$0, 2, \ldots, 24$	$25, 23, \ldots, 1$	169
1	$0, 2, \ldots, 26$	$27, 25, \ldots, 1$	196

So there are $1 + 100 + 169 + 196 = 466$ solutions satisfying the conditions. □

Exercises

7.3.1 Find the number of nonnegative integer solutions to the equation $u + v + w + x + y = 22$.

7.3.2

(a) Pat wants to buy four donuts from an ample supply of three types of donuts: glazed, chocolate, and powdered. How many different selections are possible?

(b) Pat is to select six cookies from a tray containing only chocolate chip, oatmeal, and peanut butter cookies. There are at least six of each of these three kinds of cookies on the tray. How many different assortments of six cookies can be selected?

(Source: AMC)

7.3.3 Compute the number of distinct ways in which 77 one-dollar bills can be distributed to 7 people so that no person receives less than $10. *(Source: ARML)*

7.3.4 Find the number of integer solutions to the equation $x + y + z = 10$ if x, y, and z are all less than 20.

7.3.5★ Find the number of positive integer solutions to $w + x + y + z < 25$. **Hints:** 39, 12

7.3.6★ Andrew has 10 candy bars, 10 packages of jelly beans, 10 lollipops, and 10 packs of chewing gum, and Andrew has two sisters. In how many ways can Andrew distribute the candies between his sisters, so that each sister gets 20 items total? **Hints:** 256

7.4 More Complicated Distribution Problems

Problems

Problem 7.10: A triomino game piece has three numbers on it from 1 to 9. Two pieces are considered different if they have different numbers; in other words, the order of the numbers on the piece doesn't matter. For example, the two triominos shown below are considered the same. A piece may have a number repeated, or have the same number in all three positions. How many distinct triomino pieces are there?

Problem 7.11: How many cubic polynomials $f(x)$ with positive integer coefficients are there such that $f(1) = 9$?

Problem 7.12: How many 15-digit base 4 numbers are there with eight 0's that appear in 3 groups? (For example, 230001330210000 is one such number; the 3 groups are 000, 0, and 0000.)

Problem 7.13: How many arrangements of the word PROBABILISTIC have no two I's appearing consecutively?

In most cases, a distributions problem is not going to come with a huge flashing sign that says "DISTRIBUTION!" in big neon letters. You are often going to have to look carefully to see that a more complicated problem involves distributions.

Problem 7.10: A triomino game piece has three numbers on it from 1 to 9. Two pieces are considered different if they have different numbers; in other words, the order of the numbers on the piece doesn't matter. For example, the two triominos shown below are considered the same. A piece may have a number repeated, or have the same number in all three positions. How many distinct triomino pieces are there?

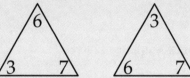

Solution for Problem 7.10: We could solve this pretty easily using casework—let's do so for practice, before seeing the solution using distributions.

Case 1: Triominos with all three numbers the same. There are 9 of these.

Case 2: Triominos with two of one number and a third different number. There are 9 choices for the numbers that appears twice, then 8 choices for the different third number, so there are 72 of these.

Case 3: Triominos with three different numbers. There are $\binom{9}{3} = 84$ of these.

So there are $9 + 72 + 84 = 165$ possible triominos.

But we can also solve this problem using distribution theory. If we let $a_1, a_2, \ldots$ represent the number of 1's, 2's, $\ldots$ on a triomino, then there is a 1-1 correspondence between triominos and nonnegative integer solutions to

$$a_1 + a_2 + \cdots + a_9 = 3.$$

We also know that these solutions are in 1-1 correspondence to *positive* integer solutions of

$$b_1 + b_2 + \cdots + b_9 = 12,$$

where $b_i = a_i + 1$. And we know that these solutions are in 1-1 correspondence with the number of ways to insert 8 dividers into 11 slots. So the answer is $\binom{11}{8} = \binom{11}{3} = 165$. □

Concept: Solving a problem via two different methods is a good way to check your answer.

The main advantage of the distribution solution to Problem 7.10 is that it scales upward easily. For example, if you were asked to count the number of hexominos (hexagons with six numbers from 1 through 9 on them), the casework would be really messy, but the distribution-based solution would be quick. (You can see for yourself in the Exercises.)

Distributions can sometimes show up in unusual places, as we see in the next problem.

Problem 7.11: How many cubic polynomials $f(x)$ with positive integer coefficients are there such that $f(1) = 9$?

Solution for Problem 7.11: We can write a such a cubic polynomial as $f(x) = ax^3 + bx^2 + cx + d$. Plugging in $x = 1$ gives us $9 = a + b + c + d$. So it's just a distribution problem! More formally, we have a 1-1 correspondence:

$$\left\{\begin{array}{l}\text{Polynomials } f(x) = ax^3 + bx^2 + cx + d \text{ with} \\ \text{positive integer coefficients such that } f(1) = 9\end{array}\right\} \leftrightarrow \{\text{Positive integer solutions to } 9 = a+b+c+d\}$$

$$\leftrightarrow \{\text{Ways to insert 3 dividers into 8 slots}\}$$

Thus there are $\binom{8}{3} = 56$ such cubic polynomials. $\square$

Problem 7.12: How many 15-digit base 4 numbers are there with eight 0's that appear in 3 groups? (For example, 230001330210000 is one such number; the 3 groups are 000, 0, and 0000.)

Solution for Problem 7.12: We think about this problem constructively: how would we build such a number? There are basically three steps:

- Decide how to break up the 0's into 3 groups;

- Decide where to place the groups of 0's within the 15-digit number;

- Choose the remaining non-0 digits.

We can count the number of choices for each step, then multiply (since the steps are independent) to get the total number of such numbers.

First, we break up the 0's into 3 groups. If the sizes of the groups are a, b, c, then we must have a, b, c all positive and $a + b + c = 8$, the total number of 0's. This is a basic distribution problem: it's the same as inserting 2 dividers into 7 possible slots, so there are $\binom{7}{2} = 21$ ways that we can break up the 0's into 3 groups.

Second, we have to place these groups into the 15-digit number. If 8 of the digits are 0, then the other 7 digits are non-0. We can think of the 3 groups of 0's as dividers that divide the non-0 digits apart. The slightly tricky thing to be aware of is that we cannot place a group of 0's at the beginning, but we can place it at the end. So there are 7 possible slots for our 3 groups (6 in the middle and 1 at the right end), hence there are $\binom{7}{3} = 35$ ways that we can insert the digits.

Finally, there are 7 non-0 digits, and each can be 1, 2, or 3. So there are $3^7 = 2187$ choices for the non-0 digits.

Combining our counts, we see that there are $(21)(35)(2187) = 1{,}607{,}445$ such numbers. $\square$

Problem 7.13: How many arrangements of the word PROBABILISTIC have no two I's appearing consecutively?

Solution for Problem 7.13: There are 3 I's and 10 other letters. We can think of placing the 3 I's into slots between 2 other letters, or at the beginning or the end of the word. Therefore, there are 11 possible positions for the I's, and 3 of them to place, so there are $\binom{11}{3} = 165$ allowed positions for the I's. Then the remaining 10 letters can be arranged in $10!/2$ ways (don't forget to divide by 2 since there are two B's!). Hence, the total number of allowed arrangements is $165(10!)/2 = 15(11!)/2$.

For practice, let's also compute this using complementary counting and PIE, to check our work. There are $13!/3!2!$ arrangements of the letters without any restrictions. There are $12!/2!$ arrangements with at least two I's together. There are $11!/2!$ arrangements with all three I's together. So, by PIE, the total number of allowed arrangements is

$$\frac{13!}{3!2!} - \frac{12!}{2!} + \frac{11!}{2!} = 11!\left(13 - 6 + \frac{1}{2}\right) = 11!\left(\frac{15}{2}\right),$$

which (of course) matches our distribution-based answer. □

Exercises

7.4.1 Calculate the number of possible hexominos: 6-sided game pieces with 6 numbers from 1 through 9 (inclusive), where a number may be repeated and where we don't care about the position of the numbers, just how many of each number are on each piece.

7.4.2 ARMLovian, the language of the fair nation of ARMLovia, consists only of words using the letters A, R, M, and L. The words can be broken up into syllables that consist of exactly one vowel, possibly surrounded by a single consonant on either or both sides. For example, LAMAR, AA, RA, MAMMAL, AMAL, LALA, MARLA, RALLAR, and AAALAAAAAMA are ARMLovian words, but MRLMRLM, MAMMMAL, MMMMM, L, ARM, ALARM, LLAMA, and MALL are not. Compute the number of 7-letter ARMLovian words. *(Source: ARML)* **Hints:** 271

7.4.3 A gardener plants 3 maple trees, 4 oak trees, and 5 birch trees in a row. He plants them in random order, each arrangement being equally likely. Find the probability that no two birch trees are next to each other. *(Source: AIME)* **Hints:** 194

7.4.4★ How many degree 6 polynomials $f(x)$ with positive integer coefficients are there such that $f(1) = 30$ and $f(-1) = 12$? **Hints:** 32, 36

7.4.5★ The Aopslandia lottery consists of randomly drawing 6 balls (without regard to order) from a bin of 44 balls numbered 1 through 44. A group of citizens is concerned that the lottery may be rigged, because they have noticed that, historically, over 50% of the drawings have resulted in a least one pair of consecutively numbered balls being drawn. Should these citizens be concerned? **Hints:** 210, 15

Extra! *The speed of communications is wondrous to behold. It is also true that speed can multiply the*
➡➡➡➡ *distribution of information that we know to be untrue. – Edward R. Murrow*

7.5 Summary

➤ Distributions are problems involving placing indistinguishable items into distinguishable boxes.

➤ If each box must contain a positive number of items, we can think of arranging the items in a row, and placing dividers between the items to divide them into the requisite number of boxes. Thinking of the problem in this way, we see that if there are n items to divide into k boxes, then we must place $k-1$ dividers among $n-1$ slots, and thus there are $\binom{n-1}{k-1}$ possible distributions.

➤ If some of the boxes may be empty, then add 1 extra item to each box, and take advantage of the 1-1 correspondences:

$$\begin{Bmatrix} \text{Distributions of } n \text{ items to } k \text{ boxes,} \\ \text{with some box(es) possibly empty} \end{Bmatrix} \leftrightarrow \begin{Bmatrix} \text{Distributions of } n+k \text{ items to } k \text{ boxes,} \\ \text{with no boxes empty} \end{Bmatrix}$$

$$\leftrightarrow \{\text{Insertion of } k-1 \text{ dividers into } n+k-1 \text{ slots}\}.$$

➤ Don't memorize the formulas. Understand where they come from. That is to say: when presented with a distribution problem, your thought process should be to think of dividers and slots, not to try to recall a formula from memory.

➤ Distribution problems with extra conditions may require casework or some other clever manipulation to convert it to a distribution problem that we know how to do.

➤ Many problems are distribution problems in disguise.

Here are some general problem-solving concepts that we saw in this chapter:

Concept:	When faced with a problem that you don't initially know how to do, try to convert it into a problem that you do know how to do.

Concept:	If you get a "nice" answer, then there's often a relatively simple combinatorial explanation for the simpler answer, and a relatively simple method of obtaining it. However, if you get an "ugly" answer, then there's probably not a simple method for getting it.

Concept:	When dealing with restrictions, it usually is best to deal with the most restrictive condition first.

Concept:	Solving a problem via two different methods is a good way to check your answer.

7.14 Determine how many triples (x, y, z) satisfy each of the following:

(a) $x + y + z = 10$ and x, y, z are positive integers.

(b) $x + y + z = 10$ and x, y, z are nonnegative integers.

(c) $x + y + z = 10$ and x, y, z are integers no less than -2.

(d) $x + y + z = 10$ and x, y, z are positive even integers.

7.15 Find the number of quintuples $(x_1, x_2, x_3, x_4, x_5)$ of positive odd integers that satisfy

$$x_1 + x_2 + x_3 + x_4 + x_5 = 2003.$$

7.16 I have 8 identical pieces of candy and 4 identical cookies to distribute to 3 children (2 boys and a girl). In how many ways can I do this, if:

(a) each child must receive at least 1 of each type of item?

(b) each child must receive exactly 4 items?

(c) the girl must receive more pieces of candy than either of the boys, and the boys must receive an equal number of cookies?

7.17 How many solutions are there in positive integers to the equation $w + x + y + z = 30$ if no variable takes on a value greater than 16?

7.18 In how many ways can the integers from 1 to 36, inclusive, be ordered such that no two multiples of 6 are adjacent?

7.19 How many terms are in the expansion of $(x + y + z)^{100}$? (For example, $(x + y + z)^2 = x^2 + y^2 + z^2 + 2xy + 2xz + 2yz$ has 6 terms.)

7.20 In how many ways can three teachers and eight students sit in the front row of 11 chairs at the auditorium if there must be at least 2 students between each pair of teachers? *(Source: AMC)*

Challenge Problems

7.21 20 chairs are set in a row. 5 people randomly sit in the chairs (no more than one person to a chair, of course!). What is the probability that nobody is sitting next to anybody else? *(Source: Mandelbrot)* **Hints:** 6

7.22 In how many ways can a word be formed with 8 A's and 5 B's if every A is next to another A and every B is next to another B? **Hints:** 121

7.23 Amy flipped a coin 20 times, and got the sequence THHTTTHTTHHTHTTTTTHH. She noticed that 3 times a heads followed a heads, 7 times a tails followed a tails, 4 times a tails followed a heads, and 5 times a heads followed a tails. How many such sequences are possible? **Hints:** 326, 261

7.24★ A bin has 10 red balls and 8 blue balls. We randomly draw out 6 balls, one at a time, without replacement. What is the probability that, at some point, we choose two consecutive balls that are red? **Hints:** 305

7.25★ In a certain lottery, 7 balls are drawn at random from n balls numbered 1 through n. If the probability that no pair of consecutive numbers is drawn equals the probability of drawing exactly one pair of consecutive numbers, find n. *(Source: Mandelbrot)* **Hints:** 61

Of course I have played outdoor games. I once played dominoes in an open air cafe in Paris. – Oscar Wilde

CHAPTER **8**

Mathematical Induction

This short chapter will introduce a proof tool called **mathematical induction**.

The principle of mathematical induction (often called just **induction**) is one of the fundamental proof methods that we use to prove statements about positive integers. The classic metaphor used to describe mathematical induction is a row of dominos all standing on end:

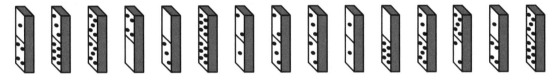

If we tip the first domino over:

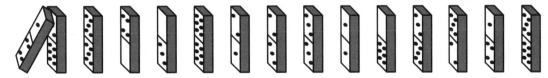

and if each domino, as it falls, knocks over the next domino in line:

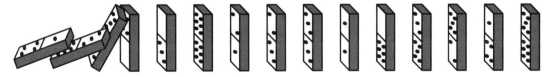

then ultimately, all of the dominos will fall over:

This is the basis of mathematical induction. Suppose that we wish to prove some mathematical statement that involves a positive integer n. Each "domino" represents the statement for a particular value of n: the first domino is the statement where $n = 1$, the second domino is the statement where $n = 2$, and so on. If we can prove that the first statement (where $n = 1$) is true, and that each statement implies the next one, then we can knock all the dominos down, and thus prove that the statement is true for all positive integers n.

In this chapter we'll formally state the principle of mathematical induction, and use it in several examples. Mathematical induction is one of the most common proof techniques that we use in counting problems, and you should master it until it becomes almost second nature.

Let's begin with a simple example:

Problem 8.1: Prove, using mathematical induction, that for any positive integer n, the sum of the first n positive integers equals $\dfrac{n(n + 1)}{2}$.

Solution for Problem 8.1: This proof, as with every proof that uses mathematical induction, consists of two parts.

- A **base case** (this is analogous to "knocking over the first domino")

- An **inductive step** (this is analogous to "proving that each domino knocks over the next one")

So let's do these two steps.

Base case: We need to prove the statement for $n = 1$. The sum of the first 1 positive integers is just 1, and indeed, this equals

$$\frac{1(1 + 1)}{2} = \frac{2}{2} = 1.$$

Inductive step: We assume that the statement is true for some positive integer k; that is, we assume that the sum of the first k positive integers is $\frac{k(k+1)}{2}$. This assumption is called the **inductive hypothesis**. Now we wish to show that the statement is also true for $k + 1$, so we compute the sum of the first $k + 1$ positive integers. Since we will want to use our inductive hypothesis, we break up the sum of the first $k + 1$ positive integers into the sum of the first k positive integers plus $k + 1$:

$$1 + 2 + \cdots + k + (k + 1) = (1 + 2 + \cdots + k) + (k + 1).$$

Now we can use our inductive hypothesis. We have assumed that the sum of the first k positive integers is $\frac{k(k+1)}{2}$, so we can substitute it into the previous equation:

$$1 + 2 + \cdots + (k + 1) = \frac{k(k + 1)}{2} + (k + 1).$$

The rest is just algebra:

$$1 + 2 + \cdots + (k + 1) = \frac{k(k + 1)}{2} + (k + 1) = \frac{k(k + 1) + 2(k + 1)}{2} = \frac{(k + 1)(k + 2)}{2} = \frac{(k + 1)((k + 1) + 1)}{2}.$$

So we see that the original statement, that the sum of the first n positive integers is equal to $\frac{n(n+1)}{2}$, is true for $n = k + 1$.

We have completed both of the necessary steps for an induction proof, so we have proved that the statement is true for all positive integers n. $\square$

The important thing to remember about a mathematical induction proof is that your proof must contain two separate parts: the verification of the base case, and the proof of the inductive step.

> **Important:** We can use **mathematical induction** when we wish to prove a statement $S(n)$ that depends on a positive integer n. The proof consists of two steps:
>
> 1. Prove $S(1)$.
> 2. Prove that $S(k)$ implies $S(k + 1)$ for any positive integer k.

For example, in Problem 8.1 above, the statement $S(n)$ was "the sum of the first n positive integers is equal to $\frac{n(n+1)}{2}$." Our base case established this for $n = 1$, and our inductive step established that the statement for $n = k$ implied the statement for $n = k + 1$.

Now that you've seen how mathematical induction works, try a few problems:

Problems

Problem 8.2: Prove that for any positive integer n,

$$1^3 + 2^3 + \cdots + n^3 = (1 + 2 + \cdots + n)^2.$$

Problem 8.3: Prove that the sum of the interior angles of a convex n-sided polygon is $180(n - 2)$ degrees. (You may assume that the sum of the interior angles of a triangle is $180°$.)

Problem 8.4: Let n be a positive integer. One square of a $2^n \times 2^n$ chessboard is removed. Prove that the remaining chessboard can be tiled with 3-square L-shaped tiles like the one shown at right.

Problem 8.5: We start with a pile of 25 stones. We divide the stones into two piles (however we wish), and write the product of the numbers of stones in the two piles on the blackboard. (For example, we might choose to divide the stones into piles of 11 and 14, in which case we would write 154 on the board.) We now choose one of the remaining piles, divide it into two smaller piles in any manner we choose, and again write the product of the numbers of stones in the two new piles on the blackboard. We repeat this process until we have 25 piles of 1 stone each. Prove that at the end of this process, no matter what choices we make along the way, the sum of the numbers written on the board will be 300.

Problem 8.6: In a large field, n people are standing so that for each person, the distances to all the other people are different. At a given signal, each person fires a water pistol and hits the person who is closest to them. When n is odd, prove that there is at least one person who is left dry.

> **Problem 8.2:** Prove that for any positive integer n,
>
> $$1^3 + 2^3 + \cdots + n^3 = (1 + 2 + \cdots + n)^2.$$

Solution for Problem 8.2: Once again, we have a statement that is claimed for all positive integers n. This is your cue to think about using induction.

> **Concept:** Whenever you have a statement that depends on a positive integer n, consider mathematical induction as a possible method of proving the statement.

Let's prove the problem statement using induction.

Base case: We see immediately that $1^3 = (1)^2$.

Inductive step: Assume that $1^3 + 2^3 + \cdots + k^3 = (1 + 2 + \cdots + k)^2$ (recall that this is called the **inductive hypothesis**). We now evaluate $1^3 + 2^3 + \cdots + k^3 + (k+1)^3$. Since we want to use the inductive hypothesis, we should try to rewrite this expression in terms of an expression present in the inductive hypothesis:

$$1^3 + 2^3 + \cdots + k^3 + (k+1)^3 = (1^3 + 2^3 + \cdots + k^3) + (k+1)^3.$$

Now we can apply the inductive hypothesis to replace $(1^3 + 2^3 + \cdots + k^3)$:

$$(1^3 + 2^3 + \cdots + k^3) + (k+1)^3 = (1 + 2 + \cdots + k)^2 + (k+1)^3.$$

We need to simplify the right side, so we apply the result from Problem 8.1 and a bit of algebra:

$$(1 + 2 + \cdots + k)^2 + (k+1)^3 = \left(\frac{k(k+1)}{2}\right)^2 + (k+1)^3 = \frac{k^2(k+1)^2 + 4(k+1)^3}{4} = \frac{k^4 + 6k^3 + 13k^2 + 12k + 4}{4}.$$

It would be nice if that numerator factored. Usually, factoring a degree 4 polynomial is a bit of a chore. However, here we have a huge clue. Remember that at the end of the day, we want the right side to equal $(1 + 2 + \cdots + (k+1))^2$, and that this is equal to $\left(\frac{(k+1)(k+2)}{2}\right)^2$. So we hope that the numerator factors as $(k+1)^2(k+2)^2$. Indeed, it does:

$$\frac{k^4 + 6k^3 + 13k^2 + 12k + 4}{4} = \frac{(k^2 + 2k + 1)(k^2 + 4k + 4)}{4} = \left(\frac{(k+1)(k+2)}{2}\right)^2.$$

Finally, we use Problem 8.1 again to finish:

$$1^3 + 2^3 + \cdots + k^3 + (k+1)^3 = \left(\frac{(k+1)(k+2)}{2}\right)^2 = (1 + 2 + \cdots + k + (k+1))^2.$$

We have proven our base case and our inductive step, so the result is true for all positive integers n. $\square$

Induction can be used in geometric problems too! Anything that depends on a positive integer n is potentially a candidate for mathematical induction.

Problem 8.3: Prove that the sum of the interior angles of a convex n-sided polygon is $180(n-2)$ degrees. (You may assume that the sum of the interior angles of a triangle is $180°$.)

Solution for Problem 8.3: We already have the base case $n = 3$, since the problem told us we can assume that the sum of a triangle's angles is $180 = 180(3-2)$. (Note: we'll omit the degree symbol and the word "degrees" in our solution; you can assume that all angle quantities are in degrees.) So we can proceed directly to the inductive step.

We assume that for some integer $k \geq 3$, the sum of the interior angles of a k-sided polygon is $180(k-2)$. Now we want to compute the sum of the interior angles of a $(k+1)$-sided polygon. How are we going to get a k-sided polygon from a $(k+1)$-sided polygon?

We can lop off a vertex!

For example, in the first picture in Figure 8.1 below, we have a 7-sided polygon. We can draw a diagonal between vertices that are two sides apart, as shown in the middle diagram. This splits the polygon into a triangle and a 6-sided polygon, as in the right diagram below.

Figure 8.1: Splitting a 7-gon into a triangle and a 6-gon

Note that the sum of the angles of the 7-sided polygon is equal to the sum of the angles of the triangle (which is 180) plus the sum of the angles of the 6-sided polygon (which is $180(4) = 720$). Therefore, the sum of the angles in the 7-sided polygon is $180 + 720 = 900 = 180(5)$.

We can do this for any $(k+1)$-sided polygon. We draw a diagonal between two vertices that are 2 sides apart on the polygon, splitting the $(k+1)$-sided polygon into a triangle and a k-sided polygon. (The dashed lines in Figure 8.2 below indicate that the polygon has an arbitrary number of sides.)

Figure 8.2: Splitting a $(k+1)$-gon into a triangle and a k-gon

We know that the sum of the angles of the triangle is 180, and by the inductive hypothesis, we know that the sum of the angles of the k-sided polygon is $180(k-2)$. Therefore, the sum of the angles in the $(k+1)$-sided polygon is the sum of these, which is

$$180 + 180(k-2) = 180(k-1) = 180((k+1)-2),$$

thus proving the assertion. □

Problem 8.4: Let n be a positive integer. One square of a $2^n \times 2^n$ chessboard is removed. Prove that the remaining chessboard can be tiled with 3-square L-shaped tiles like the one shown at right.

Solution for Problem 8.4: As we've seen before, we have a statement that we must prove to be true for all positive integers. This makes it a good candidate for induction.

Base case: If we take a $2^1 \times 2^1$ chessboard—that is, a 2×2 chessboard—and remove a square, what's left will be exactly the same shape as an L-shaped tile (although perhaps rotated). So we can simply place a single tile to cover the board, and we're done.

Inductive step: Assume that we can tile any $2^k \times 2^k$ chessboard with any square removed, where $k \geq 1$ is a positive integer. We'll consider a $2^{k+1} \times 2^{k+1}$ chessboard with a square removed, and see if we can tile it.

We need to reduce the $2^{k+1} \times 2^{k+1}$ problem down to the $2^k \times 2^k$ problem, which we assume (by the inductive hypothesis) that we can solve. What's the easiest way to do that? It seems like the easiest way is to divide our $2^{k+1} \times 2^{k+1}$ board into 4 boards of size $2^k \times 2^k$ by slicing it horizontally and vertically through the middle:

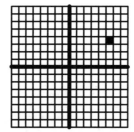

But only one of our four $2^k \times 2^k$ boards is going to have a square missing; the other three will be complete. We can use the inductive hypothesis to tile whichever quadrant contains the missing square, but what do we do about the other three? If only each of them had a square missing.... Is there a systematic way we can remove a square from each of the three complete $2^k \times 2^k$ boards?

Yes! "Remove" the squares closest to the center of the original $2^{k+1} \times 2^{k+1}$ board by covering them with a tile, as shown in the picture to the right. Now we're set! We place that tile in the center, and what's left is four boards of size $2^k \times 2^k$, each with a square missing. We can tile them, using the inductive hypothesis, and when we do so the result is a tiling of the original $2^{k+1} \times 2^{k+1}$ board.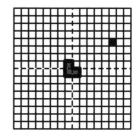

Therefore, by induction, for all positive integers n, we can tile a $2^n \times 2^n$ board with any square removed. □

Sometimes the result for k is not enough to prove the result for $k + 1$. We say that we are using **strong induction** when we must assume the result for all $1 \leq i \leq k$ in order to prove the result for $k + 1$. The informal way to think of strong induction is that it takes the combined strength of all of the first k dominos in order to knock over the $(k + 1)^{\text{st}}$ domino. Strong induction is not that common, but as we will see in the next problem, it is occasionally necessary.

Problem 8.5: We start with a pile of 25 stones. We divide the stones into two piles (however we wish), and write the product of the numbers of stones in the two piles on the blackboard. (For example, we might choose to divide the stones into piles of 11 and 14, in which case we would write 154 on the board.) We now choose one of the remaining piles, divide it into two smaller piles in any manner we choose, and again write the product of the numbers of stones in the two new piles on the blackboard. We repeat this process until we have 25 piles of 1 stone each. Prove that at the end of this process, no matter what choices we make along the way, the sum of the numbers written on the board will be 300.

Solution for Problem 8.5: We might first try to figure out where that "300" comes from. To this end, we can experiment a bit with the most "extreme" choices that we can make when dividing the stones. One extreme way of dividing the stones is to always split the stones so that one of the new piles contains just a single stone. In other words, we divide the 25 stones initially into piles of 1 and 24, then divide the 24-stone pile into piles of 1 and 23, and so on. If we do this, then the numbers that get written on the blackboard are $24, 23, \ldots, 1$, and their sum is

$$24 + 23 + \cdots + 1 = \frac{(25)(24)}{2} = \binom{25}{2} = 300.$$

Aha, there's the 300.

The difficulty is that there doesn't seem to be any obvious direct way to approach this problem. (It turns out that there's a *really* clever counting argument that proves that the sum must be 300. But it's fairly difficult to see. We'll present it at the end of the solution.) But perhaps we can prove the more general version of the problem. In particular, what if we began with n stones? Then based on our experimentation above, we would expect the sum at the end to be $\binom{n}{2}$. Indeed, if we did the same extreme pile-splitting with n stones—that is, at every step, created a new pile with a single stone in it—then the numbers that get written on the blackboard would be $n - 1, n - 2, \ldots, 1$, and their sum is

$$(n - 1) + (n - 2) + (n - 3) + \cdots + 1 = \frac{(n - 1)n}{2} = \binom{n}{2}.$$

So we'd like to prove that if we start with n stones, we will end up with a sum of $\binom{n}{2}$, no matter how we go about dividing the stones at each step. Since we have a statement for all positive integers n, we might think to try induction. However, the difficulty is that the initial pile can be divided into two piles of any size. In particular, there is no way of getting from the problem of size $n - 1$ to the problem of size n. So we'll need to use a somewhat stronger version of the inductive hypothesis.

But let's begin with the base case. When $n = 1$, we have only 1 stone, and nothing to divide. So no numbers get written on the blackboard, and the sum of no numbers is $0 = \binom{1}{2}$. (We know that $\binom{1}{2} = 0$, since $\binom{1}{2}$ counts the number of ways to choose 2 items from a set of 1 item. It is impossible to choose 2 items from a set of 1 item, so there are 0 ways to make such a choice.)

For the inductive step, we assume that the result holds for all piles of size less than k, for some integer $k > 1$. This is stronger than our usual inductive hypothesis of assuming that the result holds for a pile of size $k - 1$. We now consider a pile of size k. The initial step splits the pile into a pile of size j (where $1 \leq j < k$) and a pile of size $k - j$, so that the number $j(k - j)$ is the first number written on the board.

Consider all of the numbers that we write on the board as the j-stone pile is repeatedly split until these j stones are in piles of 1 stone each. By the inductive hypothesis, the sum of the these numbers is $\binom{j}{2}$. Similarly, as we repeatedly split the $(k - j)$-stone pile into piles of 1 stone each, the sum of the numbers written is $\binom{k-j}{2}$. These numbers may get written in any order on the blackboard (since we might choose to split the piles in any order), but the order that the numbers are written on the board does not alter the sum.

Therefore, the total of the numbers on the blackboard will be

$$j(k - j) + \binom{j}{2} + \binom{k - j}{2}.$$

We can simplify this expression:

$$\begin{aligned}
j(k - j) + \binom{j}{2} + \binom{k - j}{2} &= j(k - j) + \frac{j(j - 1)}{2} + \frac{(k - j)(k - j - 1)}{2} \\
&= \frac{2jk - 2j^2 + j^2 - j + k^2 - jk - k - jk + j^2 + j}{2} \\
&= \frac{k^2 - k}{2} \\
&= \frac{k(k - 1)}{2} = \binom{k}{2}.
\end{aligned}$$

In particular, note that all of the j terms canceled, meaning that the sum doesn't depend on the initial choice of a split into two piles.

Thus, by induction, if we start with n stones, then we will end up with a sum of $\binom{n}{2}$. In particular, when $n = 25$, the sum is $\binom{25}{2} = 300$. $\square$

The previous problem is an example of **recursion**, which we will see more of in Chapter 10.

As we mentioned, there is a very clever counting solution to the problem that does not require induction. Again, suppose we start with a pile of n stones. We connect every pair of stones by a string. Since there are $\binom{n}{2}$ pairs of stones, we will need $\binom{n}{2}$ strings.

Suppose we divide the n stones into a pile of j stones and a pile of $n - j$ stones, for some integer $0 < j < n$. We then cut all of the strings between stones in different piles. Notice that $j(n - j)$ strings get cut, which is exactly the number that gets written on the blackboard.

We continue splitting piles and cutting strings between stones in different piles. Again, note that at each split, the number of strings being cut always equals the number written on the blackboard.

At the end of the process, each stone is in its own pile, so all of the $\binom{n}{2}$ strings have been cut. But the total number of strings that have been cut equals the sum of the numbers on the blackboard! So the sum of the numbers on the blackboard must be $\binom{n}{2}$.

Problem 8.6: In a large field, n people are standing so that for each person, the distances to all the other people are different. At a given signal, each person fires a water pistol and hits the person who is closest to them. When n is odd, prove that there is at least one person who is left dry.

Solution for Problem 8.6: We are trying to prove a statement for all odd positive integers, so we'll consider using induction.

The result is clearly true for the base case $n = 1$: if there is only one person, then there is nobody else to shoot at him, so he must remain dry.

Since we want to prove this result only for odd positive integers, we'll slightly modify our inductive step. We will assume that the result is true for some odd positive integer $n = k$, and try to prove the result for $n = k + 2$. This basically skips over all the cases where n is an even positive integer, and instead shows that the result for one odd integer implies the result for the next odd integer.

Consider a group of $k + 2$ people with water pistols. There are two people that we know will definitely not stay dry: let A and B be the pair of people who are closest to each other, so A and B end up firing at each other.

We now have two possibilities: either one of the remaining k people also fires at A or B, or none of them do. If one of the remaining k people also fires at A or B, then this leaves at most $k - 1$ shooting at those k people, therefore one of the k people will be left dry.

On the other hand, if none of the k people fire at A or B, then we have a group of k people, all of whom fire at each other. But now we can use our inductive hypothesis! By the induction hypothesis, one of these k people must be left dry.

In either case, we have a person who is left dry. Therefore, the result is true for $n = k + 2$, and by induction, it is true for all odd positive integers n. $\square$

Summary

➤ We can use **mathematical induction** when we wish to prove a statement $S(n)$ that depends on a positive integer n. The proof consists of two steps:

1. Prove $S(1)$.
2. Prove that $S(k)$ implies $S(k + 1)$ for any positive integer k.

➤ Whenever you have a statement that depends on a positive integer n, consider mathematical induction as a possible method of proving the statement.

REVIEW PROBLEMS

8.7 Prove that the sum of the first n positive odd integers is n^2.

8.8 Prove, by induction, the formula for a geometric series:

$$a + ar + ar^2 + \cdots + ar^{n-1} = a\frac{r^n - 1}{r - 1}.$$

8.9 Prove that

$$1^2 + 2^2 + \cdots + n^2 = \frac{n(n+1)(2n+1)}{6}$$

for all positive integers n.

8.10 Prove that $7^n - 1$ is a multiple of 6 for all positive integers n.

8.11 Prove that

$$2! \cdot 4! \cdots \cdot (2n)! \geq ((n+1)!)^n$$

for all positive integers n.

8.12 $2n$ points (where $n > 1$ is a positive integer) are given in space, such that no 3 of them are on a line. We draw $n^2 + 1$ line segments connecting pairs of these points. Prove that there must exist a triangle whose vertices are 3 of the points and whose sides are 3 of the drawn segments.

8.13 A plane is divided into regions by a finite number of lines. Show that it is possible to color the resulting regions with two colors, white and black, so that any two bordering regions are of opposite colors. (Two regions border if and only if their common boundary is a line segment; in particular, regions that only meet at a single point do not border.)

Challenge Problems

8.14 Prove the **generalized triangle inequality**: if $x_1, x_2, \ldots, x_n$ are real numbers, then

$$|x_1 + x_2 + \cdots + x_n| \leq |x_1| + |x_2| + \cdots + |x_n|.$$

Hints: 40, 2

8.15 Prove that $2^{2^n} + 3^{2^n} + 5^{2^n}$ is divisible by 19 for all positive integers n. **Hints:** 174

8.16 For any set T whose elements are positive integers, define $f(T)$ to be the square of the product of the elements of T. For example, if $T = \{1, 3, 6\}$, then $f(T) = (1 \cdot 3 \cdot 6)^2 = 18^2 = 324$.

For any positive integer n, consider all nonempty subsets S of $\{1, 2, \ldots, n\}$ that do not contain two consecutive integers. Prove that the sum of all the $f(S)$'s of these subsets is $(n+1)! - 1$. **Hints:** 268

8.17★ An international conference consists of n representatives from each of n different countries. Prove that the n^2 people can be seated around a large round table such that, if A and B are two distinct representatives from the same country, then the people sitting to the immediate left of A and to the immediate left of B are from different countries. **Hints:** 335, 38

8.18★ Suppose there are n identical cars at different points on a circular track, and that each car needs exactly 1 gallon of gas to make it around the track. Initially, the total amount of gas in all of the cars' fuel tanks is exactly 1 gallon. Show that there is one car that can make it counterclockwise around the track by collecting all of the gasoline from each car that it passes as it moves. **Hints:** 71, 41

8.19★ **Pick's Theorem** states that given a polygon in the coordinate plane, whose vertices are all *lattice points* (that is, points with integer coordinates), the area of the polygon is given by the formula

$$I + \frac{B}{2} - 1,$$

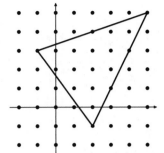

where I is the number of lattice points on the interior of the polygon, and B is the number of lattice points on the boundary of the polygon. For example, in the picture to the right, the triangle has 13 lattice points in its interior, and 6 lattice points on the boundary, so the area is $13 + 6/2 - 1 = 15$.

Assume that Pick's Theorem is true for triangles whose vertices are lattice points. (For an extra challenge, you can try to prove this, but the proof is messy and involves a lot of cases.) Prove that Pick's Theorem is true for all polygons whose vertices are lattice points, by induction on the number of vertices of the polygon. **Hints:** 122, 228

8.20★ Let S_{2002} be a set with 2002 elements, and let N be an integer with $0 \le N \le 2^{2002}$. Prove that it is possible to color every subset of S either black or white so that the following conditions hold:

1. the union of any two white subsets is white;

2. the union of any two black subsets is black;

3. there are exactly N white subsets.

(Source: USAMO) **Hints:** 336, 315

8.21★ An $n \times n$ matrix whose entries come from the set $S = \{1, 2, \ldots, 2n-1\}$ is called a **silver matrix** if, for each $i = 1, 2, \ldots, n$, the ith row and the ith column together contain all elements of S. For example, the matrix to the right is a silver matrix for $n = 4$. Show that silver matrices exist for infinitely many values of n. *(Source: IMO)* **Hints:** 237, 9, 291

$$\begin{pmatrix} 1 & 2 & 5 & 6 \\ 3 & 1 & 7 & 5 \\ 4 & 6 & 1 & 2 \\ 7 & 4 & 3 & 1 \end{pmatrix}$$

The good, of course, is always beautiful, and the beautiful never lacks proportion. – Plato

CHAPTER **9**

_____ **Fibonacci Numbers**

9.1   Introduction

$$1, 1, 2, 3, 5, 8, 13, 21, 34, 55, 89, 144, 233, 377, 610, 987, \ldots$$

What is special about the above sequence of numbers? This sequence is one of the most famous sequences of positive integers in all of mathematics. If you've seen it before, you probably know how these numbers are generated, but if not, I don't want to spoil the surprise!

In this chapter, we'll explore some of the fascinating properties of this sequence.

9.2   A Motivating Problem

Problem 9.1: Mike is climbing a flight of 10 stairs. With each step, he will climb either 1 or 2 stairs. In how many different ways can he climb the flight of stairs?

(a) Try it for 1, 2, 3, or 4 stairs. You results should appear in the list of numbers above.

(b) Solve the problem by casework, where the cases depend on the number of 2-stair steps that Mike takes.

(c) Let $f(n)$ be the number of ways to climb n steps. How is $f(n)$ related to $f(n-1)$ and $f(n-2)$?

(d) Solve the problem by writing an expression for $f(n)$ in terms of $f(n-1)$ and $f(n-2)$, and using this expression, together with your values for $f(1)$ and $f(2)$ from part (a), to compute $f(10)$.

We'll start our investigation of the special sequence listed on the previous page by looking at a problem whose answer is related to the sequence.

Problem 9.1: Mike is climbing a flight of 10 stairs. With each step, he will climb either 1 or 2 stairs. In how many different ways can he climb the flight of stairs?

Solution for Problem 9.1: We could count this directly with some casework. Let's do that first for practice.

In order to climb 10 stairs, Mike can take 0, 1, 2, 3, 4, or 5 2-stair steps (with the rest being 1-stair steps).

If he takes 0 2-stair steps and 10 1-stair steps, there is $\binom{10}{0} = 1$ way to reach the top.

If he takes 1 2-stair step and 8 1-stair steps, then we have 9 total steps to arrange, and must choose 1 of them to be the 2-stair step. Therefore, there are $\binom{9}{1} = 9$ ways to reach the top.

If he takes 2 2-stair steps and 6 1-stair steps, then we have 8 steps to arrange, and must choose 2 of them to be the 2-stair steps. Therefore, there are $\binom{8}{2} = 28$ ways to reach the top.

If he takes 3 2-stair steps and 4 1-stair steps, then there are $\binom{7}{3} = 35$ ways to reach the top.

If he takes 4 2-stair steps and 2 1-stair steps, then there are $\binom{6}{4} = 15$ ways to reach the top.

If he takes 5 2-stair steps and 0 1-stair steps, then there is $\binom{5}{5} = 1$ way to reach the top.

Therefore, there are $1 + 9 + 28 + 35 + 15 + 1 = 89$ ways in which Mike can climb the stairs.

That was simple enough. But there's another way that we can think about the problem.

Consider Mike's final step. Either Mike took a 1-stair step from the 9th stair, or he took a 2-stair step from the 8th stair. This means that we have a 1-1 correspondence:

$$\{\text{Ways to climb 10 stairs}\} \leftrightarrow \{\text{Ways to climb 9 stairs}\} \cup \{\text{Ways to climb 8 stairs}\}.$$

Therefore, when we count, we get

(Number of ways to climb 10 stairs) = (Number of ways to climb 9 stairs)

$$+ \text{(Number of ways to climb 8 stairs)}.$$

If we let $f(n)$ denote the number of ways to climb n stairs, then this means that $f(10) = f(9) + f(8)$.

This works for any number of stairs (as long as there are more than 2). In general, for any positive integer $n > 2$, we have the equation $f(n) = f(n-1) + f(n-2)$. In this way, we can build up to the number of ways to climb 10 stairs by counting the number of ways to climb a smaller number of stairs.

We start by noting that $f(1) = 1$ and $f(2) = 2$. If there is only 1 stair, then the only way to climb it is via a 1-stair step. If there are 2 stairs, then we have 2 choices: Mike can take two 1-stair steps, or he can take a single 2-stair step.

Extra! *You don't have to see the whole staircase, just take the first step.* – Martin Luther King, Jr.
▮▶ ▮▶ ▮▶ ▮▶

Now we can count the number of ways to climb larger staircases:

$$
\begin{aligned}
f(3) &= f(2) + f(1) &= 2 + 1 &= 3, \\
f(4) &= f(3) + f(2) &= 3 + 2 &= 5, \\
f(5) &= f(4) + f(3) &= 5 + 3 &= 8, \\
f(6) &= f(5) + f(4) &= 8 + 5 &= 13, \\
f(7) &= f(6) + f(5) &= 13 + 8 &= 21, \\
f(8) &= f(7) + f(6) &= 21 + 13 &= 34, \\
f(9) &= f(8) + f(7) &= 34 + 21 &= 55, \\
f(10) &= f(9) + f(8) &= 55 + 34 &= 89.
\end{aligned}
$$

Once again, we see that there are 89 ways to climb a 10-stair staircase. □

If we list $f(1), f(2), f(3), \ldots$, then we (almost) get the sequence that opened this chapter:

$$1, 2, 3, 5, 8, 13, 21, 34, 55, 89, \ldots.$$

Each number in the sequence (after the first two) is the sum of the two numbers immediately before it: $2 = 1 + 1, 3 = 2 + 1, 5 = 3 + 2$, and so on.

The only thing that's different from the sequence on page 172 is that the sequence on page 172 has an extra 1 at the start. We can add this 1 to our example, though, without any difficulty. If there are 0 stairs, then there's only 1 way to climb this staircase: do nothing! So $f(0) = 1$ makes sense. Also note that $f(2) = f(1) + f(0) = 1 + 1 = 2$, so that our equation $f(n) = f(n-1) + f(n-2)$ holds for all positive integers $n \geq 2$.

When we add this first 1 to our sequence, we get the sequence that we first saw at the start of the chapter:

$$1, 1, 2, 3, 5, 8, 13, 21, 34, 55, 89, \ldots.$$

These numbers are called the **Fibonacci numbers**, in honor of the Italian mathematician Leonardo of Pisa, whose nickname was "Fibonacci," and who first published the sequence of numbers in his *Liber abaci* in 1202.

We've seen that each Fibonacci number is the sum of the previous two Fibonacci numbers. We can write this using a more formal definition.

We typically denote sequences using a variable with a subscript, such as

$$a_1, a_2, a_3, \ldots.$$

Thus, a_1 is the first number in the sequence, a_2 is the second number in the sequence, and so on. Sometimes we'll start our lists with a_0 instead of a_1, so that our sequence would be

$$a_0, a_1, a_2, \ldots.$$

The **Fibonacci numbers** are defined by $F_1 = 1$, $F_2 = 1$, and $F_n = F_{n-1} + F_{n-2}$ for all positive integers $n > 2$. So, for example,

$$F_3 = F_2 + F_1 = 1 + 1 = 2,$$
$$F_4 = F_3 + F_2 = 2 + 1 = 3,$$
$$F_5 = F_4 + F_3 = 3 + 2 = 5,$$
$$F_6 = F_5 + F_4 = 5 + 3 = 8,$$
$$F_7 = F_6 + F_5 = 8 + 5 = 13,$$

etc.

Sometimes, it is convenient to have a value of F_0, so we also define $F_0 = 0$. Note that $1 = F_2 = F_1 + F_0 = 1 + 0$, so this definition is consistent.

Relating this notation to our solution to Problem 9.1, we see that the number of ways that Mike can climb n steps is F_{n+1} (*not* F_n).

9.3   Some Fibonacci Problems

Problems

Problem 9.2: An adult pair of rabbits is in an enclosed yard. Every month, every adult pair of rabbits produces a pair of offspring, which grows to adulthood in 2 months and then begins itself to produce offspring. (Assume that rabbits never die.)

(a) Make a table of the number of adult pairs of rabbits, the number of child pairs of rabbits, and the total number of pairs of rabbits, for each month of the year.

(b) Do you see any patterns in your table from part (a)?

(c) How does your table relate to Fibonacci numbers?

(d) Prove your results.

Problem 9.3: How many subsets of $\{1, 2, \ldots, n\}$ have no two consecutive integers as elements? (For example, the subset $\{1, 3, 4, 8\}$ has the consecutive integers 3 and 4 as elements, but the subset $\{2, 4, 7, 9\}$ has no two consecutive integers as elements.)

(a) List and count all of the valid subsets for $n = 1, 2, 3, 4$.

(b) Do you notice the pattern (again)?

(c) Establish a 1-1 correspondence:

$$\left\{ \begin{matrix} \text{Subsets of } \{1, 2, \ldots, n\} \text{ with no} \\ \text{two consecutive elements} \end{matrix} \right\}$$

$$\updownarrow$$

$$\left(\left\{ \begin{matrix} \text{Subsets of } \{1, 2, \ldots, n-1\} \text{ with} \\ \text{no two consecutive elements} \end{matrix} \right\} \cup \left\{ \begin{matrix} \text{Subsets of } \{1, 2, \ldots, n-2\} \text{ with} \\ \text{no two consecutive elements} \end{matrix} \right\} \right).$$

(d) Use this correspondence to answer the original problem.

Problem 9.4: In how many ways can we tile a $2 \times n$ checkerboard with n tiles of size 1×2, such that each tile covers exactly two squares? An example with $n = 6$ is shown below.

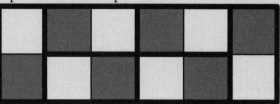

Problem 9.5: Prove that for any $k \geq 1$,

$$F_1 + F_3 + F_5 + \cdots + F_{2k+1} = F_{2k+2}.$$

Problem 9.6: Prove that for all positive integers $n \geq 1$,

$$F_1^2 + F_2^2 + \cdots + F_n^2 = F_n F_{n+1}.$$

As we will see in this section, the Fibonacci numbers pop up in a lot of different contexts in counting problems. We'll start with a classic example.

Problem 9.2: An adult pair of rabbits is in an enclosed yard. Every month, every adult pair of rabbits produces a pair of offspring, which grows to adulthood in 2 months and then begins itself to produce offspring. How many pairs of rabbits will there be after a year? (Assume that rabbits never die.)

Solution for Problem 9.2: We can experiment and make a table. We use the facts that every month:

- Every adult pair has a pair of newborn offspring;

- Every pair of newborns from last month becomes a pair of child rabbits this month;

- Every pair of child rabbits from last month becomes a pair of adult rabbits this month.

Using these facts, we can generate a table:

Extra!　　　　　　　　　　**Leonardo of Pisa a.k.a. Fibonacci**

➠➠➠➠　It is not universally agreed upon how Leonardo of Pisa got the nickname "Fibonacci." It is generally thought to be a shortening of "filius Bonacci," which means "son of Bonacci," although it is unclear whether Bonacci was a family name or simply a nickname (meaning perhaps "lucky" or "good natured"). The Fibonacci numbers were not named after Fibonacci until long after his death, by the 19th-century French mathematician Edouard Lucas (who also has a series of numbers named after him, as you will see in an Exercise later in the chapter). See the book's links page on page vi for links to sources for the above and more complete biographical information on Fibonacci.

Month	# of adult pairs	# of child pairs	# of newborn pairs	Total
0	1	0	0	1
1	1	0	1	2
2	1	1	1	3
3	2	1	2	5
4	3	2	3	8
5	5	3	5	13
6	8	5	8	21
7	13	8	13	34
8	21	13	21	55
9	34	21	34	89
10	55	34	55	144
11	89	55	89	233
12	144	89	144	377

So we see that there will be 377 pairs of rabbits after a year. (I hope it's a big yard!)

Note that our table has Fibonacci numbers all over the place! There ought to be a reason why.

Since the rabbits never die, all of the rabbits alive in month $n - 1$ will live on to month n. But new rabbits are also born in month n. Therefore,

(# of pairs of rabbits in month n) = (# of pairs rabbits in month $n - 1$)

+ (# of newborn pairs of rabbits in month n).

How many pairs of rabbits are born? All adult pairs produce offspring. The adults are precisely those rabbits that were alive two months ago. So we can write

(# of pairs of rabbits in month n) = (# of pairs of rabbits in month $n - 1$)

+ (# of pairs of rabbits in month $n - 2$).

So if we let r_n denote the number of pairs of rabbits in month n, we have the equation

$$r_n = r_{n-1} + r_{n-2}.$$

That looks like Fibonacci numbers!

We could very easily stumble into a common mistake here.

> **Bogus Solution:** Our above work shows that the number of pairs of rabbits in month n is the nth Fibonacci number. Therefore, after a year, we get the 12th Fibonacci number F_{12}.

This doesn't work because the Fibonacci numbers are defined starting with $F_1 = F_2 = 1$. Here, we're starting with $r_0 = 1$ and $r_1 = 2$. So our number of pairs of rabbits are the Fibonacci numbers, but they're all shifted. To be exact, $r_n = F_{n+2}$. This means that the answer to our rabbit problem is the 14th Fibonacci number, F_{14}, which is 377. $\square$

Sidenote: Problem 9.2 is exactly the problem that Fibonacci himself first studied in his work *Liber abaci* in 1202. One translation of the original problem from *Liber abaci* reads:

> A certain man put a pair of rabbits in a place surrounded on all sides by a wall. How many pairs of rabbits can be produced from that pair in a year if it is supposed that every month each pair begets a new pair which from the second month on becomes productive?

(Source: MacTutor History of Mathematics)

Problem 9.3: How many subsets of $\{1, 2, \ldots, n\}$ have no two consecutive integers as elements?

Solution for Problem 9.3: One idea might be to do a nasty PIE calculation to compute the number of subsets that *do* have at least two consecutive elements. But try it for about 5 minutes and you'll see that it's pretty ugly.

Another idea to try some small values of n, and look for a pattern.

Concept: When trying to find or prove something for all positive integers n, try some small values of n, and look for a pattern.

If $n = 1$, then there are 2 valid subsets: $\emptyset$ and $\{1\}$. (Don't forget about $\emptyset$!)

If $n = 2$, then there are 3 valid subsets: $\emptyset$, $\{1\}$, and $\{2\}$.

If $n = 3$, then there are 5 valid subsets: $\emptyset$, $\{1\}$, $\{2\}$, $\{3\}$, and $\{1,3\}$.

If $n = 4$, then there are 8 valid subsets: $\emptyset$, $\{1\}$, $\{2\}$, $\{3\}$, $\{4\}$, $\{1,3\}$, $\{1,4\}$, and $\{2,4\}$.

So far, for $n = 1, 2, 3, 4, \ldots$, we have $2, 3, 5, 8 \ldots$—those are Fibonacci numbers!

Let s_n denote what we're trying to count: the number of subsets of $\{1, 2, \ldots, n\}$ with no two consecutive elements. It sure looks like the s_n are the Fibonacci numbers, meaning that we want to try to show that

$$s_n = s_{n-1} + s_{n-2}$$

for all $n > 2$.

So let's think about it a little bit more carefully. Is there any way that we can build the valid subsets of $\{1, 2, \ldots, n\}$ from the valid subsets of $\{1, 2, \ldots, n-1\}$ and $\{1, 2, \ldots, n-2\}$?

Let's go back to our example and list the subsets for $n = 2, 3, 4$:

n	Valid subsets of $\{1, 2, \ldots, n\}$
2	$\emptyset, \{1\}, \{2\}$
3	$\emptyset, \{1\}, \{2\}, \{3\}, \{1,3\}$
4	$\emptyset, \{1\}, \{2\}, \{3\}, \{4\}, \{1,3\}, \{1,4\}, \{2,4\}$

Notice that all of the subsets in the $n = 3$ list are also in the $n = 4$ list. The leftovers in the $n = 4$ list (that aren't in the $n = 3$ list) are $\{4\}$, $\{1, 4\}$, and $\{2, 4\}$. Compare these with the subsets in the $n = 2$ list. They're the same subsets, but with a 4 added.

This observation can be generalized to any $n \geq 3$, as follows. Every subset of $\{1, 2, \ldots, n\}$ either contains n or doesn't contain n. A subset that doesn't contain n is a subset of $\{1, 2, \ldots, n - 1\}$. A subset that does contain n is not allowed to contain $n - 1$ (since it can't contain two consecutive elements), so the remaining elements form a subset of $\{1, 2, \ldots, n - 2\}$.

Therefore, we've established a 1-1 correspondence:
$$\left\{\begin{matrix} \text{Subsets of } \{1, 2, \ldots, n\} \text{ with no} \\ \text{two consecutive elements} \end{matrix}\right\} \leftrightarrow \left(\left\{\begin{matrix} \text{Subsets of } \{1, 2, \ldots, n-1\} \text{ with} \\ \text{no two consecutive elements} \end{matrix}\right\} \cup \left\{\begin{matrix} \text{Subsets of } \{1, 2, \ldots, n-2\} \text{ with} \\ \text{no two consecutive elements} \end{matrix}\right\}\right).$$
The above correspondence means that
$$s_n = s_{n-1} + s_{n-2}$$
for all $n > 2$. Look familiar?

Beware! As in Problem 9.2, don't jump to the conclusion that the answer is the nth Fibonacci number. We still need to check the initial conditions. We already listed the subsets for $n = 1$ and $n = 2$ and found that $s_1 = 2$ and $s_2 = 3$. Therefore, we see that $s_n = F_{n+2}$, so the answer is the $(n + 2)^{\text{nd}}$ Fibonacci number. □

> **Sidenote:** Since the answers to Problems 9.2 and 9.3 were the same, can you find a natural 1-1 correspondence between
>
> $$\{\text{Pairs of rabbits after } n \text{ months}\} \leftrightarrow \left\{\begin{matrix} \text{Subsets of } \{1, 2, \ldots, n\} \text{ with no} \\ \text{two consecutive elements} \end{matrix}\right\}?$$
>
> The answer is on page 191.

Problem 9.4: In how many ways can we tile a $2 \times n$ checkerboard with n tiles of size 1×2, such that each tile covers exactly two squares? An example with $n = 6$ is shown below.

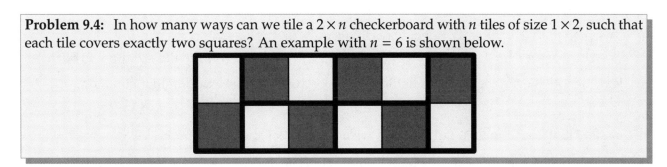

Solution for Problem 9.4: As we often do, let's start with a few small examples to get an idea of what the solution might be.

There's only 1 way to tile a 2×1 board:

There are 2 ways to tile a 2×2 board:

There are 3 ways to tile a 2×3 board:

There are 5 ways to tile a 2×4 board:

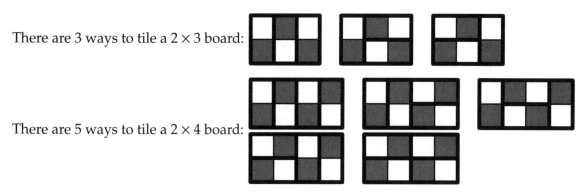

Hmmm... the first four values of n give us $1, 2, 3, 5$ possible tilings. So the answer might be the Fibonacci numbers. With this goal in mind, we try to think about how tilings of the $2 \times n$ board can be built from tilings of the $2 \times (n-1)$ and $2 \times (n-2)$ boards.

Suppose $n > 2$, and think about how we might place the tile which covers the lower-left square. There are two cases.

Case 1: Place the tile vertically. Then we must tile the remaining $2 \times (n-1)$ board.

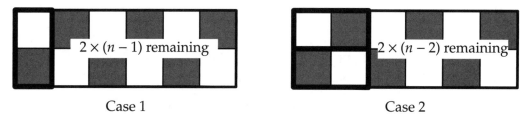

Case 1   Case 2

Case 2: Place the tile horizontally. Then we must also place a horizontal tile covering the upper-left square (since that's the only way to cover that square). What's left is a $2 \times (n-2)$ board that we must tile.

So we see that

(# of ways to tile a $2 \times n$ board) = (# of ways to tile a $2 \times (n-1)$ board)

+ (# of ways to tile a $2 \times (n-2)$ board).

If we let t_n denote the number of ways to tile a $2 \times n$ board, then we see the familiar equation

$$t_n = t_{n-1} + t_{n-2}.$$

Once again, Fibonacci numbers! But as usual, we need to check the initial conditions.

There's only one way to tile a 2×1 board, so $t_1 = 1$. There are two ways to tile a 2×2 board (either place both tiles horizontally or both tiles vertically), so $t_2 = 2$.

Since $t_1 = F_2$, $t_2 = F_3$, and $t_n = t_{n-1} + t_{n-2}$, we see that in general, $t_n = F_{n+1}$, which is the $(n+1)^{\text{st}}$ Fibonacci number. $\square$

The Fibonacci numbers satisfy lots and lots of identities. Here's a fairly simple one:

Problem 9.5: Prove that for any $k \geq 1$,

$$F_1 + F_3 + F_5 + \cdots + F_{2k+1} = F_{2k+2}.$$

Solution for Problem 9.5: Just to convince ourselves that the result is actually true, let's check it for $k = 4$:

$$F_1 + F_3 + F_5 + F_7 + F_9 = 1 + 2 + 5 + 13 + 34 = 55 = F_{10}.$$

Let's play with our example and see if we can determine why it works.

> **Concept:** Before tackling the general case of an identity or equation, it often helps to play with a simple example or two, in order to get a feel for what's going on.

If we start collapsing the sum from left-to-right, something interesting happens:

$$1 + 2 + 5 + 13 + 34 = 3 + 5 + 13 + 34$$
$$= 8 + 13 + 34$$
$$= 21 + 34$$
$$= 55.$$

Notice that the first terms of each intermediate sum are the "missing" Fibonacci numbers. For example, $3 = F_4$, $8 = F_6$, and $21 = F_8$.

Now let's look at the general case. To get the ball rolling, we need to use the fact that $F_1 = F_2 = 1$. Then the sum collapses, just like in our example.

$$
\begin{aligned}
F_1 + F_3 + F_5 + F_7 + \cdots + F_{2k+1} &= F_2 + F_3 + F_5 + F_7 + \cdots + F_{2k+1} \\
&= F_4 + F_5 + F_7 + \cdots + F_{2k+1} \\
&= F_6 + F_7 + \cdots + F_{2k+1} \\
&\vdots \\
&= F_{2k} + F_{2k+1} \\
&= F_{2k+2}.
\end{aligned}
$$

We can also prove the result more formally using mathematical induction.

Base case: If $k = 1$, then
$$F_1 + F_3 = 1 + 2 = 3 = F_4.$$

So the result holds for $k = 1$.

Inductive step: Assume that the result holds for some positive integer k. We will attempt to prove the result for $k + 1$:

$$
\begin{aligned}
F_1 + F_3 + \cdots + F_{2(k+1)+1} &= (F_1 + F_3 + \cdots + F_{2k+1}) + F_{2k+3} \\
&= F_{2k+2} + F_{2k+3} \\
&= F_{2k+4},
\end{aligned}
$$

where the last step $F_{2k+2} + F_{2k+3} = F_{2k+4}$ comes from the definition of the Fibonacci numbers.

Therefore, the result is true by induction. $\square$

Let's see a more complicated Fibonacci identity.

Problem 9.6: Prove that for all positive integers $n \geq 1$,

$$F_1^2 + F_2^2 + \cdots + F_n^2 = F_n F_{n+1}.$$

Solution for Problem 9.6: There are two basic ways we can approach this: we can try to prove it algebraically with an induction proof, or we can try to look for a clever counting argument. Let's do the induction proof here.

Base case: If we let $n = 1$, we get $F_1^2 = 1^2 = 1 = (1)(1) = F_1 F_2$, so the base case holds.

Inductive step: Assume that the identity is true for some positive integer k. We will write the left side of the identity for $k + 1$:

$$F_1^2 + F_2^2 + \cdots + F_k^2 + F_{k+1}^2.$$

All but the final term of this expression can be replaced by $F_k F_{k+1}$ using the inductive hypothesis, so we have:

$$F_1^2 + \cdots + F_k^2 + F_{k+1}^2 = F_k F_{k+1} + F_{k+1}^2.$$

The right side of the above can be factored as $(F_k + F_{k+1})F_{k+1}$. And what is that expression inside the parentheses? It's $F_k + F_{k+1}$, which by the definition of the Fibonacci numbers is equal to F_{k+2}. Therefore, the whole expression is simply $F_{k+2}F_{k+1}$, and thus we have proved that

$$F_1^2 + \cdots + F_{k+1}^2 = F_{k+1}F_{k+2}.$$

Hence, by induction, the result is proved.

Although the algebraic induction proof was relatively straightforward, it didn't really provide any insight into *why* the identity is true. For a deeper understanding, we can try to find a counting argument that proves the identity. We'll leave this for you as an Exercise. □

You'll get a chance to see more Fibonacci identities in the Exercises and Challenge Problems. We'll also revisit identities in general in Chapter 12.

Exercises

9.3.1 Norman wishes to buy a can of soda costing 75 cents from a vending machine. He has an unlimited supply of identical nickels (worth 5 cents each) and dimes (worth 10 cents each). In how many different orders can he insert coins into the machine to pay for his soda?

9.3.2 Three sheets of glass are arranged in parallel. A light ray passes through the top sheet, reflects a number of times between the sheets, and exits the glass. For example, the diagram below shows the possible paths that the ray can take, if it reflects exactly 3 times:

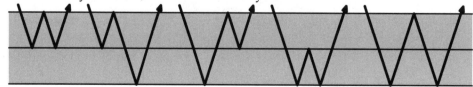

How many different paths can the ray take if it reflects exactly 9 times?

9.3.3 Suppose you toss a fair coin 10 times. What is the probability that you do not get two heads in a row? **Hints:** 45, 147

9.3.4 Prove that F_n is a multiple of 3 if and only if n is a multiple of 4. **Hints:** 304

9.3.5 Prove that for any $k \geq 1$,

$$1 + F_2 + F_4 + F_6 + \cdots + F_{2k} = F_{2k+1}.$$

9.3.6 A subset S of $\{1, 2, \ldots, n\}$ is called *selfish* if $\#(S) \in S$. A selfish subset is called *minimal* if none of its proper subsets are selfish. Find the number of minimal selfish subsets of $\{1, 2, \ldots, n\}$. *(Source: Putnam)* **Hints:** 310

9.3.7★ Find a counting argument to prove the identity from Problem 9.6. **Hints:** 142, 199

9.4 A Formula for the Fibonacci Numbers

Note: the problems are this section are somewhat algebraically intense and can be a bit abstract. If you are having difficulty with these problems, you may wish to skip ahead to Chapter 10 and then come back to this section after you've learned more about the general theory of recursions.

Problems

Problem 9.7: In this problem, we work out a formula for the Fibonacci numbers.

(a) List the first several Fibonacci numbers, and compute the ratios between successive Fibonacci numbers. Do you notice these ratios approaching a common value?

(b) Suppose that $a_n = c^n$ satisfies the recursion $a_n = a_{n-1} + a_{n-2}$ for some constant c. Find the possible values of c.

(c) Show that $a_n = \lambda_1 c_1^n + \lambda_2 c_2^n$ satisfies the recursion from part (b), where c_1 and c_2 are the values you found in part (b), and λ_1 and λ_2 are arbitrary constants.

(d) Find a formula for the nth Fibonacci number, by using the known values of a_1 and a_2 to solve for λ_1 and λ_2 in part (c).

Problem 9.8: Is there a simpler formula for the Fibonacci numbers? (Here, you are allowed to use the nearest integer function: $[x]$ is the integer closest to x, so that for example $[1.2] = 1$ and $[1.7] = 2$. By convention, we always round towards even numbers, so that $[1.5] = [2.5] = 2$.)

Problem 9.9: Prove that

$$F_{n-1}F_{n+1} - F_n^2 = (-1)^n,$$

where F_n is the nth Fibonacci number.

As we have seen, the Fibonacci numbers are given by the recursion

$$F_n = F_{n-1} + F_{n-2},$$

for all $n > 2$, where $F_1 = 1$ and $F_2 = 1$. Although we can now go ahead and list arbitrarily many Fibonacci numbers using this recursion, it's a bit of a pain to calculate, say, the 100th Fibonacci number, because we'd have to list out the first 99 Fibonacci numbers in order to find the 100th. It'd be awfully nice if there were just a formula for F_n that we could plug $n = 100$ into and get the 100th Fibonacci number right away.

The good news is that there is such a formula! The bad news is that, on first glance, it's *really* messy. Let's work it out.

Problem 9.7: Find a formula for F_n, the nth Fibonacci number, where $F_1 = F_2 = 1$ and $F_n = F_{n-1} + F_{n-2}$ for all $n > 2$.

Solution for Problem 9.7: Let's list the first twelve Fibonacci numbers:

$$1, 1, 2, 3, 5, 8, 13, 21, 34, 55, 89, 144, \ldots$$

Where can we even begin our search for a general formula?

We can start by drawing a graph of the Fibonacci numbers F_n as a function of n; the graph is shown at the right. Note that the y-axis is scaled differently than the x-axis, since otherwise the graph wouldn't fit. We see that F_n grows fast. Really fast. And the fastest-growing "nice" functions that we know of are exponential functions (hence the phrase "exponential growth"). So our strategy is to suppose that the Fibonacci numbers satisfy an exponential expression, and hope that we get lucky. Looking at the ratio of successive Fibonacci numbers (rounded to three decimal places), we suspect that we will get lucky:

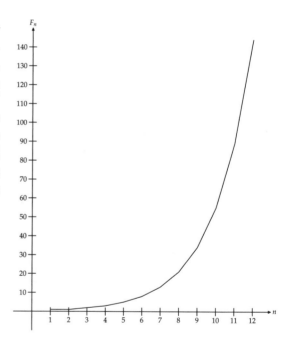

n	F_n/F_{n-1}	Decimal
2	1/1	1.000
3	2/1	2.000
4	3/2	1.500
5	5/3	1.667
6	8/5	1.600
7	13/8	1.625
8	21/13	1.615
9	34/21	1.619
10	55/34	1.618
11	89/55	1.618
12	144/89	1.618

We see that the ratio between consecutive terms of the Fibonacci sequence appears to converge to about 1.618. This means that the Fibonacci sequence appears to converge toward a geometric sequence with ratio about 1.618. Let's see if we can make this statement more precise.

Specifically, we let $a_n = c^n$ for some unknown constant c. If we plug this into our recurrence relation for the Fibonacci numbers, we get

$$a_n = a_{n-1} + a_{n-2} \quad \Rightarrow \quad c^n = c^{n-1} + c^{n-2}.$$

Since $c \neq 0$ (otherwise it would be a pretty stupid formula), we can divide by c^{n-2} to get $c^2 = c + 1$.

Good, we know how to solve this! If we rewrite it as $c^2 - c - 1 = 0$, we can use the quadratic formula:

$$c = \frac{1 \pm \sqrt{5}}{2}.$$

This gives us two possible candidates for c:

$$c_1 = \frac{1 + \sqrt{5}}{2}, \qquad c_2 = \frac{1 - \sqrt{5}}{2}.$$

Note that $c_1 \approx 1.618$ and $c_2 \approx -0.618$; in particular, c_1 seems to match the ratio of consecutive Fibonacci numbers that we found in our earlier experimentation.

How do we know which value, c_1 or c_2, to use, or if either one works? Based on our calculations above, we know that $a_n = c_1^n$ and $a_n = c_2^n$ each satisfy the relation $a_n = a_{n-1} + a_{n-2}$. Unfortunately, neither satisfies the initial conditions $a_1 = a_2 = 1$. So how do we fix it?

We observe that any a_n of the form

$$a_n = \lambda_1 c_1^n + \lambda_2 c_2^n,$$

where λ_1 and λ_2 are constants, satisfies the recurrence relation. (λ is the greek letter *lambda*, which is often used to represent unknown constants.) To see this, we simply plug it in for a_n, a_{n-1}, and a_{n-2} in our recurrence relation, and verify that it holds:

$$
\begin{aligned}
a_n = \lambda_1 c_1^n + \lambda_2 c_2^n &= \lambda_1 (c_1^{n-1} + c_1^{n-2}) + \lambda_2 (c_2^{n-1} + c_2^{n-2}) \\
&= \lambda_1 c_1^{n-1} + \lambda_2 c_2^{n-1} + \lambda_1 c_1^{n-2} + \lambda_2 c_2^{n-2} \\
&= a_{n-1} + a_{n-2}.
\end{aligned}
$$

Now we have a plan: use the above expression for a_n, plug in the initial conditions, and solve for the unknown constants λ_1 and λ_2.

$$
\begin{aligned}
1 &= \lambda_1 c_1 + \lambda_2 c_2, \\
1 &= \lambda_1 c_1^2 + \lambda_2 c_2^2.
\end{aligned}
$$

We're going to have to get our hands dirty a bit, and plug in c_1 and c_2. Let's clear the denominators while we're at it.

$$
\begin{aligned}
2 &= \lambda_1 (1 + \sqrt{5}) + \lambda_2 (1 - \sqrt{5}), \\
4 &= \lambda_1 (1 + \sqrt{5})^2 + \lambda_2 (1 - \sqrt{5})^2.
\end{aligned}
$$

This is probably not the prettiest system of linear equations that you've ever seen.

Fortunately, we can use a little trick to make the system a lot simpler. Extend the sequence back one term to $a_0 = 0$. Then our system (using $a_0 = 0$ and $a_1 = 1$) becomes:

$$0 = \lambda_1 c_1^0 + \lambda_2 c_2^0,$$
$$1 = \lambda_1 c_1 + \lambda_2 c_2.$$

This simplifies quite nicely:

$$0 = \lambda_1 + \lambda_2,$$
$$2 = \lambda_1(1 + \sqrt{5}) + \lambda_2(1 - \sqrt{5}).$$

Now solving is fairly simple (if a bit messy). The first equation gives us $\lambda_2 = -\lambda_1$, and substituting this into the second equation gives:

$$2 = \lambda_1(-(1 - \sqrt{5}) + (1 + \sqrt{5})).$$

This simplifies to

$$\lambda_1 = \frac{2}{2\sqrt{5}} = \frac{1}{\sqrt{5}}.$$

Then plugging λ_1 back into $0 = \lambda_1 + \lambda_2$, we see that

$$\lambda_2 = -\frac{1}{\sqrt{5}}.$$

Thus we conclude that

$$a_n = \frac{1}{\sqrt{5}}\left(\frac{1 + \sqrt{5}}{2}\right)^n - \frac{1}{\sqrt{5}}\left(\frac{1 - \sqrt{5}}{2}\right)^n.$$

We can make it a little bit nicer by factoring, and we leave it in the final (relatively) nice form:

$$a_n = \frac{(1 + \sqrt{5})^n - (1 - \sqrt{5})^n}{2^n\sqrt{5}}.$$

□

Important: A closed formula for the nth Fibonacci number F_n is

$$F_n = \frac{(1 + \sqrt{5})^n - (1 - \sqrt{5})^n}{2^n\sqrt{5}}.$$

This formula is known as **Binet's formula** for the Fibonacci numbers, as it was derived by the French mathematician Jacques Binet in the mid-1800s (although it was certainly known earlier).

For example, let's compute F_4 and verify that indeed we get $F_4 = 3$. If nothing else, this is good practice in using the Binomial Theorem.

$$F_4 = \frac{(1 + \sqrt{5})^4 - (1 - \sqrt{5})^4}{2^4\sqrt{5}} = \frac{(1 + 4\sqrt{5} + 30 + 20\sqrt{5} + 25) - (1 - 4\sqrt{5} + 30 - 20\sqrt{5} + 25)}{16\sqrt{5}} = \frac{48\sqrt{5}}{16\sqrt{5}} = 3.$$

This may seem like a very weird formula for a sequence of integers! But look carefully—when we compute $(1 + \sqrt{5})^n - (1 - \sqrt{5})^n$, all of the terms without a $\sqrt{5}$ in them will cancel out. All of the remaining terms will have odd powers of $\sqrt{5}$, and thus the $\sqrt{5}$ will get cancelled out by the $\sqrt{5}$ term in the denominator, so the final result will be a rational number. (In fact it will be an integer: the powers of 2 will cancel as well, though this is a lot harder to see right away.)

Although this formula works, it's still messy to use in practice. For example, this is not a practical way to calculate F_{100}, since we'd have to expand $(1 + \sqrt{5})^{100}$.

Problem 9.8: Is there a simpler formula for the Fibonacci numbers?

Solution for Problem 9.8: To be fair, the answer to this question depends on how we define "simpler."

Let's take another look at our formula:

$$F_n = \frac{(1 + \sqrt{5})^n - (1 - \sqrt{5})^n}{2^n \sqrt{5}}.$$

Let's break it up into its two terms, and put the 2^n inside the numerator's exponentials:

$$F_n = \frac{1}{\sqrt{5}} \left(\frac{1 + \sqrt{5}}{2} \right)^n - \frac{1}{\sqrt{5}} \left(\frac{1 - \sqrt{5}}{2} \right)^n.$$

Take a look at that second term. Not very big, is it?

In fact, when n is large, it's really tiny. Note that $1 - \sqrt{5} \approx -1.24$, so $(1 - \sqrt{5})/2 \approx -0.62$. So we have

$$F_n \approx \frac{1}{\sqrt{5}} \left(\frac{1 + \sqrt{5}}{2} \right)^n - \frac{(-0.62)^n}{\sqrt{5}}.$$

That term at the right end always has absolute value less than $\frac{1}{2}$. It becomes much, much smaller as n gets larger, but even where $n = 0$, it's $1/\sqrt{5} \approx 0.447$, which is less than 0.5. Therefore, since we know that F_n is an integer, we conclude that F_n is whatever integer is closest to the left term, since the right term only alters the value of F_n by at most $\frac{1}{2}$.

Let $[x]$ denote the closest integer to x. For example, $[1.2] = 1$ and $[1.8] = 2$. Basically, we're rounding x to the nearest integer. By convention, if we're exactly halfway between two integers, we always round towards the even integer, so for example, $[1.5] = [2.5] = 2$. Then our discussion above establishes that

$$F_n = \left[\frac{1}{\sqrt{5}} \left(\frac{1 + \sqrt{5}}{2} \right)^n \right].$$

□

Sidenote: **The golden ratio**

The number

$$\phi = \frac{1 + \sqrt{5}}{2} \approx 1.618033989\ldots$$

is a very special irrational number. It is called the **golden ratio**.

continued on next page...

Sidenote: ... *continued from previous page*

♪ The golden ratio many nice properties. We have already seen that the golden ratio is the value to which the ratio of successive Fibonacci numbers converges, and also that the golden ratio is the positive root of $x^2 - x - 1 = 0$. The golden ratio can also be written as a continued square root as

$$\phi = \sqrt{1 + \sqrt{1 + \sqrt{1 + \sqrt{1 + \cdots}}}}.$$

The ratio is also considered an aesthetically-pleasing ratio of side lengths of a rectangle. For example, the large rectangle at right has the golden ratio as the ratio of its side lengths. Note how we can remove a square from the figure and get a smaller rectangle with the same ratio of side lengths—can you figure out why this is so?

Now at least we can use our calculator or computer to determine large Fibonacci numbers. But the real power of Binet's formula is in proving some nontrivial Fibonacci identities. Here's an example:

Problem 9.9: Prove that
$$F_{n-1}F_{n+1} - F_n^2 = (-1)^n,$$
where F_n is the nth Fibonacci number.

Solution for Problem 9.9: In order to simplify the algebra, let $\nu = 1 + \sqrt{5}$ and $\mu = 1 - \sqrt{5}$.

Concept: If the same "ugly" term pops up a lot in an algebraic computation, consider assigning it a variable name so that your algebra is easier to read and write.

Now we can calculate:

$$F_{n-1}F_{n+1} - F_n^2 = \left(\frac{\nu^{n-1} - \mu^{n-1}}{2^{n-1}\sqrt{5}}\right)\left(\frac{\nu^{n+1} - \mu^{n+1}}{2^{n+1}\sqrt{5}}\right) - \left(\frac{\nu^n - \mu^n}{2^n\sqrt{5}}\right)^2$$

$$= \frac{(\nu^{n-1} - \mu^{n-1})(\nu^{n+1} - \mu^{n+1}) - (\nu^n - \mu^n)^2}{2^{2n}(5)}$$

$$= \frac{(\nu^{2n} - \nu^{n-1}\mu^{n+1} - \nu^{n+1}\mu^{n-1} + \mu^{2n}) - (\nu^{2n} - 2\nu^n\mu^n + \mu^{2n})}{2^{2n}(5)}$$

$$= \frac{\nu^{n-1}\mu^{n-1}(-\nu^2 - \mu^2 + 2\nu\mu)}{2^{2n}(5)}$$

$$= -\frac{\nu^{n-1}\mu^{n-1}(\nu - \mu)^2}{2^{2n}(5)}.$$

Now we note that $\nu\mu = (1 + \sqrt{5})(1 - \sqrt{5}) = -4$ and $\nu - \mu = (1 + \sqrt{5}) - (1 - \sqrt{5}) = 2\sqrt{5}$. Therefore we have

$$F_{n-1}F_{n+1} - F_n^2 = -\frac{(-4)^{n-1}(2\sqrt{5})^2}{2^{2n}(5)} = -\frac{(-1)^{n-1}2^{2n}(5)}{2^{2n}(5)} = (-1)^n.$$

Certainly we could prove this identity via other means, but the formula makes it fairly straightforward to prove by just plugging-and-chugging with algebra. $\square$

Exercises

9.4.1 Prove, using Binet's formula, that for all $n > 0$,

$$F_{2n-1} = F_n^2 + F_{n-1}^2.$$

9.4.2 Use Binet's formula to find a simple approximation for the number of digits in F_n.

9.4.3★ The **Lucas numbers** are defined by $L_0 = 2$, $L_1 = 1$, and $L_n = L_{n-1} + L_{n-2}$ for all $n \geq 2$.

(a) Use a procedure similar to Problem 9.7 to find a formula for L_n.

(b) Prove that $F_{2n} = F_n L_n$. **Hints:** 269

(c) Prove that $L_n^2 - 5F_n^2 = 4(-1)^n$. **Hints:** 20

(d)★ Prove that, except for 1, 2, and 3, no positive integer is both a Fibonacci number and a Lucas number. **Hints:** 51, 267

9.5 Summary

➤ The Fibonacci numbers are the sequence of numbers

$$0, 1, 1, 2, 3, 5, 8, 13, 21, 34, 55, 89, 144, 233, 377, 610, 987, \ldots$$

where each number (after the first two) is the sum of the two preceding numbers.

➤ A more formal definition is that $F_0 = 0$, $F_1 = 1$, and $F_n = F_{n-1} + F_{n-2}$ for all $n \geq 2$.

➤ Many counting problems have the Fibonacci numbers as their solution, especially those in which we can relate a problem of size n to the same problem of size $n - 1$ and $n - 2$.

➤ A closed formula for the nth Fibonacci number F_n is

$$F_n = \frac{(1 + \sqrt{5})^n - (1 - \sqrt{5})^n}{2^n \sqrt{5}}.$$

Here are some general problem-solving concepts:

> **Concept:** When trying to find or prove something for all positive integers n, try some small values of n, and look for a pattern.

> **Concept:** Before tackling the general case of an identity or equation, it often helps to play with a simple example or two, in order to get a feel for what's going on.

> **Concept:** If the same "ugly" term pops up a lot in an algebraic computation, consider assigning it a variable name so that your algebra is easier to read and write.

REVIEW PROBLEMS

9.10 The parking lot outside our building has 12 parking spaces. Compact cars can easily fit within a single space, but SUVs take up 2 spaces. In how many different ways can the lot be filled?

9.11 For any positive integer n, determine the number of ordered sums of positive integers greater than 1 summing to n. (For example, if $n = 6$, then the sums are 6, 4 + 2, 2 + 4, 3 + 3, and 2 + 2 + 2.)

9.12

(a) Fourteen people sit in a row of 14 chairs, one person per chair. At the sound of a bell, they all are allowed to change seats, but each person is permitted to move no farther than one seat from her original chair. Each person is not required to move, and there must be one person per chair in the rearrangement. The bell sounds; how many rearrangements can the people form?

(b) What is the answer if the 14 people are sitting in chairs around a round table?

9.13 On the planet Venus, female Venusians have a mother and a father, but male Venusians have only a mother. For any positive integer n, how many n-generation ancestors does a male Venusian have? (1-generation ancestors are parents, 2-generation ancestors are grandparents, 3-generation ancestors are great grandparents, and so on.)

9.14 Prove that for any $k \geq 1$,

$$F_1 + F_2 + F_3 + F_4 + \cdots + F_k = F_{k+2} - 1.$$

9.15 How many n-digit base-4 numbers are there that start with the digit 3 and in which each digit is exactly one more or one less than the previous digit? (For example, 321010121 is such a 9-digit number.)

9.16 How many paths are there from hex A to hex B in the diagram below, if each step of a path must be to a hex immediately adjacent on the right? (A sample such path is shown.)

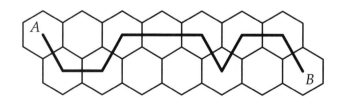

9.17 A large elementary school class goes on a field trip to see a play. The front row of the theater has 11 seats. No boy wants to sit between 2 girls (or sit at the end of the row next to a girl) and no girl wants to sit between 2 boys (or sit at the end of the row next to a boy). In how many ways can the row of seats be assigned to boys and girls? **Hints:** 251

9.18 In Problem 9.1, we saw using casework that the answer to the 10-stair problem was

$$\binom{10}{0} + \binom{9}{1} + \binom{8}{2} + \binom{7}{3} + \binom{6}{4} + \binom{5}{5} = 89,$$

and that this was also equal to the Fibonacci number F_{11}. Can you generalize and prove this result? **Hints:** 156

9.19 Simplify the product

$$\prod_{k=2}^{100} \left(\frac{F_k}{F_{k-1}} - \frac{F_k}{F_{k+1}} \right).$$

(Source: Mandelbrot) **Hints:** 23

9.20 Find the value of the infinite sum

$$\frac{1}{3} + \frac{1}{9} + \frac{2}{27} + \cdots + \frac{F_n}{3^n} + \cdots.$$

(Source: Mandelbrot) **Hints:** 188

9.21 For any nonnegative real number x, let $\langle x \rangle$ denote the fractional part of x; that is, $\langle x \rangle = x - \lfloor x \rfloor$, where $\lfloor x \rfloor$ denotes the greatest integer less than or equal to x. Suppose that a is a positive real number with $2 < a^2 < 3$ such that $\langle a^{-1} \rangle = \langle a^2 \rangle$. Find $a^{12} - 144a^{-1}$. *(Source: AIME)* **Hints:** 213, 35

> **Sidenote:** Recall the problem on page 179. The answer is that a subset of $\{1, 2, \ldots, n\}$ represents a particular pair's "family tree." For example, if $n = 12$ and the subset is $\{2, 4, 7, 11\}$, then the pair corresponding to this subset was born in month 11, its parents were born in month 7, its grandparents were born in month 4, and its great grandparents were born in month 2 (to the original pair). Note that, in particular, the subset $\emptyset$ corresponds to the original pair of rabbits. We leave it to you to fill in the missing details of this correspondence, and also to determine why the condition "no two consecutive elements" on the subset is necessary.

To understand recursion, you must first understand recursion. – Anonymous

CHAPTER **10**

_____

Recursion

10.1 Introduction

The Fibonacci numbers from the last chapter are an example of **recursion**. Whenever we have a sequence of numbers in which the next number in the sequence is derived from previous numbers, we have a recursion. For example, with the Fibonacci numbers, we have the **recurrence relation**

$$F_n = F_{n-1} + F_{n-2}, \ F_0 = 0, \ F_1 = 1.$$

Each Fibonacci number (after the first two) depends on the two previous Fibonacci numbers.

In this chapter, we'll look at several problems in which we can use recursion as a solution method, and discuss more generally the types of problems for which recursion may be useful. In brief, whenever we have a problem in which we have to compute some quantity that can be expressed in terms of a positive integer n, and we can replace that computation with a computation of the same quantity but for smaller values (such as $n - 1$, $n - 2$, etc.), we may be able to use recursion to solve the problem. This can be a little confusing as an abstract concept, but should become more clear as we work through several examples.

Later in the chapter, we will explore a special sequence of numbers called the **Catalan numbers**. As we will see, these numbers pop up in a surprising number of different counting situations. The eminent combinatorist Richard Stanley, in his book *Enumerative Combinatorics: Volume 2* and on its accompanying website, lists (as of this writing) 149 different mathematical objects that are counted by the Catalan numbers. The items on Stanley's list come from many different branches of mathematics, and several of these items have deep mathematical significance.

We will examine several problems whose answers involve the Catalan numbers. We will use our results from these problems to determine both the recursive definition of the Catalan numbers and also a fairly simple closed-form formula for the numbers.

10.2 Examples of Recursions

Problem 10.1: The diagram below shows 3 pegs and 8 rings of different sizes. All of the rings start on the left peg. The goal is to move all of the rings to the right peg. On each move, we can remove the top ring on any peg and place it on any other peg, provided that we do not place a ring on top of a smaller ring.

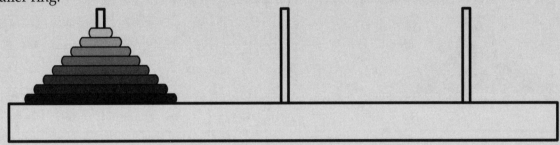

(a) Try the problem for 1, 2, or 3 rings. What is the minimum number of moves necessary to move all of the rings to the right peg?

(b) What is the minimum number of moves necessary in the full 8-ring version of the game?

(c) Suppose there were n rings, where n is an arbitrary positive integer. What is the minimum number of moves necessary?

Problem 10.2: I work for a valet parking company. Each of our customers drives either a Cadillac, a Continental, or a Porsche. My boss told me I have to reserve spaces in our parking lot and mark them as being for a Cadillac, a Continental, or a Porsche. Cadillacs and Continentals each take 2 spaces while Porsches only require 1.

(a) Suppose that our parking lot has 3 spaces. In how many ways can I allocate the spaces?

(b) What if the lot has 12 spaces?

(c)★ What if the lot has 500 spaces?

Problem 10.3: Find the number of 10-digit ternary sequences (that is, sequences with digits 0, 1, or 2) such that the sequence does not contain two consecutive zeros.

Problem 10.4: 6 sprinters are in the 100-meter dash. Ties are allowed in the final standings, so that, for example, one possible order of finish is:

Runner #6 wins; #2 and #5 tie for 2nd; and #1, #3 and #4 all tie for last.

How many different finishing orders are possible?

The Fibonacci numbers are just one example of recursion. We'll look at some more recursion problems in this section. The common theme is that we break down each problem into smaller versions of the same problem.

> **Concept:** Whenever you can take a problem and express its solution in terms of smaller versions of the same problem, that problem is a good candidate for recursion.

Problem 10.1: The diagram below shows 3 pegs and 8 rings of different sizes. All of the rings start on the left peg. The goal is to move all of the rings to the right peg. On each move, we can remove the top ring on any peg and place it on any other peg, provided that we do not place a ring on top of a smaller ring.

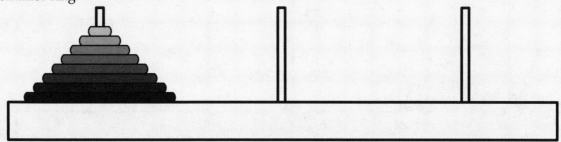

(a) What is the minimum number of moves necessary to win the game?

(b) Suppose there were n rings, where n is an arbitrary positive integer. What is the minimum number of moves necessary?

Solution for Problem 10.1:

(a) Since we don't really know what the best strategy is, let's look at some smaller examples.

> **Concept:** Experiment with smaller examples in order to get a feel for a hard problem.

If there's only 1 ring, then we only need 1 move: we just move it to the right peg and we win!

If there are 2 rings, it's still pretty easy to win. We move the smaller ring to the middle peg, then move the larger ring to the right peg, and finally move the smaller ring to the right peg. We need 3 moves to win, and it's pretty clear that there's no way to win in fewer moves.

Experiment with 3 rings before reading further. How many moves are necessary?

It turns out that the best way to win is via the following sequence of moves:

- Start position

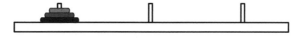

- Move smallest ring to right peg.

- Move middle ring to center peg.

- Move smallest ring to center peg.

- Move largest ring to right peg.

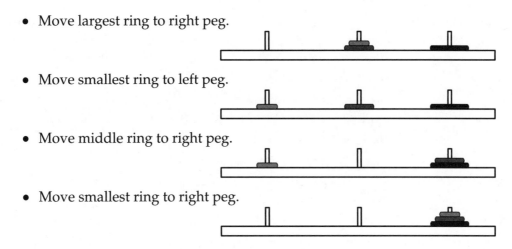

- Move smallest ring to left peg.

- Move middle ring to right peg.

- Move smallest ring to right peg.

So we can win in 7 moves. But how can we be sure that this is the least possible number of moves, and that we're not overlooking a better strategy that might allow us to win in 6 or fewer moves?

We can break up the necessary steps for solving the 3-ring version as follows:

- Move the smaller two rings to the middle peg.
- Move the largest ring to the right peg.
- Move the smaller two rings to the right peg.

Note that this must be the optimal strategy. We cannot move the largest ring until the smaller 2 rings have been moved off of it. We can't put the largest ring on the right peg unless it's empty. So there's no other possible course of action except for that which is listed above.

We know that moving the two smaller rings to the middle peg takes 3 moves, since that's just the 2-ring problem. Then our 4th move is moving the largest ring. Finally, we know that it takes 3 more moves to move the two smaller rings onto the right peg. So 7 moves is the best we can do.

We have also gained the necessary insight to solve the 8-ring case. The general strategy is

- Move the smaller 7 rings to the middle peg.
- Move the largest ring to the right peg.
- Move the smaller 7 rings to the right peg.

Therefore, we can say that

$$(\text{\# of moves to win with 8 rings}) = 2(\text{\# of moves to win with 7 rings}) + 1.$$

If we let h_n denote the number of moves necessary to solve the problem with n rings, then we have $h_n = 2h_{n-1} + 1$. This type of formula, in which the elements of a sequence are defined in terms of previous element(s) of the sequence, is called a **recurrence relation** or simply a **recurrence**.

Extra! This problem is known as the **Towers of Hanoi** and was invented by the French mathematician Edouard Lucas in 1883. An animated version of the Towers of Hanoi problem (and its solution) was prominently featured on the website for the 2007 International Mathematical Olympiad, which was held in Hanoi, Vietnam.

We can now work our way up to h_8 by starting at $h_1 = 1$ and using the above recurrence relation:

$$h_2 = 2h_1 + 1 = 2(1) + 1 = 3,$$
$$h_3 = 2h_2 + 1 = 2(3) + 1 = 7,$$
$$h_4 = 2h_3 + 1 = 2(7) + 1 = 15,$$
$$h_5 = 2h_4 + 1 = 2(15) + 1 = 31,$$
$$h_6 = 2h_5 + 1 = 2(31) + 1 = 63,$$
$$h_7 = 2h_6 + 1 = 2(63) + 1 = 127,$$
$$h_8 = 2h_7 + 1 = 2(127) + 1 = 255.$$

So we need 255 moves to win the game.

(b) Hopefully, the pattern of the above numbers is pretty clear. Each number is 1 less than a power of 2, so it looks like $h_n = 2^n - 1$. Observing a pattern is not a proof, though. We need to prove that this works. We do so by induction, checking that it works for the initial condition $n = 1$ and that it satisfies the recurrence relation.

First, we see that $h_1 = 2^1 - 1 = 2 - 1 = 1$. Now we show that our formula for h_n satisfies the recurrence relation, by induction. Assume that the formula works for h_{n-1}; that is, assume that $h_{n-1} = 2^{n-1} - 1$. Then by our recurrence relation:

$$h_n = 2h_{n-1} + 1 = 2(2^{n-1} - 1) + 1 = 2^n - 2 + 1 = 2^n - 1.$$

So, by induction, $h_n = 2^n - 1$ for all positive integers n.

□

A little later, we'll see how you might have come up with the formula if you didn't see a pattern. In fact, we see a bit of this approach in the next problem.

Problem 10.2: I work for a valet parking company. Each of our customers drives either a Cadillac, a Continental, or a Porsche. My boss told me I have to reserve spaces in our parking lot and mark them as being for a Cadillac, a Continental, or a Porsche. Cadillacs and Continentals each take 2 spaces while Porsches only require 1.

(a) The lot has 12 spaces. In how many ways can I allocate the spaces?

(b) What if the lot has 500 spaces?

Solution for Problem 10.2: We see that this problem easily scales, meaning that there's nothing particularly special about the number "12" in the problem: it would be essentially the same problem regardless of the number. We also see that, after we allocate the first space(s) for the first car, we have the same problem (with a smaller number of spots) as we started with. This makes the problem a good candidate for a recursive solution.

(a) Let p_n denote the number of ways to allocate a parking lot with n spots. If we allocate the first spot to a Porsche, then we can allocate the remaining $n - 1$ spots in p_{n-1} ways. If we allocate the first two spots to a Cadillac, then we can allocate the remaining $n - 2$ spots in p_{n-2} ways. Similarly, if we allocate the first two spots to a Continental, then we can allocate the remaining $n - 2$ spots in

p_{n-2} ways. Therefore, our recurrence relation is

$$p_n = p_{n-1} + 2p_{n-2}.$$

We start with $p_1 = 1$ (since if there's only one spot, we have only one choice, which is to allocate it to a Porsche) and $p_2 = 3$ (since our choices are a Cadillac, a Continental, or two Porsches). Then we can use the recurrence relation to work our way up to p_{12}:

$$p_3 = p_2 + 2p_1 = 3 + 2(1) = 5$$
$$p_4 = p_3 + 2p_2 = 5 + 2(3) = 11$$
$$p_5 = p_4 + 2p_3 = 11 + 2(5) = 21$$
$$p_6 = p_5 + 2p_4 = 21 + 2(11) = 43$$
$$p_7 = p_6 + 2p_5 = 43 + 2(21) = 85$$
$$p_8 = p_7 + 2p_6 = 85 + 2(43) = 171$$
$$p_9 = p_8 + 2p_7 = 171 + 2(85) = 341$$
$$p_{10} = p_9 + 2p_8 = 341 + 2(171) = 683$$
$$p_{11} = p_{10} + 2p_9 = 683 + 2(341) = 1365$$
$$p_{12} = p_{11} + 2p_{10} = 1365 + 2(683) = 2731$$

So there are 2731 ways in which we can allocate the 12 spaces.

(b) We don't want to keep listing out terms all the way to p_{500}. Instead, we'd like a nice formula that we can simply plug $n = 500$ into and get the answer. But the pattern, if there even is a pattern, is not so obvious here.

One pattern that you might notice is that each number in the above list is almost twice the number immediately before it:

$$11 = 2(5) + 1, \ 21 = 2(11) - 1, \ 43 = 2(21) + 1, \ 85 = 2(43) - 1, \ \ldots$$

But it's not at all clear how to turn this into a formula.

However, the "almost-doubling" of the numbers makes us think about exponential growth, so we can try the same tactic that we used to find the Binet formula for the Fibonacci numbers. We look for solutions to the recurrence of the form c^n for some constant c. If we substitute this into the recurrence, we get
$$c^n = c^{n-1} + 2c^{n-2}.$$

We can divide through by c^{n-2} and we're left with $c^2 = c + 2$, which gives $c^2 - c - 2 = 0$. This factors as $(c - 2)(c + 1) = 0$, so we see that $c = 2$ or $c = -1$.

Then, just as we did when finding the Binet formula, we have that $p_n = \lambda_1 2^n + \lambda_2(-1)^n$, where λ_1 and λ_2 are unknown constants, and we use the initial conditions $p_1 = 1$ and $p_2 = 3$ to solve for the constants. This gives us a system of linear equations:

$$p_1 = 1 = 2\lambda_1 - \lambda_2,$$
$$p_2 = 3 = 4\lambda_1 + \lambda_2.$$

Adding them together gives $4 = 6\lambda_1$, so $\lambda_1 = \frac{2}{3}$, which then gives us $\lambda_2 = \frac{1}{3}$. Therefore we have

$$p_n = \frac{2}{3}2^n + \frac{1}{3}(-1)^n.$$

We can now quickly compute p_n for any value of n; in particular

$$p_{500} = \frac{2}{3}2^{500} + \frac{1}{3} = \frac{2^{501} + 1}{3}.$$

$\square$

At this point, you may have a pretty good idea how to approach finding a closed-form formula for many recursions. We "guess" that the formula will somehow involve c^n, and then solve for c. We'll come back to this idea in the next section.

Problem 10.3: Find the number of 10-digit ternary sequences (that is, sequences with digits 0, 1, or 2) such that the sequence does not contain two consecutive zeros.

Solution for Problem 10.3: We can do casework based on the first digit of the sequence. If the first digit is a 0, then the next digit must be a 1 or a 2, and the remaining $n - 2$ digits must not have two consecutive zeros. If the first digit is a 1 or a 2, then the remaining $n - 1$ digits must not have two consecutive zeros.

So if a_n is the number of n-digit ternary sequences with no two consecutive zeros, then a_n satisfies the recurrence relation

$$a_n = 2a_{n-1} + 2a_{n-2}.$$

This looks a lot like the Fibonacci sequence, except that each term is *double* the sum of the previous two terms.

Noting that $a_1 = 3$ and $a_2 = 8$ (any two-digit sequence except 00 is allowed), we can generate the sequence:

$$3, 8, 22, 60, 164, 448, 1224, 3344, 9136, 24960, \ldots$$

So the answer is 24960. $\square$

Problem 10.4: 6 sprinters are in the 100-meter dash. Ties are allowed in the final standings, so that, for example, one possible order of finish is:
Runner #6 wins; #2 and #5 tie for 2nd; and #1, #3 and #4 all tie for last.
How many different finishing orders are possible?

Solution for Problem 10.4: What makes this problem difficult is that any number of people could tie for any position, including all 6 of them tying for first place. So trying to attack this problem using a direct method like casework seems very difficult, as there are a large number of cases to consider: we could have all 6 tied, or 5 could be tied with the other alone (ahead or behind), or it might go 4-1-1 (or 1-4-1 or 1-1-4), or 3-2-1 or 4-2 or 3-3 or 3-1-1-1 or.... We could try to list all of the possible cases and count the number of possibilities for each, but it would be long and messy, and it'd be easy to overlook a case and get the wrong answer.

So instead, let's try to look at some smaller version of the same problem to try to get a handle for what's going on. The easiest way to simplify the problem is to reduce the number of runners.

If there's only 1 runner, then obviously there's only 1 possible order of finish.

If there are two runners A and B, then either A could beat B, or B could beat A, or they could tie. So there are 3 possible orders.

Let's go to 3 runners. Already it's starting to get complicated. If none of them tie, then there are 3! orders of finish. Or they could all finish in a 3-way tie. If 2 of them tie, then there are $\binom{3}{2} = 3$ choices for which two tie, and then 2 choices for the order between the two tied runners and the 3rd non-tied runner, for a total of 6 possible orders when 2 of them tie. So there's a total of $6 + 1 + 6 = 13$ possible finishing orders for the 3-runner race.

Going to 4 runners leads to even nastier casework, and no clear pattern emerged from the numbers that we've gotten so far: 1, 3, 13.

So let's look at the problem another way.

> **Concept:** If a solution method doesn't seem to be getting you anywhere, don't be afraid to start over and try a completely different approach.

We can look at the runner(s) that come in first. If k runners tie for first, then there are $\binom{6}{k}$ ways to choose those runners. Then, we have $6 - k$ runners left to arrange. This suggests a recursive approach, since the "$6 - k$ runners left to arrange" part is a smaller version of the same problem.

So let a_n be the number of possible orders of finish for an n-runner race. As above, we can break up the problem into cases based on how many runners tie for first. If k runners tie for first, then we have $\binom{n}{k}$ ways to choose them, and a_{n-k} ways that the remaining $n - k$ runners can finish.

This gives us our recursion:

$$a_n = \sum_{k=1}^{n} \binom{n}{k} a_{n-k}.$$

Note that a_n depends on all of the previous terms of the sequence of a's.

Now we just crunch the numbers, noting that $a_0 = 1$: if there are no runners to arrange, then there are no choices to make.

$$a_1 = a_0 = 1$$
$$a_2 = 2a_1 + a_0 = 1 + 2 = 3$$
$$a_3 = 3a_2 + 3a_1 + a_0 = 9 + 3 + 1 = 13$$
$$a_4 = 4a_3 + 6a_2 + 4a_1 + a_0 = 52 + 18 + 4 + 1 = 75$$
$$a_5 = 5a_4 + 10a_3 + 10a_2 + 5a_1 + a_0 = 375 + 130 + 30 + 5 + 1 = 541$$

Finally, we get our answer:

$$a_6 = 6a_5 + 15a_4 + 20a_3 + 15a_2 + 6a_1 + a_0$$
$$= 3246 + 1125 + 260 + 45 + 6 + 1$$
$$= 4683.$$

The numbers that we found:

$$1, 3, 13, 75, 541, 4683, \ldots,$$

don't seem to fit any nice pattern, so there doesn't seem to be any obvious closed-form formula. □

Exercises

10.2.1 I have a 10-foot flagpole. I have 3 different types of 2-foot flags and 2 different types of 1-foot flags. I have billions of each of the types of flags. Find the number of ways I can arrange flags to exactly fill the 10-foot flagpole (where the orders of the flags matter, and while flags of the same type are indistinguishable, flags of different types are distinguishable).

10.2.2 Let $a_1, a_2, \ldots$ be a sequence with the following properties:
(i) $a_1 = 1$, and
(ii) $a_{2n} = n(a_n)$ for any positive integer n.
What is the value of $a_{2^{100}}$?

10.2.3 How many 6-digit base-4 numbers have the property that they contain at least one 1 and that the first 1 is to the left of the first 0 (or there is no 0)?

10.2.4 Let $a_1 = p$ and $a_2 = q$, where p and q are positive integers, and let $a_n = a_{n-1}a_{n-2}$ for all $n \geq 3$. Find a formula for a_n in terms of n, p, and q. **Hints:** 197

10.2.5 Call a set of integers *spacy* if it contains no more than one out of any three consecutive integers. How many subsets of $\{1, 2, 3, \ldots, 12\}$ are spacy? *(Source: AMC)* **Hints:** 231

10.2.6★ Find the number of 10-digit binary sequences that have exactly one pair of consecutive 0's. **Hints:** 59

10.3 Linear Recurrences

Problems

Problem 10.5: We wish to find a closed-form formula for the recurrence where $a_1 = 1$, $a_2 = 2$, and $a_n = a_{n-1} + 6a_{n-2}$ for all $n > 2$.

(a) Ignore the initial conditions for a_1 and a_2. Let $a_n = c^n$ for some unknown constant c. Plug this into the recurrence relation and solve for c.

(b) You should have found in part (a) that there is more than one value of c that works. How can we combine the different values of c into a common solution?

(c) In (a) and (b), you did not use the conditions $a_1 = 1$ and $a_2 = 2$. Use these conditions to finish the problem.

Problem 10.6: Find a closed-form formula for the recurrence $a_n = 4a_{n-1} - 4a_{n-2}$ for all $n > 2$, with initial conditions $a_1 = 1$ and $a_2 = 3$.

In Problem 10.2, we saw a small piece of the general theory for finding formulas for recurrences. In this section, we'll explore this theory in a bit more detail, and introduce some terminology.

> **Important:** It is not necessary to memorize this terminology. It is more important to understand the underlying concepts.

We're not going to try to cover completely the general theory of recurrence relations—doing so would require more algebra than we wish to get into in this book. Instead, we'll focus on a couple of key examples.

> **Problem 10.5:** Find a closed formula for the recurrence where $a_1 = 1$, $a_2 = 2$, and $a_n = a_{n-1} + 6a_{n-2}$ for all $n > 2$.

Solution for Problem 10.5: As we did when we found the closed-form formula for the recurrence in Problem 10.2, we start this problem by trying a solution of the form c^n for some unknown constant c. We plug this into the recurrence to get

$$c^n = c^{n-1} + 6c^{n-2}.$$

Dividing by c^{n-2} and bringing all the terms to one side gives

$$c^2 - c - 6 = 0.$$

Factoring gives us $(c - 3)(c + 2) = 0$, so the roots are $c = 3$ and $c = -2$.

Therefore, we know that for any constants λ_1 and λ_2,

$$a_n = \lambda_1 3^n + \lambda_2 (-2)^n$$

satisfies the recurrence. To find a solution to the recurrence that also satisfies the initial conditions $a_1 = 1$ and $a_2 = 2$, we substitute $n = 1$ and $n = 2$ into the above equation to get the system of linear equations:

$$1 = 3\lambda_1 - 2\lambda_2,$$
$$2 = 9\lambda_1 + 4\lambda_2.$$

Multiplying the top equation by 2 and adding gives $4 = 15\lambda_1$, so $\lambda_1 = \frac{4}{15}$, and plugging this back in gives $\lambda_2 = -\frac{1}{10}$.

Therefore our solution is

$$a_n = \frac{4}{15}(3)^n - \frac{1}{10}(-2)^n.$$

As a quick check, we can compute $a_3 = a_2 + 6a_1 = 8$, and check that

$$\frac{4}{15}(3)^3 - \frac{1}{10}(-2)^3 = \frac{4}{15}(27) - \frac{1}{10}(-8) = \frac{36}{5} + \frac{4}{5} = \frac{40}{5} = 8.$$

$\square$

We can use this same method as a general procedure to solve recurrences of the form $a_n = pa_{n-1} + qa_{n-2}$, where p and q are constants. Such recurrences are called **linear** recurrences, since each term of the recurrence is an element of the sequence (such as a_n or a_{n-1}) multiplied by a constant.

We first look for solutions that are exponentials of the form c^n, which we can plug into the recurrence to get

$$c^n = pc^{n-1} + qc^{n-2}.$$

We divide by c^{n-2} and move all terms to one side to get

$$c^2 - pc - q = 0.$$

This is called the **characteristic equation** of the recurrence. We solve this to get roots c_1 and c_2. Then, if $c_1 \neq c_2$, we have that

$$a_n = \lambda_1 c_1^n + \lambda_2 c_2^n$$

for some constants λ_1 and λ_2. This is called the **general solution** of the recurrence. Finally, we use the initial conditions on a_1 and a_2 to solve for the λ's.

The next problem explores what we do if the roots of the characteristic equation are equal; that is, if $c_1 = c_2$ in the above discussion.

Problem 10.6: Find a closed-form formula for the recursion $a_n = 4a_{n-1} - 4a_{n-2}$ for all $n > 2$, with initial conditions $a_1 = 1$ and $a_2 = 3$.

Solution for Problem 10.6: It's a linear recurrence relation, so we start by forming the characteristic polynomial:

$$c^2 - 4c + 4 = 0.$$

This factors as $(c - 2)^2 = 0$, so we have a double root at $c = 2$. What do we do now?

Bogus Solution: If we try to simply set

$$a_n = \lambda 2^n,$$

we plug in $a_1 = 1$ and see that $\lambda = \frac{1}{2}$, so it looks like $a_n = \frac{1}{2}(2^n) = 2^{n-1}$.

But this doesn't work, since $a_2 = 3 \neq 2^1$.

Let's compute a few terms of the sequence, to see if there's an easily-observed pattern:

$$a_1 = 1,$$
$$a_2 = 3,$$
$$a_3 = 4(3 - 1) = 8,$$
$$a_4 = 4(8 - 3) = 20,$$
$$a_5 = 4(20 - 8) = 48,$$
$$a_6 = 4(48 - 20) = 112,$$
$$\vdots$$

We suspect that 2^n might somehow be involved, so let's divide each term by 2^n:

$$\frac{a_1}{2} = \frac{1}{2}, \quad \frac{a_2}{4} = \frac{3}{4}, \quad \frac{a_3}{8} = 1, \quad \frac{a_4}{16} = \frac{5}{4}, \quad \frac{a_5}{32} = \frac{3}{2}, \quad \frac{a_6}{64} = \frac{7}{4}, \quad \cdots$$

It appears that every time n increases by 1, $a_n/2^n$ increases by $\frac{1}{4}$. In particular, we can conjecture that

$$\frac{a_n}{2^n} = \frac{n+1}{4} \qquad \Rightarrow \qquad a_n = \frac{n+1}{4}(2^n).$$

We see that there is an $n2^n$ term in our solution. In fact, this is what happens in general. If we have a double root of our characteristic equation, then we will get a term with a factor of n in the general solution. In this problem, our general solution is

$$a_n = \lambda_1 2^n + \lambda_2 n2^n.$$

Now we can plug in our initial values and get a system of linear equations:

$$1 = 2\lambda_1 + 2\lambda_2,$$
$$3 = 4\lambda_1 + 8\lambda_2.$$

Solving this system gives $\lambda_1 = \frac{1}{4}$ and $\lambda_2 = \frac{1}{4}$, so our solution is

$$a_n = \frac{1}{4}2^n + \frac{1}{4}n2^n = \frac{1}{4}(1+n)2^n.$$

$\square$

Here is the general statement of what we observed in the previous problem:

> **Important:** If c is a double root of the characteristic equation, then the general solution will have terms of the form
> $$\lambda_1 c^n + \lambda_2 nc^n.$$

Of course, we have not proved that this always works. It is not terribly difficult to algebraically show that this is always the general solution of a recurrence with a double root, but we will leave the details of the proof as an Exercise.

Exercises

10.3.1 Find a formula for a_n where $a_n = -2a_{n-1} + 15a_{n-2}$, with initial conditions $a_0 = 0$ and $a_1 = 1$.

10.3.2 Find a formula for a_n where $a_n = 4a_{n-1} - 3a_{n-2}$, with initial conditions $a_0 = 1$ and $a_1 = 1$.

10.3.3 Find a formula for a_n where $a_n = 6a_{n-1} - 9a_{n-2}$, with initial conditions $a_0 = 1$ and $a_1 = 4$.

10.3.4 Prove that, if $a_n = pa_{n-1} + qa_{n-2}$ is a recurrence such that the characteristic equation has a double root c, then

$$a_n = \lambda_1 c^n + \lambda_2 nc^n$$

is a solution to the recurrence. **Hints:** 4

10.3.5 Find a formula for a_n where $a_n = 2a_{n-1} - 5a_{n-2}$, with initial conditions $a_1 = 2$ and $a_2 = 1$. **Hints:** 252

10.3.6★ Find a closed-form formula for the recurrence $a_n = 2a_{n-1} + a_{n-2} - 2a_{n-3}$ with initial conditions $a_0 = 0$, $a_1 = 1$, and $a_2 = 3$. **Hints:** 165

10.4 A Hard Recursion Problem

Problem 10.7: I have 2048 letters numbered 1 to 2048. I have to address every single one. Originally they're stacked in order with #1 on top. To make the task a bit less mind-numbing, I address every other one, starting with #1. When I address a letter, I put it in my outbox. The ones I skip I stack as I skip them (so #2 is on the bottom of the stack after my first pass). After I finish my first pass, I have 1024 letters which are not addressed; #2048 is on top, #2 is on the bottom. I then repeat my procedure over and over until there's only one letter left. What number is that letter? *(Source: AIME)*

In the real world, we're not usually just given a simple recurrence relation and told to solve it. Recursion is usually hidden within a problem, and the recursion may not be so easy to see or write down. Let's do an example of a problem in which a recursion is clearly present, but precisely describing that recursion is somewhat difficult.

Problem 10.7: I have 2048 letters numbered 1 to 2048. I have to address every single one. Originally they're stacked in order with #1 on top. To make the task a bit less mind-numbing, I address every other one, starting with #1. When I address a letter, I put it in my outbox. The ones I skip I stack as I skip them (so #2 is on the bottom of the stack after my first pass). After I finish my first pass, I have 1024 letters which are not addressed; #2048 is on top, #2 is on the bottom. I then repeat my procedure over and over until there's only one letter left. What number is that letter? *(Source: AIME)*

Solution for Problem 10.7: We could try to brute-force it, writing out the stack at each step. However, that sounds like a very long process, and the potential for error is huge. So instead, as we often do, we can try to get a feel for the problem by playing with smaller cases. Since the number of letters get halved on every iteration of the procedure, it makes sense to just look at stacks with sizes that are powers of 2.

If we start with 1 letter, then obviously #1 is the last letter remaining. Let's denote by a_n the number of the last letter remaining if we start with a stack of 2^n letters numbered from #1 to #2^n. So we've established that $a_0 = 1$, and we're trying to find a_{11}.

If we start with 2 letters, then I address #1, and #2 is left over. So $a_1 = 2$.

If we start with 4 letters, on the first pass I address #1 and #3 and am left with a stack with #4 on top and #2 on the bottom. On the second pass, I address #4, and #2 is the last letter remaining. So $a_2 = 2$.

For a stack with 8 letters, I had better start using a chart, shown at right. In each stack, I cross off the letters that I sign, and then the next stack is the uncrossed letters, in reverse order. We see from the chart at the right that letter #6 is the last letter remaining, so $a_3 = 6$.

#1̶
#2
#3̶   #8̶
#4   #6
#5̶ $\Rightarrow$ #4̶ $\Rightarrow$ #2̶ $\Rightarrow$ #6
#6   #2   #6
#7̶
#8

I can try to keep going to 16 letters, but now's a good time to start looking for a shortcut. Let's just list the stack after doing the first pass for 16 letters. It will be a stack with just the even-numbered letters remaining, with #16 at the top and #2 at the bottom, as shown on the left below.

#16
#14
#12
#10
#8
#6
#4
#2

I don't need to keep going, because I can take advantage of recursion! I know that in the 8-letter problem, the 6th letter from the top is the last one remaining. So I know that in my stack at left, the 6th letter from the top is the one last remaining! This just happens to be letter #6, so we know that $a_4 = 6$.

As n increases, we could keep listing the stacks, but as the stacks get larger it quickly gets tedious to do so. However, because we have an iterative process (in which every step of the process is essentially a smaller version of the previous step), we think of using recursion. Can we determine a formula for a_n in terms of a_{n-1}?

In our exploratory work above, we saw that a_4 was the number of the letter that was in the 6th position after the first pass. But $a_3 = 6$, so we can also say that a_4 was the number of the letter that was in the $(a_3)^{rd}$ position after the first pass. Extending this logic to the general case, we know that a_n will be the number of the letter in the $(a_{n-1})^{th}$ position after we do our first pass.

> **Concept:** Try to use your experimentation as a guide for creating a description or formula for the general problem. This is what scientists and mathematicians do: they perform specific experiments to try to understand a general theory.

We still need to convert the phrase "$(a_{n-1})^{th}$ position" into a formula for the number of that letter.

One observation that should be apparent from working out the smaller cases is that the letters remaining after the first pass through the stack are exactly the even-numbered letters, and they are stacked in reverse order, with letter #2^n at the top and letter #2 at the bottom. We want the $(a_{n-1})^{th}$ letter in this stack. This is also the letter that is $2^{n-1} - a_{n-1} + 1$ from the bottom (make sure you see why we need to add 1 here). Therefore, it is letter number $2(2^{n-1} - a_{n-1} + 1)$.

We have just established our recurrence relation—it's

$$a_n = 2(2^{n-1} - a_{n-1} + 1),$$

with initial condition $a_0 = 1$. We can now plug-and-chug with this recurrence to get our value of a_{11}. For practice—and to check our work—we'll start at a_0, even though we've already computed the first few values of a.

$$a_0 = 1$$
$$a_1 = 2(2^0 - a_0 + 1) = 2(1 - 1 + 1) = 2$$
$$a_2 = 2(2^1 - a_1 + 1) = 2(2 - 2 + 1) = 2$$
$$a_3 = 2(2^2 - a_2 + 1) = 2(4 - 2 + 1) = 6$$
$$a_4 = 2(2^3 - a_3 + 1) = 2(8 - 6 + 1) = 6$$
$$a_5 = 2(2^4 - a_4 + 1) = 2(16 - 6 + 1) = 22$$

$$a_6 = 2(2^5 - a_5 + 1) = 2(32 - 22 + 1) = 22$$
$$a_7 = 2(2^6 - a_6 + 1) = 2(64 - 22 + 1) = 86$$
$$a_8 = 2(2^7 - a_7 + 1) = 2(128 - 86 + 1) = 86$$
$$a_9 = 2(2^8 - a_8 + 1) = 2(256 - 86 + 1) = 342$$
$$a_{10} = 2(2^9 - a_9 + 1) = 2(512 - 342 + 1) = 342$$
$$a_{11} = 2(2^{10} - a_{10} + 1) = 2(1024 - 342 + 1) = 1366$$

Therefore, envelope #1366 will be the last one remaining. $\square$

Exercises

10.4.1★ Surely you noticed the pattern that $a_k = a_{k+1}$ whenever k is odd. Can you come up with an explanation for that pattern? **Hints:** 57

10.4.2★ Can you find a closed-form formula for a_n? **Hints:** 255, 99

10.5 Problems Involving Catalan Numbers

Problems

Problem 10.8: In how many ways can 10 people sitting around a circular table simultaneously shake hands (so that there are 5 handshakes going on), such that no two people cross arms? For example, the handshake arrangement on the left side below is valid, but the arrangement on the right side is invalid.

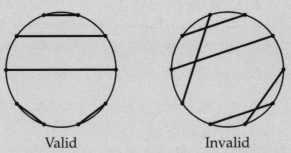

Valid   Invalid

(a) Compute by hand the number of handshake arrangements for 2, 4, or 6 people sitting around a table.

(b) It's a bit hard to do it by hand for 8 people (you can try if you like), so we'll look for a more clever approach. Pick one person (out of the 8); how many people can he shake hands with?

(c) For each possible handshake for the first person in (b), in how many ways can the rest of the table shake hands?

(d) Use your answers from (b) and (c) to count the number of 8-person handshake arrangements.

(e) Can you extend your reasoning from (b)-(d) above to solve the 10-person problem?

Problem 10.9: How many ways are there to arrange 5 open parentheses "(" and 5 closed parentheses ")" such that the parentheses "balance," meaning that, as we read left-to-right, there are never more)'s than ('s? For example, the arrangement ((()())()) is valid, but the arrangement (()())(() is invalid.

(a) Compute by hand the number of arrangements for 1, 2, and 3 pairs of parentheses. Do your answers look familiar?

(b) Try to find a 1-1 correspondence between arrangements of n pairs of parentheses and handshake arrangements of $2n$ people (from Problem 10.8).

Problem 10.10: How many 10-step paths are there from $(0,0)$ to $(5,5)$ on the grid below?

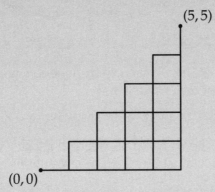

(a) It may be tempting to answer $\frac{1}{2}\binom{10}{5} = 126$. Explain why this is incorrect.

(b) Compute by hand the number of paths on the half-grid to $(1,1)$, $(2,2)$, and $(3,3)$. Notice anything familiar?

(c) Try to find a 1-1 correspondence between solutions to this problem and solutions to one of the two previous problems.

Problem 10.11: In how many ways can a convex heptagon (a 7-sided polygon) be triangulated? (To **triangulate** a polygon means to draw enough diagonals to divide it into a bunch of triangles, as in the example shown at right.)

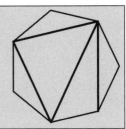

We'll explore several problems that look very different on the surface, but that actually all have the same underlying structure. As we work through these problems, try to keep them all in the back of your mind, with an eye towards the features in the various problems that are similar.

Extra!	Eugène Catalan 1814–1894
⇒⇒⇒⇒	The Catalan numbers (which we will be exploring in this section) are named after the 19th-century mathematician Eugène Catalan. He is also known for his conjecture (made in 1844) that 8 and 9 are the only consecutive positive integers that are perfect powers ($8 = 2^3$ and $9 = 3^2$). This conjecture remained unproven until 2002, when it was proved by Preda Mihăilescu.

Problem 10.8: In how many ways can 10 people sitting around a circular table simultaneously shake hands (so that there are 5 handshakes going on), such that no two people cross arms? For example, the handshake arrangement on the left side below is valid, but the arrangement on the right side is invalid.

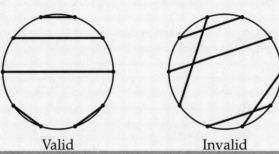

Valid   Invalid

Solution for Problem 10.8: As we often do, we can experiment on smaller versions of the same problem, in order to get some idea for what's going on in general.

If there are 2 people, then there is obviously only one way for them to shake hands.

If there are 3 people, then there's no way that they can all shake hands, because there will always be an odd person left out. In general, we must always have an even number of people.

If there are 4 people, then there are 3 ways for them to shake hands (pick one of the people, and choose one of the other 3 people to shake hands with him; the other two people are then forced to shake with each other). But one of these ways is illegal: the pairs of people sitting across from each other cannot shake hands, since their arms would cross. So there are only 2 legal handshake configurations. In the figure below, we see the two legal handshake configurations on the left, and the 3rd (illegal) configuration on the right.

This person
can't shake
with anybody

If there are 6 people, then things get a bit more complicated. The first thing to note is that no one can shake hands with the person sitting 2 positions away from them on the left or on the right, because if they did, they'd "cut off" a person who would not be able to shake hands with anyone, as in the figure on the left.

This leaves us with two cases.

Case 1: Some pair of people who are directly across from each other shake hands. There can only be one such pair, since two or more such pairs would cross each other at the center of the table. There are 3 choices for a pair of opposite people, and once we have chosen such a pair, the rest of the handshakes are fixed (the two people on each side of the central handshake must shake with each other). These three cases are shown in the figure below:

  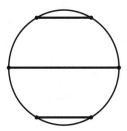

Case 2: Everybody shakes hands with one of his/her neighbors. There are two possibilities, depending on whether a specific person shakes hands to the left or to the right, as shown in the figure below:

So there are a total of $3 + 2 = 5$ ways for 6 people to legally shake hands.

When we get up to 8 people, it's starting to get too complicated to list all the configurations. So let's look at it from a particular person's point-of-view. As in the 6-person case above, we cannot leave an odd number of people on either side of this person's handshake. So our initial person cannot shake hands with anybody that is an even number of people away. In the figure to the right, we show a circled initial person, and his allowed handshakes are shown by dashed lines. Note that each of these handshakes ends at a person who is an odd number of people away from our initial (circled) person.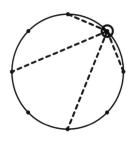

If the initial person shakes hands with a neighbor, we can think of the remaining 6 people as being on a smaller circle, as in the figure below:

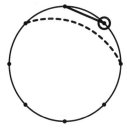

These 6 remaining people have 5 ways to shake, just as in the 6-person problem. Since the initial person has 2 neighbors with whom to shake, this means that there are $2(5) = 10$ handshake arrangements that start with our initial person shaking hands with a neighbor.

Otherwise, our initial person has to shake with a person who is 3 positions to his left or right. Once this is done, the two people who are "cut off" from the rest must shake with each other, and the other 4 people form a 4-person mini-table that can shake in 2 ways:

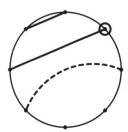

This gives another 2(2) = 4 handshake arrangements, since there are 2 choices of the person that is 3 away from the original person, and then 2 choices to finish the handshaking at the 4-person mini-table.

Therefore, there are 10 + 4 = 14 ways for 8 people to shake hands.

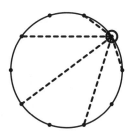

Finally, we can use this same strategy for 10 people. Choose an initial person. This person has 5 choices for whom to shake hands with, as shown in the picture to the right. If he shakes with one of his neighbors (2 choices), then the remaining 8 people form a mini-table that can shake in 14 ways. If he shakes with a person 3 positions away (2 choices), then 2 people are cut off (and must shake), and the other 6 people form a mini-table that can shake in 5 ways. If he shakes with the person directly opposite (1 choice), then each side of the table has a group of 4 people, each of which can shake in 2 ways.

Therefore, the number of handshake arrangements for 10 people is

$$2(14) + 2(5) + 1(2)(2) = 28 + 10 + 4 = 42.$$

□

Before we go on, let's list the numbers that we found while working through the previous problem:

Number of people	2	4	6	8	10
Number of handshake configurations	1	2	5	14	42

Keep these numbers in mind as we continue through this section.

Problem 10.9: How many ways are there to arrange 5 open parentheses "(" and 5 closed parentheses ")" such that the parentheses "balance," meaning that, as we read left-to-right, there are never more)'s than ('s? For example, the arrangement ((()())() is valid, but the arrangement (()()))(() is invalid.

Solution for Problem 10.9: As we often do, let's experiment with small values.

If we have 1 set of parentheses, then we only have one possibility: ().

If we have 2 sets of parentheses, we can either nest them as (()), or we can list both pairs one after the other as ()(). So there are 2 possibilities.

If we have 3 sets of parentheses, then a little experimentation will show that there are 5 possibilities:

$$()()(), (())(), ()(()), (()()), ((())).$$

Hmmm..., 1, 2, 5, ... Do you recognize these numbers? They are the same numbers that we got for the number of non-crossing handshakes of people sitting at a round table in Problem 10.8. Perhaps there is a connection between the two problems.

> **Concept:** When you see the same answer for two different problems, look for a connection, or better yet, for a 1-1 correspondence between them.

Since the parentheses come in pairs, it's natural to think that in any 1-1 correspondence between parenthesis-arrangements and valid handshakes around a table, each set of parentheses will represent two people shaking hands. The fact that the parentheses must be properly nested will somehow correspond to the condition that handshakes cannot cross.

For instance, we can list all of the arrangements of 3 pairs of parentheses, and their corresponding handshake arrangements. We'll label both the parentheses and the people with the letters A through F, and note how each pair of parentheses corresponds to a pair of people that are shaking hands.

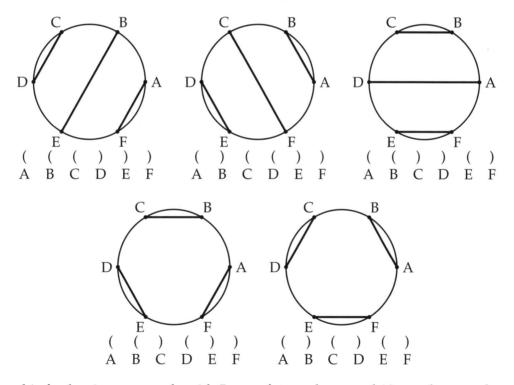

Let's see this further in an example with 5 sets of parentheses and 10 people around a table. We'll label the people around the table A through J, and the parentheses will also be labeled with A through J as we read from left to right. Each matching pair of parentheses corresponds to a handshake. A sample correspondence is shown on the next page.

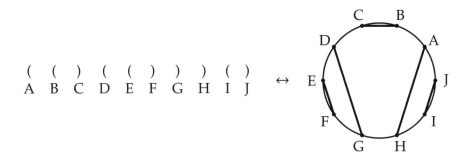

This leads to a 1-1 correspondence

$$\left\{\begin{array}{l}\text{parenthesis arrangements of}\\ n\text{ pairs of parentheses}\end{array}\right\} \leftrightarrow \left\{\begin{array}{l}\text{handshake arrangements of}\\ 2n\text{ people around a table}\end{array}\right\}.$$

Thus the answer to our problem is the same as the number of handshake arrangements of 10 people, which is 42. □

That's two problems so far involving the sequence 1,2,5,14,42,.... You should therefore not be surprised by what you will find in the next problem.

Problem 10.10: How many 10-step paths are there from $(0,0)$ to $(5,5)$ on the grid below?

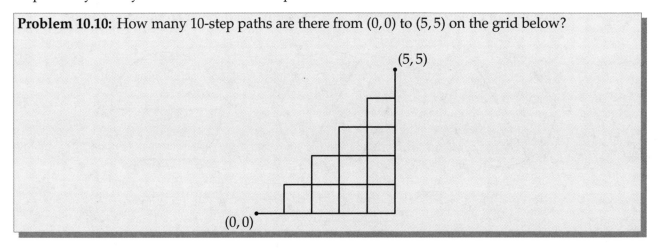

Solution for Problem 10.10: The first thing that we notice is that the grid shown is exactly the part of the full 5×5 grid that is below the main diagonal, as shown below:

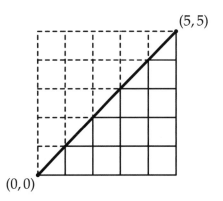

This might suggest the following quick "solution":

> **Bogus Solution:** We know that there are $\binom{10}{5} = 252$ paths on the full grid. Since we only have the lower-half of the grid to work with, that means that we have $\frac{252}{2} = 126$ paths on the lower-half of the grid.

This is of course absurd, as there are many paths that pass through both halves of the grid, like the one shown below:

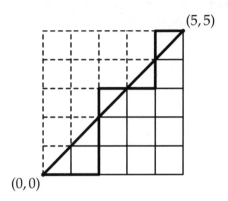

So how can we count the paths that only go below the main diagonal?

Once again, let's count the paths in some smaller cases.

If the half-grid is 1×1, then there's only one path:

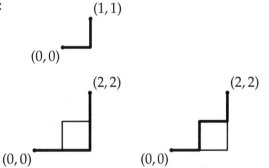

If the half-grid is 2×2, then there are 2 paths:

If the half-grid is 3×3, then there are 5 paths:

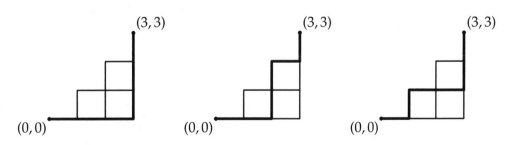

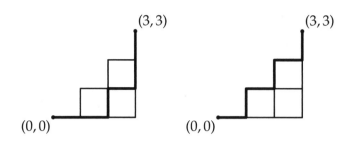

There are those numbers again: 1,2,5,.... So we'll once again look for a 1-1 correspondence between this problem and one of the previous problems. Since each path from $(0,0)$ to (n,n) consists of n moves up and n moves to the right, we think to try to find a correspondence between these paths and lists of n "("s and n ")"s.

Indeed, we can make a 1-1 correspondence

{balanced expressions with n pairs of parentheses} ↔ {paths on an $n \times n$ grid below the diagonal},

by letting each "(" represent a move to the right and each ")" represent a move up. As long as there are more "("s than ")"s, there will be more rights than ups, and the path will never cross above the main diagonal of the $n \times n$ grid.

Therefore, there are 42 paths on the 5×5 half-grid, since there are 42 possible nested expressions with 5 pairs of parentheses. □

You can probably guess ahead of time what the answer to the next problem will be!

Problem 10.11: In how many ways can a convex heptagon (a 7-sided polygon) be triangulated? (To **triangulate** a polygon means to draw enough diagonals to divide it into a bunch of triangles, as in the example shown at right.)

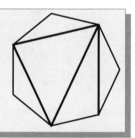

Solution for Problem 10.11:   As we often do, we can build up to a solution by looking at smaller cases. There's only 1 way to triangulate a triangle: do nothing, because it's already triangulated! There are 2 ways to triangulate a convex quadrilateral: draw in either diagonal.

Triangulating a pentagon is the first tricky case. The easiest way to think about it is to pick an edge, and think about the possibilities for that edge. For example, we look at the bottom edge in the regular pentagon shown below, and we see that it can be a part of one of three possible triangles in some triangulation:

  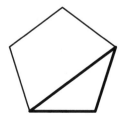

In the middle pentagon shown above, we see that the triangulation is already finished. In the left and right pentagons, we still need to draw in a diagonal of the remaining quadrilateral region, so there are 2 choices for how to finish each of those triangulations. Therefore, there are a total of $2 + 1 + 2 = 5$ possible triangulations of the pentagon.

Hmmm... 1, 2, and 5 ways to triangulate triangles, quadrilaterals, and pentagons. We should not be surprised if we find 14 ways to triangulate a hexagon! Given a hexagon, we fix an edge and draw the possible triangulations:

Figure 10.1: Four possibilities to triangulate a hexagon

In the left and right hexagons in Figure 10.1, we have 5 ways to finish the triangulation by triangulating the pentagon region that remains. In the middle two hexagons in Figure 10.1, we have 2 ways to finish the triangulation by triangulating the quadrilateral region that remains. Therefore there are $5 + 2 + 2 + 5 = 14$ ways to triangulate a hexagon. (No surprise!)

Finally, there are five ways to start the triangulation of a heptagon from a fixed edge:

We see that there are $14 + 5 + 4 + 5 + 14 = 42$ ways to finish the triangulation (note that the center heptagon above has 4 ways since each of the two remaining quadrilaterals must be triangulated in one of 2 ways). □

We saw the sequence

$$1, 2, 5, 14, 42, \ldots$$

in each of the four problems in this section. These are the **Catalan numbers**.

Exercises

10.5.1 In how many ways can an octagon be triangulated?

10.5.2 Compute the number of ways to place 5 indistinguishable balls into 5 distinguishable boxes $B_1, B_2, \ldots, B_5$ such that boxes B_1 through B_i have a total of no more than i balls (for all $1 \le i \le 5$).

10.5.3 How many 5-digit numbers are such that the digits, as read left-to-right, are nondecreasing, and that the ith digit from the left is at most i? (For example, 12235 is such a number.) **Hints:** 263

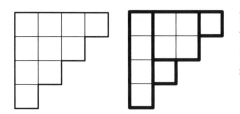

10.5.4★ In how many ways can the shape at the far left be tiled with 4 rectangular tiles, such that each tile has integer side lengths (where a side length of 1 corresponds to a side of one of the small squares)? A sample such tiling is shown at left. **Hints:** 284, 328

10.5.5★ Determine the number of paths from $(0,0)$ to $(6,6)$ in the grid at right, in which every step is either up or to the right, that pass through none of the points $(1,1)$, $(3,3)$, or $(5,5)$ (these points are marked with large X's in the grid). **Hints:** 90, 55

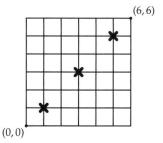

10.6 Formulas for the Catalan Numbers

Problem 10.12: Can you write a recurrence relation for the Catalan numbers?

Problem 10.13: Compare the nth Catalan number with the binomial coefficient $\binom{2n}{n}$. Do you notice any pattern?

Problem 10.14: Find a 1-1 correspondence between:

$$\left\{\begin{matrix} \text{paths from } (0,0) \text{ to } (n,n) \text{ that go above the} \\ \text{main diagonal} \end{matrix}\right\} \quad \leftrightarrow \quad \{\text{paths from } (0,0) \text{ to } (n-1, n+1)\}.$$

Problem 10.15: Find a formula for the nth Catalan number.

As we've seen in the problems in the previous section, the nth Catalan number can be defined as:

- the number of ways that $2n$ people sitting around a table can shake hands, so that no two handshakes cross arms;

- the number of ways to write n ('s and n)'s such that the parentheses are balanced;

- the number of $2n$-step paths on a rectangular grid from $(0,0)$ to (n,n) that do not cross above the main diagonal;

- the number of ways to triangulate a convex $(n+2)$-gon.

It would be nice if we could easily compute the Catalan numbers. For now, let's focus on the recursive definition.

Problem 10.12: What is the recurrence relation for the Catalan numbers?

Solution for Problem 10.12: We've actually already seen it in the problems in the previous section. For each of the problems in the previous section, we can break down the problem of size n into cases, where each case is composed of two smaller problems whose sizes add to $n - 1$.

For instance, in Problem 10.8, we start with a table with $2n$ people. Once we place the initial handshake, we are left with two smaller tables with $2(n - 1)$ people combined.

In Problem 10.9, we can look at the first parenthesis on the left and its corresponding closing parenthesis. This splits the rest of the parentheses into two groups: those that are inside this first pair, and those that are to the right of this pair. For example, in the following 10-pair nesting, the first set of parentheses (in bold) splits the rest of the parentheses into a 6-pair group (inside the bold parentheses) and a 3-pair group (to the right of the bold parentheses):

$$\big((()\,())(()\,()\,()\big)(())\,()$$

The first set of parentheses will always split the remaining $n - 1$ pairs into two groups of balanced parentheses, although one of the groups may be empty.

We can use the 1-1 correspondence between Problem 10.9 and Problem 10.10 to see how to set up the recursion for the paths on the half-grid from $(0, 0)$ to (n, n). The idea is that the end of the *first* complete set of parentheses corresponds to the place where the path *first* touches the diagonal after leaving $(0, 0)$. For example, we show a 5-parentheses nesting and its corresponding path in the figure below. The first set of parentheses is shown in bold, and it corresponds to the path's first touching of the main diagonal at $(2, 2)$.

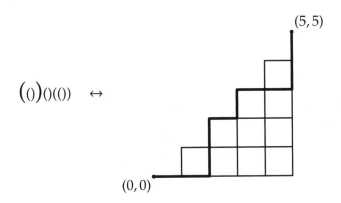

The path is now broken into 2 paths on 2 smaller half-grids.

In all of these problems, the solution is the nth Catalan number C_n, and we arrive at the solution by breaking up the problem into a sum of two smaller problems. Specifically, we see that C_n is the sum of all possible products of the form $C_k C_l$ where $k + l = n - 1$. That is,

$$C_n = C_0 C_{n-1} + C_1 C_{n-2} + \cdots + C_{n-1} C_0 = \sum_{k=0}^{n-1} C_k C_{n-1-k}.$$

The sequence starts at $C_0 = 1$. $\square$

We can once again verify this recursion for the numbers that we've already computed:

$$C_0 = 1,$$
$$C_1 = C_0 C_0 = 1,$$
$$C_2 = C_0 C_1 + C_1 C_0 = 1 + 1 = 2,$$
$$C_3 = C_0 C_2 + C_1 C_1 + C_2 C_0 = 2 + 1 + 2 = 5,$$
$$C_4 = C_0 C_3 + C_1 C_2 + C_2 C_1 + C_3 C_0 = 5 + 2 + 2 + 5 = 14,$$
$$C_5 = C_0 C_4 + C_1 C_3 + C_2 C_2 + C_3 C_1 + C_4 C_0 = 14 + 5 + 4 + 5 + 14 = 42.$$

Let's continue and compute the next couple of Catalan numbers:

$$C_6 = C_0 C_5 + C_1 C_4 + C_2 C_3 + C_3 C_2 + C_4 C_1 + C_5 C_0 = 42 + 14 + 10 + 10 + 14 + 42 = 132,$$
$$C_7 = C_0 C_6 + C_1 C_5 + C_2 C_4 + C_3 C_3 + C_4 C_2 + C_5 C_1 + C_6 C_0 = 132 + 42 + 28 + 25 + 28 + 42 + 132 = 429.$$

So now we have a recursive formula for the Catalan numbers. However, it is somewhat unsatisfying. Not only it is recursive, but each Catalan number depends on *all* of the preceding Catalan numbers, not just the one or two immediately prior. It would be much nicer to have a closed-form formula into which we could plug some value of n and have C_n just pop out. But where can we begin to find such a formula?

Problem 10.10 looks most promising, as it's most related to a problem that we feel like we understand well and know how to find a formula for, namely paths on a grid from $(0,0)$ to (n,n). We know that, without any restrictions, there are $\binom{2n}{n}$ such paths. So that's a good place to start.

Problem 10.13: Compare the nth Catalan number with the binomial coefficient $\binom{2n}{n}$. Do you notice any pattern?

Solution for Problem 10.13: Let's list the first 7 Catalan numbers and the first 7 values of $\binom{2n}{n}$ and see if we notice anything.

n	1	2	3	4	5	6	7
C_n	1	2	5	14	42	132	429
$\binom{2n}{n}$	2	6	20	70	252	924	3432

It's not too clear how to find a pattern between these two rows of numbers, but the one column that might jump out at you is the $n = 4$ column with the numbers 14 and 70, since $70 = 5(14)$. This might cause you to notice that $\binom{2n}{n}$ always appears to be a multiple of C_n. Let's expand our chart:

n	1	2	3	4	5	6	7
C_n	1	2	5	14	42	132	429
$\binom{2n}{n}$	2	6	20	70	252	924	3432
$\dfrac{\binom{2n}{n}}{C_n}$	2	3	4	5	6	7	8

Now we have strong experimental evidence that $C_n = \dfrac{1}{n+1}\dbinom{2n}{n}$. $\square$

Of course, observing a pattern is not a proof. Let's further examine the "paths on a grid" problem and see what else we might determine. We know that $\binom{2n}{n}$ counts the number of paths from $(0,0)$ to (n,n) on a rectangular grid. We also know that C_n counts the number of these paths that don't go above the diagonal. So $\binom{2n}{n} - C_n$ counts the number of paths that *do* go above the diagonal. Since we suspect that $C_n = \frac{1}{n+1}\binom{2n}{n}$, we suspect that the number of paths that go above the diagonal should be:

$$\binom{2n}{n} - \frac{1}{n+1}\binom{2n}{n} = \frac{n}{n+1}\binom{2n}{n}.$$

We do a bit of algebraic manipulation with this quantity:

$$\frac{n}{n+1}\binom{2n}{n} = \frac{n(2n)!}{(n+1)n!n!} = \frac{(2n)!}{(n+1)!(n-1)!} = \binom{2n}{n-1}.$$

This last quantity gives us an idea for a 1-1 correspondence:

Problem 10.14: Find a 1-1 correspondence between:

$$\left\{\begin{matrix}\text{paths from } (0,0) \text{ to } (n,n) \text{ that go above the}\\ \text{main diagonal}\end{matrix}\right\} \quad \leftrightarrow \quad \{\text{paths from } (0,0) \text{ to } (n-1,n+1)\}.$$

Solution for Problem 10.14: This can be a bit tricky to see, so let's play with the $n = 2$ case.

There are $\binom{4}{2} = 6$ paths from $(0,0)$ to $(2,2)$, and we know that $C_2 = 2$ of them stay on or below the main diagonal, so the other 4 go above the diagonal. We also know that there are $\binom{4}{1} = 4$ paths from $(0,0)$ to $(1,3)$. (Good—there are the same number of paths in each category, which is a necessity for there to be a 1-1 correspondence.)

Let's draw the 4 paths in each category, and see if we can match them up. (I'm going to help you out and list them in the order that we will match them—see if you can find the correspondence.)

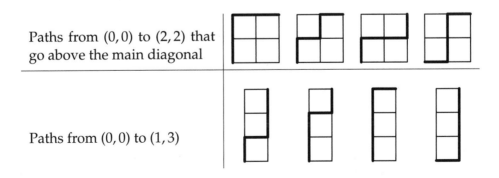

Paths from $(0,0)$ to $(2,2)$ that go above the main diagonal	
Paths from $(0,0)$ to $(1,3)$	

In each column, let's start at $(0,0)$, and let's mark (with a circle) the point on each path where the two paths differ. In other words, the path from $(0,0)$ to the circled point is the same in both paths, but after the circled point, one path goes up whereas the other goes right.

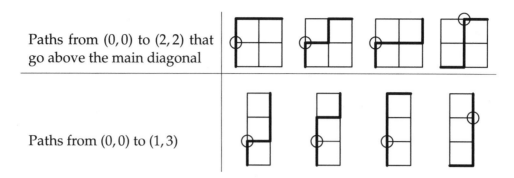

| Paths from $(0,0)$ to $(2,2)$ that go above the main diagonal | |
| Paths from $(0,0)$ to $(1,3)$ | |

We see that in each column, the path from $(0,0)$ to the circled point in both pictures is the same. However, what's more interesting is what happens after the circled point. Compare the paths after the circled point in both pictures of a column. They're mirror images of each other!

To be more precise, let's list the paths using "r" for a step to the right and "u" for a step up. We'll place in bold all of the steps after the circled point.

Paths from $(0,0)$ to $(2,2)$ that go above the main diagonal	u**urr**	ur**ur**	urr**u**	ruur
Paths from $(0,0)$ to $(1,3)$	u**ruu**	uu**ru**	uuu**r**	ruuu

Note that the unbolded parts of the paths—the parts between $(0,0)$ and the circled point—are identical, and the bolded parts of the paths—the parts between the circled point and the end—are exactly reversed.

This suggests a general strategy for finding a 1-1 correspondence. Given a path from $(0,0)$ to (n,n) that goes above the diagonal, circle the *first* point at which the path crosses above the diagonal. Then, reverse all steps past the circled point: change ups to rights and rights to ups.

Here's an example where $n = 5$. The original path is shown as solid, and the new path (after the transformation described above) is shown as dashed.

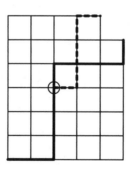

Note that the solid path, before the circled point, has one more up step than right step. After the circled point, the solid path has one more right step than up step (since the circled point lies one "up" step above the diagonal). After the reversal transformation, the dashed path has, after the circled point, one more up step than right step. Hence, starting at $(0,0)$, the combined new path has 2 more up steps than right steps. Since it still has $2n$ steps in total, it must have $n + 1$ up steps and $n - 1$ right steps, and thus the path ends at $(n - 1, n + 1)$.

This process is clearly reversible, and hence we have a 1-1 correspondence:

$$\left\{\begin{matrix}\text{paths from } (0,0) \text{ to } (n,n) \text{ that go above the}\\ \text{main diagonal}\end{matrix}\right\} \leftrightarrow \{\text{paths from } (0,0) \text{ to } (n-1, n+1)\}.$$

☐

Problem 10.15: Find a formula for the nth Catalan number.

Solution for Problem 10.15: We know that the nth Catalan number is the number of paths from $(0,0)$ to (n,n) that don't go above the diagonal. However, we know from Problem 10.14 that the paths that do go above the diagonal are in 1-1 correspondence with paths to $(n-1, n+1)$. Since there are $\binom{2n}{n}$ paths from $(0,0)$ to (n,n) and $\binom{2n}{n-1}$ paths from $(0,0)$ to $(n-1, n+1)$, we have that

$$C_n = \binom{2n}{n} - \binom{2n}{n-1}.$$

This does not exactly look like what we conjectured in Problem 10.13, so let's try to simplify it a bit. We start by writing out the expressions for the binomial coefficients:

$$C_n = \frac{(2n)!}{n!n!} - \frac{(2n)!}{(n-1)!(n+1)!}.$$

We can factor out like terms and simplify:

$$C_n = \frac{(2n)!}{(n-1)!n!}\left(\frac{1}{n} - \frac{1}{n+1}\right) = \frac{(2n)!}{(n-1)!n!} \cdot \frac{1}{n(n+1)} = \frac{(2n)!}{(n+1)!n!}.$$

Removing an $(n+1)$ term from the denominator gives us our result:

$$C_n = \frac{1}{n+1}\left(\frac{(2n)!}{n!n!}\right) = \frac{1}{n+1}\binom{2n}{n}.$$

☐

> **Important:** The formula for the nth Catalan number is
>
> $$C_n = \frac{1}{n+1}\binom{2n}{n}.$$

Exercises

10.6.1 Compute C_8 both recursively and by the closed-form formula, and verify that they match.

10.6.2 In how many ways can identical coins be placed in one or more rows on a flat surface, such that there are n coins in the bottom row, and each coin (above the bottom row) is tangent to two coins directly underneath it? The possible configurations for $n = 3$ are shown on the next page. **Hints:** 60

10.6.3 Brazil defeats Germany in a wild World Cup final by the score of 8 to 6. Assuming the 14 goals were equally likely to be scored in any order, find the probability that the score was never tied (except at 0-0). **Hints:** 110

10.6.4 We form a *rooted binary tree* by as follows. Starting with a root node, we can draw 2 branches (but not 1) from the root to new nodes. From each of these new nodes we can draw 2 branches (but not 1) to new nodes, and so on. Each node either has exactly 2 branches (in which case we call it an *internal node*) or 0 branches (in which case we call it a *leaf*). The possibilities with 3 internal nodes are shown below. Prove that the number of rooted binary trees with n internal nodes is the nth Catalan number. **Hints:** 211

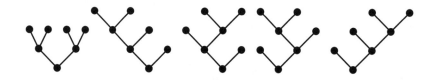

10.6.5★ Determine all values of n such that C_n is odd. **Hints:** 343, 124, 329

10.7 Summary

➤ Recursion is the name for the general concept of constructing later terms in a sequence from earlier terms.

➤ In a way, recursion is the opposite of constructive counting. In constructive counting, we think about how we would build the items that we're trying to count. In recursion, we think about how we would break up the items that we're trying to count into smaller pieces.

➤ Some simple recurrences can be solved by hand, by determining the recursive formula and then simply number-crunching to get the answer that you want.

➤ Linear recursions are solved in three steps:

1. Assume that the solution is an exponential, to get the characteristic equation.

2. Find the roots of the characteristic equation, to get the general solution as a sum of exponential terms with unknown constants.

3. Use the initial conditions to solve for the constants in the general solution.

➤ The Catalan numbers C_n are given by the following recurrence, where $C_0 = 1$:

$$C_n = C_0 C_{n-1} + C_1 C_{n-2} + \cdots + C_{n-1} C_0 = \sum_{k=0}^{n-1} C_k C_{n-1-k}.$$

They are also given by the closed-form formula $C_n = \dfrac{1}{n+1}\dbinom{2n}{n}$.

Also keep in mind the following problem-solving concepts:

> **Concept:** Experiment with smaller examples in order to get a feel for a hard problem. Try to use your experimentation as a guide for creating a description or formula for the general problem. This is what scientists and mathematicians do: they perform specific experiments to try to understand a general theory.

> **Concept:** If a solution method doesn't seem to be getting you anywhere, don't be afraid to start over and try a completely different approach.

> **Concept:** After solving a recurrence (or any problem for that matter), it is often a good idea to plug in some small values to check your work.

> **Concept:** When you see the same answer for two different problems, look for a connection, or better yet, for a 1-1 correspondence between them.

REVIEW PROBLEMS ▶

10.16 A teacher wishes to split his $2n$ students into n pairs. Use recursion to find a_n, the number of ways he can form the pairs.

10.17

(a) If we draw n lines in the plane, what is the largest number of different regions we can create (in terms of n)?

(b) If we draw n circles in the plane, what is the largest number of different regions we can create (in terms of n)?

(c) If we draw n pairs of parallel lines in the plane, what is the largest number of different regions we can create (in terms of n)?

10.18 Find a closed-form formula for the recurrence $a_n = 2a_{n-1} + a_{n-2}$ for all $n \geq 2$, with the initial conditions $a_0 = 2$ and $a_1 = 3$.

10.19 A solitaire game is played as follows. Six distinct pairs of matched tiles are placed in a bag. The player randomly draws tiles one at a time from the bag and retains them, except that matching tiles are put aside as soon as they appear in the player's hand. The game ends if the player ever holds three tiles, no two of which match; otherwise the drawing continues until the bag is empty. Find the probability that the player wins the game (by emptying the bag). *(Source: AIME)*

10.20 In how many ways can a 3×10 rectangle be tiled with tiles of size 1×2?

10.21 A collection of 8 cubes consists of one cube with edge-length k for each integer $1 \leq k \leq 8$. A tower is to be built using all 8 cubes according to the following rules:

- Any cube may be the bottom cube in the tower.

- The cube immediately on top of a cube with edge-length k must have edge-length at most $k + 2$.

How many different towers can be constructed? *(Source: AIME)*

10.22 Define a half-rectangular array of positive integers (shown below) by placing a 1 in the top row, and then letting every subsequent number be the sum of the number immediately above and the number immediately to the left. (If a number is missing, treat it as 0.) If the top row is Row 0, then Row n has $n + 1$ entries. Prove that the last entry in Row n is the nth Catalan number C_n, for all $n \geq 0$.

$$
\begin{array}{llllll}
1 \\
1 & 1 \\
1 & 2 & 2 \\
1 & 3 & 5 & 5 \\
1 & 4 & 9 & 14 & 14 \\
1 & 5 & 14 & 28 & 42 & 42
\end{array}
$$

Challenge Problems

10.23 We form a word using only A's, B's and C's. Suppose we can never have an A next to a C. Find the number of 8-letter words that can be formed. **Hints:** 81

10.24 A mail carrier delivers mail to the 19 houses on the east side of Elm Street. The carrier notices that no two adjacent houses ever get mail on the same day, but that there are never more than two houses in a row that get no mail on the same day. How many different subsets of houses that get mail on any particular day are possible? *(Source: AIME)* **Hints:** 196

10.25 We have the coins $C_1, C_2, \ldots, C_n$. For each k, C_k is biased so that, when tossed, it has probability $1/(2k + 1)$ of showing heads. If n coins are tossed, what is the probability that the number of heads is odd? *(Source: Putnam)* **Hints:** 70

10.26   For any positive integer n, prove that the number of positive integers whose digit-sums are $2n$ and whose digits are all 1, 3, or 4, is a perfect square. **Hints:** 248, 264

10.27   We are given a $2 \times n$ array of nodes, where n is a positive integer. A *valid connection* of the array is the addition of 1-unit-long horizontal and vertical edges between nodes, such that each node is connected to every other node via the edges, and there are no loops of any size. We give some examples for $n = 3$:

Let T_n denote the number of valid connections of the $2 \times n$ array. Find T_{10}. *(Source: USAMTS)* **Hints:** 330

10.28   Consider a regular decagon (10-sided polygon) with center O. We color, with one of 3 colors, each of the ten triangles formed by connecting O to adjacent vertices of the decagon. How many ways can we perform the coloring such that no two adjacent triangles are the same color? **Hints:** 166, 168

10.29★   We construct sequences of numbers as follows: start with

$$S_1 = 1, 1$$

and form each subsequent sequence by placing a new term between each term in the previous sequence equal to the sum of the two terms it's being put between:

$$S_2 = 1, 2, 1$$
$$S_3 = 1, 3, 2, 3, 1$$
$$S_4 = 1, 4, 3, 5, 2, 5, 3, 4, 1$$

and so on.

(a)   What is the largest element of S_n? **Hints:** 104, 172

(b)   Find a relationship between the number of odd terms and the number of even terms in the sequence S_n for even n. **Hints:** 285, 333

(Source: ARML)

10.30★   A *double-good* nesting of order n is an arrangement of $2n$ ")"s and n "("s such that as we read left-to-right, the number of ")"s that have appeared at any point is no more than 2 times the number of "("s that have appeared to that point. For example, the complete list of the double-good nestings of order 2 is:

$$())()$$
$$()())$$
$$(()))$$

Find the number of double-good nestings of order n. **Hints:** 148, 8, 246

10.31★ Suppose that Germany and Brazil play a soccer rematch in which there are $2n$ goals scored, and that the final score is Brazil $n + m$, Germany $n - m$. How many possible sequences of goals are there such that Germany is never more than $2m$ goals behind? **Hints:** 348, 128

10.32★ Count the number of sequences of integers a_1, a_2, a_3, a_4, a_5 with $a_i \leq 1$ for all i such that all partial sums $(a_1, a_1 + a_2, a_1 + a_2 + a_3, a_1 + a_2 + a_3 + a_4, a_1 + a_2 + a_3 + a_4 + a_5)$ are nonnegative. *(Source: HMMT)* **Hints:** 241, 72

Extra!	**General theory of linear recurrences**

⮕⮕⮕⮕ We can use methods similar to those of Section 10.3 to solve arbitrary linear recurrences of the form

$$a_n = p_1 a_{n-1} + p_2 a_{n-2} + \cdots + p_k a_{n-k},$$

where $p_1, p_2, \ldots, p_k$ are constants. We substitute $a_n = c^n$ to get the characteristic equation of the form

$$c^k - p_1 c^{k-1} - p_2 c^{k-2} - \cdots - p_k = 0.$$

Suppose that this characteristic equation has roots $r_1, r_2, \ldots, r_j$ with multiplicities $m_1, m_2, \ldots, m_j$; in other words, the characteristic equation factors as

$$(c - r_1)^{m_1} (c - r_2)^{m_2} \cdots (c - r_j)^{m_j} = 0.$$

Each root r_i contributes a term to the general solution of the form

$$(\lambda_{i1} + \lambda_{i2} n + \lambda_{i3} n^2 + \cdots + \lambda_{im_i} n^{(m_i - 1)})(r_i)^n,$$

so that the general solution is of the form

$$a_n = \sum_{i=1}^{j} \left(\sum_{s=1}^{m_i} \lambda_{is} n^{s-1} \right) (r_i)^n.$$

Note that there are k constants λ_{is} in this expression, since the multiplicities of the roots must sum to k. Given k initial values of a_n, we can always solve the resulting system of linear equations for the λ_{is}'s. This works regardless of whether the roots are real or complex.

Proving that this works for any linear recurrence requires advanced algebraic techniques and some knowledge of linear algebra.

The thing about science fiction is that it's totally wide open. But it's wide open in a conditional way.

– Octavia Butler

CHAPTER 11

Conditional Probability

11.1 Introduction

As the name suggests, conditional probability is probability with some added restriction. For many of these problems, we can use our usual method of computing probability:

$$P(\text{event}) = \frac{\text{Number of successful outcomes}}{\text{Number of possible outcomes}}.$$

The main difference with conditional probability is that the number of "possible" outcomes is usually restricted by some condition. Like many concepts in this book, it's really not as confusing as it sounds.

However, it is true that some conditional probability problems produce results that may seem rather counterintuitive. We'll see several examples of this, including a famous problem that sparked a lot of controversy in the mathematical community.

11.2 Basic Examples of Conditional Probability

Problems

Problem 11.1: A fair 6-sided die is rolled. If we know that the result is even, then what is the probability that we rolled a ⚃?

(a) In how many ways can we roll an even number?

(b) In how many ways can we roll a ⚃?

(c) What is the desired probability?

Problem 11.2: Two fair 6-sided dice are rolled. If at least one of the dice shows ⚅, then what is the probability that the sum of the two dice is 7?

(a) In how many ways can we roll the dice so that at least one is a ⚅?

(b) How many of the ways from (a) sum to 7?

(c) What is the desired probability?

Problem 11.3: Three coins are flipped.

(a) If the first coin comes up heads, then what is the probability that they all come up heads?

(b) If at least one of the coins comes up heads, then what is the probability that they all come up heads?

(c) Why are your answers to (a) and (b) different?

As we saw in the *Introduction to Counting & Probability* text, finding the probability of an event when all outcomes are equally likely is usually pretty simple:

$$P(\text{event}) = \frac{\text{Number of successful outcomes}}{\text{Number of possible outcomes}}.$$

Remember, for this to work, the outcomes must be equally likely.

Conditional probability is exactly the same. We'll start with a simple example.

Problem 11.1: A fair 6-sided die is rolled. If we know that the result is even, then what is the probability that we rolled a ⚃?

Solution for Problem 11.1: Since we know that the result is even, there are 3 equally likely possible outcomes: ⚁, ⚃, and ⚅. Only one of these, ⚅, is a "successful" outcome. Therefore, the probability is $\frac{1}{3}$. □

See how easy that was? There was nothing really new. The main thing to remember in conditional probability—in fact the *only* thing that you need to remember in conditional probability—is the same thing you need to know for pretty much *any* probability problem with equally likely outcomes:

$$P(\text{event}) = \frac{\text{Number of successful outcomes}}{\text{Number of possible outcomes}}.$$

The "condition" in conditional probability will just reduce the number of possible outcomes (and sometimes the number of successful outcomes). But it's exactly the same concept of probability that you've been using all along.

Extra!   *Luck is probability taken personally. It is the excitement of bad math.* – Penn Jillette

> **Important:** Conditional probability is not mysterious. If all events are equally likely, it's just
> $$P(\text{event}) = \frac{\text{Number of successful outcomes}}{\text{Number of possible outcomes}}.$$
> The only difference is that the condition might change the numbers of possible and successful outcomes.

The next example is a little more subtle, but it's still just the same basic concept.

Problem 11.2: Two fair 6-sided dice are rolled. If at least one of the dice shows ⚁, then what is the probability that the sum of the two dice is 7?

Solution for Problem 11.2: We'll start with a "solution" that exposes the most common pitfall in conditional probability calculations.

> **Bogus Solution:** There are 36 possible outcomes when rolling two dice. There are two ways to get a 7 that includes a ⚁, namely ⚁⚄ and ⚄⚁. Therefore, the probability is $\frac{2}{36} = \frac{1}{18}$.

This is incorrect, because there are not 36 possible outcomes! We only want to count possible outcomes that satisfy the condition that at least one of the dice shows ⚁.

The number of possible outcomes is easy to count using PIE:

$$\text{Possible outcomes} = \text{First die shows ⚁} + \text{Second die shows ⚁} - \text{Both dice show ⚁}$$
$$= 6 + 6 - 1 = 11.$$

There are 11 possible outcomes, and 2 of them are successful, so the probability is $\frac{2}{11}$. □

Once again, we just count the possible outcomes, count the successful outcomes, and take their quotient. All we need to remember is that we only count outcomes that satisfy the condition.

Here's another one:

Problem 11.3: Three coins are flipped. If at least one of them comes up heads, then what is the probability that they all come up heads?

Solution for Problem 11.3: It's tempting to take the following "shortcut":

> **Bogus Solution:** If one flip is heads, then the probability that the other two are heads is $\frac{1}{2} \times \frac{1}{2} = \frac{1}{4}$. Therefore the answer is $\frac{1}{4}$.

The error here is quite subtle. The above computation assumes that a *specific* coin comes up heads. If the question had been "If the first coin comes up heads, then what is the probability that they all come up heads?", then the above reasoning would have been perfectly correct, and the answer indeed would have been $\frac{1}{4}$.

However, that's not what the question asked. It asked "If *at least one of them* comes up heads, then what is the probability that they all come up heads?", which is a very different question. It is a much weaker condition to say that *some* coin comes up heads than it is to say that a *specific* coin comes up heads.

Fortunately, it's still easy to compute the probability, just using our basic concept of conditional probability. Since all of the outcomes are equally likely, we can calculate:

$$P(\text{All heads, given at least 1 head}) = \frac{\#(\text{ways to have all heads})}{\#(\text{ways to have at least one head})}.$$

There are 7 equally likely ways in which we can have at least one head: there are 8 ways to flip 3 coins (2^3), and we have to exclude the one outcome without any heads (namely TTT). Of the 7 equally likely outcomes, only one of them—HHH—has all three heads. Therefore, the probability is $\frac{1}{7}$. $\square$

Exercises

11.2.1 A pair of fair dice is rolled. Given that neither die shows ⚀, what is the probability that the sum of the dice is 7?

11.2.2 A fair standard die is tossed three times. Given that the sum of the first two tosses equals the third, what is the probability that at least one ⚀ is tossed? *(Source: AMC)* **Hints:** 105

11.2.3 Two fair coins are simultaneously flipped. This is done repeatedly until at least one of the coins comes up heads, at which point the process stops. What is the probability that the other coin also came up heads on this last flip? *(Source: HMMT)* **Hints:** 294

11.2.4 4 cards are drawn at random from a standard 52-card deck. At least 3 of them are ♡s. What is the probability that they are all ♡s?

11.2.5★ Let x and y be drawn (with replacement) from $\{1, 2, 3, \ldots, 99\}$ such that each ordered pair (x, y) is equally likely. Given that $x + y$ is even, determine the probability that the sum of the units digits of x and y is less than 10. **Hints:** 158

11.3 Some Definitions and Notation

Problem 11.4: Bag X has 2 red balls and 3 blue balls. Bag Y has 8 red balls and 2 blue balls. One of the two bags is chosen at random, and then a ball is chosen at random from that bag. If a blue ball is chosen, what is the probability that it came from bag X?

So far so good. If the outcomes are equally likely, we just count. But what happens if the outcomes are not equally likely? Let's see an example:

Problem 11.4: Bag X has 2 red balls and 3 blue balls. Bag Y has 8 red balls and 2 blue balls. One of the two bags is chosen at random, and then a ball is chosen at random from that bag. If a blue ball is chosen, what is the probability that it came from bag X?

Solution for Problem 11.4: We just count, right?

Bogus Solution: There are 5 blue balls. Two of them come from bag X. Therefore, the probability is $\frac{2}{5}$.

What do we always say about counting outcomes in probability problems? The outcomes have to be *equally likely*. The balls here are not all equally likely to be drawn! Because there are fewer balls in bag X than in bag Y, each individual ball in bag X is more likely to be drawn. Specifically, each of the 5 balls in bag X has probability $\frac{1}{10}$ to be drawn, whereas each of the 10 balls in bag Y has probability $\frac{1}{20}$ to be drawn. Note the the probabilities add up to 1:

$$5\left(\frac{1}{10}\right) + 10\left(\frac{1}{20}\right) = 1.$$

We have to alter our conditional probability calculation, so that instead of counting outcomes, we keep track of probabilities.

The probability of drawing a blue ball is:

$$P(\text{bag } X)P(\text{blue ball from } X) + P(\text{bag } Y)P(\text{blue ball from } Y) = \frac{1}{2} \cdot \frac{3}{5} + \frac{1}{2} \cdot \frac{1}{5} = \frac{3}{10} + \frac{1}{10} = \frac{2}{5}.$$

The probability of drawing a blue ball from bag X is:

$$P(\text{bag } X)P(\text{blue ball from } X) = \frac{1}{2} \cdot \frac{3}{5} = \frac{3}{10}.$$

Therefore,

$$P(\text{bag } X \text{ given blue ball}) = \frac{P(\text{blue ball from bag } X)}{P(\text{blue ball})} = \frac{\frac{3}{10}}{\frac{2}{5}} = \frac{3}{4}.$$

$\square$

In words, what we've done is:

$$P(\text{event given condition}) = \frac{P(\text{event and condition})}{P(\text{condition})}.$$

In Problem 11.4, the "event" was "selecting bag X" and the "condition" was "selecting a blue ball."

We can formalize this by introducing some notation.

Let A and B be two events. We denote by $P(A|B)$ the probability of A occurring given that B occurs, and we denote by $P(A \cap B)$ the probability that both A and B occur. We then have:

> **Important:**
>
> $$P(A|B) = \frac{P(A \cap B)}{P(B)}.$$

In Problem 11.4, A was "selecting bag X" and B was "selecting a blue ball."

> **Concept:** It is usually a bad idea to blindly memorize formulas like the one above.
> Instead, make sure that you understand the *process* of calculating conditional probability. Then, you won't need to memorize a formula, because you'll *know* how to apply it.

Exercises

11.3.1 Suppose there are 9 red balls and 1 white ball in Bag X and 2 white balls in Bag Y. A bag is chosen at random, then a ball is selected from the bag. Given that the ball is white, what is the probability that the bag was Bag X?

11.3.2 The probability that event A occurs is $\frac{3}{4}$, and the probability that event B occurs is $\frac{2}{3}$.

(a) What are the minimum and maximum possible values of $P(A \cap B)$?

(b) What are the minimum and maximum possible values of $P(A|B)$? Of $P(B|A)$?

(Source: AMC) **Hints:** 242

11.3.3 I have two cards. One is red on both sides. The other is red on one side, green on the other. A card is chosen at random, then placed on a table with the face pointing up chosen at random. The face that is showing is red. What is the probability the other side is also red?

11.3.4 Is it true that $P(A|B) = P(B|A)$? Why or why not? **Hints:** 79

11.3.5★ Urn A contains 4 white balls and 2 red balls. Urn B contains 3 red balls and 3 black balls. An urn is randomly selected, and then a ball inside of that urn is removed. We then repeat the process of selecting an urn and drawing out a ball, without returning the first ball. What is the probability that the first ball drawn was red, given that the second ball drawn was black? *(Source: HMMT)*

11.4 Harder Examples

Problems

> **Problem 11.5:** There are m men and n women in the Lawn Bowling Club. Two members are randomly chosen to go to the state tournament. During the press conference after the selection, a reporter asks the President of the club if at least one woman was selected, to which the President (truthfully) replies "Yes." What is the probability that both members selected are women?

Problem 11.6: The CDC has developed a new test for the Boogie-Woogie Flu. Unfortunately, the test is only 90% accurate: 10% of the time it gives the wrong result (it says a person is infected when she isn't, or it says a person is healthy when he's infected). Suppose that 5% of the general population has the flu. If a person tests positive for the flu, what is the probability that she actually has it?

Problem 11.7: Richard runs a blivet-making factory. Unfortunately, due to circumstances beyond his control, 20% of the blivets that Richard makes are defective. Defective blivets fail 10% of the time (and you don't want to be around when a blivet fails!), but good (non-defective) blivets never fail. How many times does Richard have to test a blivet that comes off his assembly line, in order for him to be 95% sure that the blivet is good?

Problem 11.8: Valentin and Naoki plan to meet at a restaurant for lunch. They agree that they will each separately show up at a random time between 12 P.M. and 2 P.M. They will each wait for 15 minutes; if the other shows up in that time period, they will have lunch together. Otherwise, they will leave. Today, Naoki showed up, waited for 15 minutes, and left because Valentin had not shown up. What is the probability that they would have had lunch together if Naoki had been willing to wait 15 more minutes?

We're ready to look at some harder conditional probability problems.

Problem 11.5: There are m men and n women in the Lawn Bowling Club. Two members are randomly chosen to go to the state tournament. During the press conference after the selection, a reporter asks the President of the club if at least one woman was selected, to which the President (truthfully) replies "Yes." What is the probability that both members selected are women?

Solution for Problem 11.5: As in Problem 11.3, beware the following "shortcut":

> **Bogus Solution:** After one woman is chosen, there are m men and $n - 1$ women remaining. The probability that a second woman is chosen is thus $\frac{n-1}{m+n-1}$, and hence the answer is $\frac{n-1}{m+n-1}$.

The problem is that we don't know if the *first* person chosen is a woman; we only know that *at least* one of them is.

However, this problem is easy if we just count outcomes that satisfy the condition.

First, we count possible outcomes. Rather than counting the number of ways that there is at least one woman, we count the number of ways that there are no women (in other words, 2 men). (Whenever you see "at least one" in a counting problem, think about counting the opposite—"none.") There are $\binom{n + m}{2}$ ways to choose 2 people from the $m + n$ person club, and $\binom{m}{2}$ ways to choose 2 men; therefore, there are $\binom{n + m}{2} - \binom{m}{2}$ ways to choose two people including at least one woman (and they're all equally likely).

To count the successful outcomes—those in which both people are women—we simply count the number of ways to choose two women from the club, which is $\binom{n}{2}$.

Therefore, the probability is

$$\frac{\binom{n}{2}}{\binom{n+m}{2} - \binom{m}{2}}.$$

$\square$

> **Concept:** If you can count *equally likely* outcomes, then that's usually the easiest way to approach conditional probability problems.

We could leave the answer in the above form, but it's probably nicer to do a little simplification. If we write out the binomial coefficients, and get rid of all the 2's in the denominators, we have

$$\frac{n(n-1)}{(n+m)(n+m-1) - m(m-1)}.$$

We expand the denominator and cancel terms, to get a relatively nice form:

$$\frac{n(n-1)}{n^2 + nm - n + nm + m^2 - m - m^2 + m} = \frac{n(n-1)}{n^2 + 2nm - n} = \frac{n-1}{n + 2m - 1}.$$

We can do a quick check of this formula by looking at the extreme cases. If the club is all women, then of course the answer to the problem is 1 (they're guaranteed to send 2 women since they're all women), and our formula gives

$$\frac{n-1}{n-1} = 1,$$

since $m = 0$. If the club has only 1 woman, then the answer to the problem is 0 (there's no way that they can send 2 women), and our formula gives

$$\frac{1-1}{1 + 2m - 1} = \frac{0}{2m} = 0,$$

since $n = 1$.

> **Concept:** Checking the extreme cases is often an easy way to do a quick check of a formula.

Problem 11.6: The CDC has developed a new test for the Boogie-Woogie Flu. Unfortunately, the test is only 90% accurate: 10% of the time it gives the wrong result (it says a person is infected when she isn't, or it says a person is healthy when he's infected). Suppose that 5% of the general population has the flu. If a person tests positive for the flu, what is the probability that she actually has it?

Solution for Problem 11.6: Here, we don't have items that we can count. So instead we will compute:

$$P(\text{event given condition}) = \frac{P(\text{event and condition})}{P(\text{condition})}.$$

In this problem, the event is "person has the flu" and the condition is "person tests positive."

The probability of having the flu and testing positive for it is $(0.05)(0.90) = 0.045$.

The probability of testing positive for the flu contains two cases: either the person has the flu and correctly tests positive, or the person is healthy and incorrectly tests positive. Thus, the probability is $(0.05)(0.90) + (0.95)(0.10) = 0.045 + 0.095 = 0.140$.

Therefore, the probability that a person who tests positive for the flu is actually infected is

$$\frac{0.045}{0.140} = \frac{45}{140} = \frac{9}{28} \approx 32.1\%.$$

□

Problem 11.6 illustrates a very important point. Many people, in their daily lives, would draw the following conclusion:

> **Bogus Solution:** The flu test is 90% accurate. Therefore, if a person tests positive for the flu, there is a 90% probability that that person has the flu.

As we saw above, this is a completely bogus conclusion.

The next problem has a similar theme.

> **Problem 11.7:** Richard runs a blivet-making factory. Unfortunately, due to circumstances beyond his control, 20% of the blivets that Richard makes are defective. Defective blivets fail 10% of the time (and you don't want to be around when a blivet fails!), but good (non-defective) blivets never fail. How many times does Richard have to test a blivet that comes off his assembly line, in order for him to be 95% sure that the blivet is good?

Solution for Problem 11.7: Here's a common—but incorrect—way to solve the problem.

> **Bogus Solution:** A defective blivet will pass n tests with probability $(0.90)^n$. Since we want to be 95% sure that a blivet is good, we need $(0.90)^n \leq 0.05$. Solving for n, we see that $n \geq \log_{0.90} 0.05 \approx 28.4$, so we need to test each blivet 29 times.

This is the type of solution than can result if you just start manipulating numbers in a way that "seems right" rather than *thinking* about what you're doing. What you have actually calculated by $(0.90)^n$ is the probability that a defective blivet will pass n tests. This is not the same thing as the probability that a blivet that passes n tests is defective.

In fact, what we need to compute is

$$P(\text{a blivet is good} \mid \text{the blivet passes } n \text{ tests}).$$

We start by computing the probability of the condition. A good blivet, which occurs with probability 0.80, will always pass n tests. A bad blivet, which occurs with probability 0.20, will pass n tests with probability $(0.90)^n$. Therefore, the probability that a random blivet passes n tests is $(0.80) + (0.20)(0.90)^n$. This will be the denominator of our conditional probability expression.

Hence, the probability that a blivet that passes n tests is good is

$$\frac{0.80}{0.80 + (0.20)(0.90)^n}.$$

This is the quantity that we need to be at least 0.95, so we need to solve the inequality:

$$\frac{0.8}{0.8 + (0.2)(0.9)^n} \geq 0.95.$$

This simplifies to give:

$$0.8 \geq 0.95(0.8 + (0.2)(0.9)^n),$$
$$\Leftrightarrow \quad 0.8 \geq 0.76 + (0.19)(0.9)^n,$$
$$\Leftrightarrow \quad 0.04 \geq (0.19)(0.9)^n,$$
$$\Leftrightarrow \quad \tfrac{4}{19} \geq (0.9)^n.$$

Therefore $n \geq \log_{0.9}(4/19) \approx 14.8$, so we need to test the blivet 15 times. $\square$

Concept: In conditional probability problems, make sure that you properly identify what is the condition and what is the successful outcome. Don't just blindly jump into calculations until you've first determined what it is you need to calculate.

Conditional probability is not just for discrete probability problems. We can use it on continuous probability problems too. This usually means using geometry, but it's the same basic principle. To compute a conditional probability, we compute

$$P(\text{event given condition}) = \frac{\text{area(event with condition)}}{\text{area(condition)}},$$

where we can replace "area" with "length" or "volume" as appropriate.

Here is a basic example of conditional probability in a continuous probability problem:

Problem 11.8: Valentin and Naoki plan to meet at a restaurant for lunch. They agree that they will each separately show up at a random time between 12 P.M. and 2 P.M. They will each wait for 15 minutes; if the other shows up in that time period, they will have lunch together. Otherwise, they will leave. Today, Naoki showed up, waited for 15 minutes, and left because Valentin had not shown up. What is the probability that they would have had lunch together if Naoki had been willing to wait 15 more minutes?

Solution for Problem 11.8: Since the possible outcomes—that is, the times at which Naoki and Valentin show up at the restaurant—form a continuous set, we cannot simply count the possible outcomes, since there are infinitely many. So we will have to use a geometric approach.

We know that Valentin and Naoki miss each other, so their arrival times must have differed by at least 15 minutes.

If we plot Naoki's arrival time on the x-axis and Valentin's arrival time on the y-axis, then the region of possible outcomes is shaded in the diagram at right. Note that this is the region above the line $y = x + 15$ (which signifies that Valentin arrived at least 15 minutes after Naoki) and the region below the line $y = x - 15$ (which signifies that Valentin arrived at least 15 minutes before Naoki). Together, the two halves of the "possible outcomes" region form a square with side length 105 (where the units are measured in minutes), so it has area $(105)^2 = 11025$.

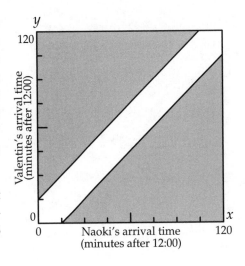

Had Naoki waited 15 more minutes, then he would have met Valentin provided that Valentin doesn't arrive more than 30 minutes after Naoki. This is the region between the lines $y = x + 15$ and $y = x + 30$, as shown in the figure on the left below.

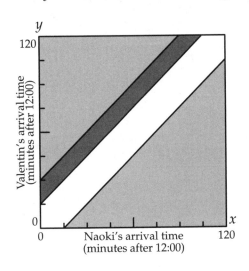

The area of the darker "successful outcomes" region can be calculated in many ways. Possibly the simplest is to notice that is it the area of an isosceles right triangle with side length 105 minus the area of an isosceles right triangle with side length 90. Therefore, its area is $(105)^2/2 - (90)^2/2 = 5512.5 - 4050 = 1462.5$.

Thus, the probability that Naoki would have met Valentin, had Naoki waited another 15 minutes, is

$$\frac{1462.5}{11025} = \frac{2925}{22050} = \frac{13}{98}.$$

□

Exercises

11.4.1 A bag of popcorn has $\frac{2}{3}$ white kernels and $\frac{1}{3}$ yellow kernels. Only half the white kernels will pop, but $\frac{2}{3}$ of the yellow kernels will. If a kernel is taken at random from the bag and is popped, what is the probability that it was a yellow kernel?

11.4.2 What fraction of all permutations of the numbers $1, 2, 3, 4, 5, 6$ such that the first term is not 1 has third term 3? *(Source: AMC)*

11.4.3 Two real numbers r and s are chosen at random between 0 and 1. If we know that $|r - s| < \frac{1}{4}$, what is the probability that $r < \frac{1}{2} < s$? **Hints:** 327

11.4.4 The inhabitants of Liarland tell the truth 40% of the time. One-third of the inhabitants have six toes on their right foot. You ask 4 of the inhabitants of Liarland if the King of Liarland has six toes on his right foot. They've all seen the King's feet (since the King always goes barefoot) and they all say "Yes." What is the probability that the King's right foot has 6 toes?

11.4.5 In Problem 11.7, suppose that quality control has improved to the point where only 15% of the blivets produced are defective. How many times does a blivet need to be tested so that Richard can be 95% sure that the blivet is good?

11.4.6 You are being tested for arachnophobia. It is known that 30% of people have arachnophobia. Those who have arachnophobia shiver $\frac{9}{10}$ of the time they are shown a picture of a black widow spider. Those who do not have arachnophobia shiver $\frac{1}{5}$ of the time they are shown a picture of a black widow.

(a) You are shown a picture of a black widow three times and never shiver. What is the probability you have arachnophobia?

(b) You are shown a picture of a black widow three times and always shiver. What is the probability you have arachnophobia?

(c)★ You are shown a picture of a black widow three times and shiver exactly twice. What is the probability you have arachnophobia?

11.5 Let's Make a Deal!

Problems

Problem 11.9: A popular game show is played as follows. A stage has three giant curtains. Behind one of the three curtains (chosen at random by the show's producers) is a brand-new car. Behind each of the other two curtains is a goat. The host knows which curtain is hiding the car. The contestant naturally wants to win the car and not a goat. The contestant chooses one of the curtains. The host then opens one of the *other* curtains, exposing one of the goats. The contestant is now given the opportunity to switch to the other unopened curtain, or keep the one that he originally chose. Should he switch?

Problem 11.10: You and two other prisoners are in jail. Tomorrow, two of you will be sentenced to life in prison, but the third will be set free; any one of the three of you is equally likely to be the one set free. The jailer is sitting just outside the bars. You ask the jailer if you are the one to be set free.

(a) The jailer says he knows who will be let free, but he won't tell you if you are the one to be freed. He does, however, point at one of the other two and tells you that that person will not be released. What is the probability that you will be set free?

(b) The jailer says that he can't give out that information, but he does have the names of the two to be sentenced to life written on two scraps of paper in his pocket. When the jailer turns away from you, you're able to sneak a hand through the bars and snatch one piece of paper out of his pocket. It doesn't have your name on it. What is the probability that you will be set free?

(c) Your probability of being freed is different in these two cases. Why?

In this section we'll explore one of the more famous problems involving conditional probability. This problem has a lot of history, due to its arguably counterintuitive answer. The problem gained notoriety when professional mathematicians argued about the result in the national press!

This problem is loosely based on the old TV game show *Let's Make a Deal*, and it is thus often known as the **Monty Hall problem** after the host of the show, Monty Hall.

> **Problem 11.9:** A popular game show is played as follows. A stage has three giant curtains. Behind one of the three curtains (chosen at random by the show's producers) is a brand-new car. The host knows which curtain is hiding the car. Behind each of the other two curtains is a goat. The contestant naturally wants to win the car and not a goat. The contestant chooses one of the curtains. The host then opens one of the *other* curtains, exposing one of the goats. The contestant is now given the opportunity to switch to the other unopened curtain, or keep the one that he originally chose. Should he switch?

Solution for Problem 11.9: A *very* common "solution" is the following:

> **Bogus Solution:** There are two unopened curtains. The contestant has no information about which curtain has the car and which curtain has the goat. Therefore, each has probability $\frac{1}{2}$ of having the car, and it doesn't matter if the contestant switches or not—the chance of winning the car is $\frac{1}{2}$ either way.

As we alluded to above, there are some professional mathematicians (not most, fortunately) that would argue that the above "solution" is correct! It seems to make a lot of intuitive sense.

However, it's bogus.

Certainly everyone agrees that before the host does anything, the contestant has a $\frac{1}{3}$ probability of picking the curtain with the car behind it. At the start of the game, the contestant *really* does not have any information about what's behind the curtains, so it's just a random guess. Since there are 3 curtains, and the car is equally likely to be behind any one of them, the probability that the initial guess is correct is $\frac{1}{3}$.

Now the host reveals a goat, which is *something that he can always do!* The two unpicked curtains either have two goats, or a car and a goat, so the host can always expose a goat behind an unchosen curtain. Since we know that the host can always expose a goat, regardless of the contestant's initial choice, why should the contestant's probability of winning magically increase from $\frac{1}{3}$ to $\frac{1}{2}$?

In fact, it doesn't. The probability that the contestant's initial choice was the car is still $\frac{1}{3}$. Therefore, if the contestant switches to the only other remaining curtain, he has a $\frac{2}{3}$ chance of switching to the car.

Perhaps you don't buy that argument—after all, it was pretty informal. It's good that you're skeptical! Let's prove it more rigorously.

Let's label the curtains A, B, and C, and let's suppose that the contestant picks A and the host reveals a goat. Here is a table showing the four possibilities and the probabilities that they occur. The boldface indicates the exposed goat.

Curtain A	Curtain B	Curtain C	Probability	Switching
Car	**Goat**	Goat	1/6	Loses
Car	Goat	**Goat**	1/6	Loses
Goat	Car	**Goat**	1/3	Wins
Goat	**Goat**	Car	1/3	Wins

Therefore, we see that if the contestant switches, he has a probability $\frac{1}{3} + \frac{1}{3} = \frac{2}{3}$ of winning, and probability $\frac{1}{6} + \frac{1}{6} = \frac{1}{3}$ of losing. $\square$

> **Sidenote:** The controversy surrounding the publication of the Monty Hall problem in the popular Sunday newspaper supplement *Parade* was detailed in a July 21, 1991 front-page story in *The New York Times*.

Problem 11.10: You and two other prisoners are in jail. Tomorrow, two of you will be sentenced to life in prison, but the third will be set free; any one of the three of you is equally likely to be the one set free. The jailer is sitting just outside the bars. You ask the jailer if you are the one to be set free.

(a) The jailer says he knows who will be let free, but he won't tell you if you are the one to be freed. He does, however, point at one of the other two and tells you that that person will not be released. What is the probability that you will be set free?

(b) The jailer says that he can't give out that information, but he does have the names of the two to be sentenced to life written on two scraps of paper in his pocket. When the jailer turns away from you, you're able to sneak a hand through the bars and snatch one piece of paper out of his pocket. It doesn't have your name on it. What is the probability that you will be set free?

(c) Your probability of being freed is different in these two cases. Why?

Solution for Problem 11.10: We first note that before the jailer responds, the probability that you will be set free is $\frac{1}{3}$, since you are told that the three prisoners are all equally likely to be freed.

(a) The jailer, unfortunately, is not giving you any additional information. You already knew that one of your two cellmates was going to be sentenced to life. So your probability of being freed is still $\frac{1}{3}$.

(b) You will be set free if and only if your name is not in the jailer's pocket. If your name was in his pocket, which occurs with probability $\frac{2}{3}$, then you would draw somebody else's name with probability $\frac{1}{2}$. If your name was not in his pocket, which occurs with probability $\frac{1}{3}$, then you would draw somebody else's name with probability 1. Therefore, the probability of drawing someone else's name is

$$\frac{2}{3} \cdot \frac{1}{2} + \frac{1}{3} \cdot 1 = \frac{2}{3}.$$

Hence, the probability that you will be freed (which is the same as the probability that your name is not in the jailer's pocket), given that you drew someone else's name, is

$$\frac{P(\text{you're freed and you draw another name})}{P(\text{you draw another name})} = \frac{\frac{1}{3}}{\frac{2}{3}} = \frac{1}{2}.$$

(c) Why the difference? It's subtle. In part (a), you got no information: whether you were going to be freed or not, the jailer can still give you someone else's name. But in part (b), when you reached into the jailer's pocket, there was always a chance that you might have drawn your own name. But you didn't, and that's *information* that you can use.

$\square$

━━━

Exercises

11.5.1 Suppose there are 10 curtains with 1 car and 9 goats. After the initial selection, the host reveals 8 of the goats. What's the probability of winning if the contestant switches?

11.5.2★ Suppose there are 4 curtains with 1 car and 3 goats. The host reveals one goat and gives the contestant the opportunity to switch. Then the host reveals a second goat and again gives the contestant the opportunity to switch. What's the best strategy, and what's the probability of winning the car if it is followed? **Hints: 7**

11.6 Summary

➤ Conditional probability is the probability of an event occurring, subject to some extra additional condition.

➤ If all events are equally likely, it's just

$$P(\text{event}) = \frac{\text{Number of successful outcomes}}{\text{Number of possible outcomes}}.$$

The only difference is that all of the outcomes must satisfy the condition.

➤ More formally, if A is the outcome and B is the condition, then the probability of A given that B occurs, denoted $P(A|B)$, is:

$$P(A|B) = \frac{P(A \cap B)}{P(B)}.$$

➤ Conditional probability problems sometimes lead to counterintuitive answers, so check your work carefully! Make sure that you properly identify what is the condition and what is the successful outcome.

Here are some general problem-solving concepts:

Concept: ⚷	Checking the extreme cases is often an easy way to do a quick check of a formula.

Concept: ⚷	Don't just blindly jump into calculations until you've first determined what it is you need to calculate.

━━━

REVIEW PROBLEMS

11.11 An integer is randomly chosen from 1 to 20, inclusive. Given that the number is prime, what is the probability it is even?

11.12   Bag A has three white and two black marbles. Bag B has four white and three black marbles. A marble is taken at random from Bag A and placed in Bag B. A marble is then taken at random from Bag B. Given that the marble taken from Bag B is white, what is the probability that the marble taken from Bag A was white?

11.13   A jar has 499 fair pennies and one penny with heads on both sides. A penny is chosen from the jar at random and flipped 9 times. It comes up heads every time. What is the probability that the coin is the two-headed coin?

11.14   On average, one in five Martians is a compulsive liar, and the rest always tell the truth. It rains 30% of the time on Mars. If three randomly-chosen Martians tell Astronaut Sue that it is raining, then what is the probability that it is actually raining? *(Source: Mandelbrot)*

11.15   A point on a circular table of radius 5 cm is chosen, and a quarter (of radius 1 cm) is placed on the table with its center at the chosen point. If no part of the quarter hangs over the edge of the table, then what is the probability that part of the quarter overlaps the center of the table?

11.16   Louise has a 75% probability of attending the annual Mathematical Association of America (MAA) convention. Thelma has an 80% chance of attending if Louise also attends; otherwise she has a 50% chance of attending. If I go to the convention and see Themla there, then what is the probability that Louise is also there? *(Source: Mandelbrot)*

Challenge Problems

11.17   The Royals and the Cubs play in the World Series while you fly to the moon and back. You find out when you return that the series lasted 6 games. You know that the Royals had a $\frac{2}{3}$ probability of winning each game. What is the probability that the Royals won the World Series? (Note that the World Series is a series of games played between two teams until one team has won 4 games.) **Hints:** 205

11.18   Roger and Stacy each go to the county fair on the same day. They each separately show up at a random time between 12:00 and 6:00. Roger stays for an hour, whereas Stacy stays for 2 hours. If we know that at some time they were both at the fair simultaneously, what is the probability that they were both there at exactly 3:00? **Hints:** 254

11.19   A coin is flipped 20 times in a row. Given that exactly 14 heads appeared, find the probability that no two consecutive coin flips were both tails. **Hints:** 64, 85

11.20★   At the Lucky Losers' Casino, the craps game is played with a set of 5 dice. The pit boss notices one of the customers cheating by replacing one of the dice with a loaded die that rolls ⚁ with probability $\frac{1}{2}$ (each of the other numbers on the loaded die show up with equal probability). Unfortunately, the boss can't tell which die is the loaded die (they all look identical), so he chooses one at random and rolls it 10 times. How many ⚁'s would he have to roll in order to be at least 90% sure that the chosen die is the loaded die? **Hints:** 257

11.21★   3 points A, B, C are randomly chosen on the circumference of a circle. If A, B, C all lie on a semicircle, then what is the probability that all of the angles of triangle ABC are less than 120°? **Hints:** 192

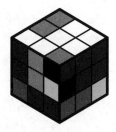

To be idle requires a strong sense of personal identity. – Robert Louis Stevenson

CHAPTER **12**

_____ **Combinatorial Identities**

12.1 Introduction

In this chapter, we'll discuss some combinatorial identities. Some of these you should hopefully already know, and some will likely be new to you.

While more advanced combinatorial identities aren't often used to solve problems (though sometimes they *are* the problem), mastering combinatorial identities requires a very deep understanding of what our common counting tools really mean in terms of counting, as opposed to just knowing how to perform computations.

12.2 Basic Identities

Problem 12.1: Prove that

$$\binom{n}{r} = \binom{n}{n-r}$$

for all positive integers n and all integers $0 \le r \le n$, using

(a) a committee-forming argument: show that both sides of the identity count, in a different way, the number of ways to form a certain committee.

(b) a block-walking argument: show that both sides count, in a different way, the number of certain types of paths on a grid or on Pascal's Triangle.

(c) algebra: show algebraically that the two sides are equal.

Problem 12.2: Prove that

$$\binom{n}{r} + \binom{n}{r+1} = \binom{n+1}{r+1}$$

for all positive integers n and all integers $0 \le r < n$, using

(a) a committee-forming argument.

(b) a block-walking argument.

(c) algebra.

Problem 12.3: Prove that for any positive integer n,

$$(x+y)^n = \binom{n}{0}x^n + \binom{n}{1}x^{n-1}y + \binom{n}{2}x^{n-2}y^2 + \cdots + \binom{n}{n}y^n.$$

Problem 12.4: Prove that

$$\binom{n}{0} + \binom{n}{1} + \binom{n}{2} + \cdots + \binom{n}{n} = 2^n$$

using

(a) algebra.

(b) a committee-forming argument.

(c) a block-walking argument.

(d) the identity from Problem 12.2 and mathematical induction.

Problem 12.5: Use Problems 12.3 and 12.4 to compute the sum

$$\binom{n}{0} + \binom{n}{2} + \binom{n}{4} + \cdots$$

for any positive integer n.

Problem 12.6: Prove that

$$\binom{k}{k} + \binom{k+1}{k} + \binom{k+2}{k} + \cdots + \binom{n}{k} = \binom{n+1}{k+1}$$

using

(a) a block-walking argument.

(b) algebra, specifically Pascal's identity and induction.

(c) a committee-forming argument.

All of the identities in this section were covered in the book *Introduction to Counting & Probability*. However, even if you've read that book, we advise you to work through the identities again here.

We will pay special attention to the fact that there is usually more than one way to prove a given identity. Indeed, there are three major techniques.

First, we can use a committee-forming argument, or more generally, show that the quantity on the left side of the equation counts the same thing, but in a different way, as the quantity on the right side of the equation. Often this will be two different ways to form a committee (or committees) with certain properties.

Second, we can look at our old friend Pascal's Triangle, and show that both sides of an identity count the number of paths to the same point, but in different ways.

Finally, we can use the definition $\binom{n}{k} = \dfrac{n!}{k!(n-k)!}$ and use algebraic manipulation to prove identities. Also, there is the closely related method of using identities that we have already proven to prove more complicated identities.

In many of the problems below, we'll see all three methods in action. Also note that in some of these methods, we will need to use mathematical induction to make the proof precise. This is most common when using the algebraic method.

A quick note about notation: in identities, it is very common to use the abbreviations **LHS** for "left hand side" and **RHS** for "right hand side". For example, in the identity

$$(a + b)^2 = a^2 + 2ab + b^2,$$

the LHS is $(a + b)^2$ and the RHS is $a^2 + 2ab + b^2$. Of course, in identities, we are trying to show that the LHS and the RHS are equal.

We'll start with what is probably the most basic identity involving combinations.

Problem 12.1: Prove that

$$\binom{n}{r} = \binom{n}{n-r}$$

for all positive integers n and all integers $0 \le r \le n$, using

(a) a committee-forming argument.

(b) a block-walking argument.

(c) algebra.

Solution for Problem 12.1:

(a) Both sides count the number of ways to form a committee of r people from a group of n people. The LHS chooses the r people who are on the committee. The RHS chooses the $n - r$ people who are *not* on the committee.

(b) A path to the rth point of the nth row of Pascal's Triangle requires r steps down-and-to-the-right and $n - r$ steps down-and-to-the-left. The LHS of the identity chooses r steps (out of an n-step path) to go right, and the RHS of the identity chooses $n - r$ steps (out of an n-step path) to go left.

(c) We can use the algebraic definition: $\binom{n}{r} = \dfrac{n!}{r!(n-r)!} = \dfrac{n!}{(n-r)!(n-(n-r))!} = \binom{n}{n-r}$.

$\square$

The next identity is just as fundamental and is frequently used to prove other, more complicated, identities.

Problem 12.2: Prove that
$$\binom{n}{r} + \binom{n}{r+1} = \binom{n+1}{r+1}$$

for all positive integers n and all integers $0 \le r < n$, using

(a) a committee-forming argument.

(b) a block-walking argument.

(c) algebra.

This identity is known as **Pascal's identity**.

Solution for Problem 12.2:

(a) Usually, when trying to form a counting argument to prove an identity, we'll start with the simpler side of the identity first. In this case, the RHS is simpler; in fact, it's just a single binomial coefficient. Therefore, it makes sense to start by imagining a club with $n + 1$ members. This way, the RHS of the identity simply counts the number of ways to form an $(r + 1)$-person committee from the club members. The LHS is a sum of two binomial coefficients, so we will try to break up this count into two cases.

> **Concept:** When trying to prove identities by committee-forming arguments:
>
> - Usually try to find a counting explanation for the simpler side first.
>
> - A single binomial coefficient of the form $\binom{n}{r}$ means that a good place to start is choosing r items from a set of n items.
>
> - A sum of binomial coefficients may mean trying a casework argument.

Suppose that one of the club members is named Katie. Our two cases will be whether or not Katie is on the committee. If Katie is on the committee, then we must choose r people from the n remaining club members to complete the committee, which can be done in $\binom{n}{r}$ ways. If Katie is not on the committee, then the entire $(r + 1)$-person committee must be chosen from the remaining n members, which can be done in $\binom{n}{r+1}$ ways. Since these two cases are exclusive (either Katie is on the committee or she isn't), we sum the counts of the two cases to get the total number of committees, thus establishing the identity.

(b) The RHS counts the number of paths on Pascal's Triangle to the $(r + 1)^{st}$ point of the $(n + 1)^{st}$ row. Every such path must pass through exactly one of the two points immediately above: these are the rth and $(r + 1)^{st}$ points of the nth row. There are $\binom{n}{r}$ paths to the first point and $\binom{n}{r+1}$ paths to the second point. Summing them gives the number of paths to our desired point, thus establishing the identity.

(c) We can apply the definition of binomial coefficients and use algebra. We start by writing out the definitions for the terms on the LHS:

$$\binom{n}{r} + \binom{n}{r+1} = \frac{n!}{r!(n-r)!} + \frac{n!}{(r+1)!(n-r-1)!}.$$

Factor out all the common terms:

$$\frac{n!}{r!(n-r)!} + \frac{n!}{(r+1)!(n-r-1)!} = \frac{n!}{r!(n-r-1)!}\left(\frac{1}{n-r} + \frac{1}{r+1}\right),$$

$$= \frac{n!}{r!(n-r-1)!}\left(\frac{n+1}{(n-r)(r+1)}\right).$$

Now recombine the terms into factorials:

$$\frac{n!}{r!(n-r-1)!}\left(\frac{n+1}{(n-r)(r+1)}\right) = \frac{n!(n+1)}{r!(r+1)(n-r-1)!(n-r)},$$

$$= \frac{(n+1)!}{(r+1)!(n-r)!}.$$

Finally, we see that this last expression is exactly what we want on the RHS of the identity:

$$\frac{(n+1)!}{(r+1)!(n-r)!} = \binom{n+1}{r+1}.$$

$\square$

You should recognize the next identity as the **Binomial Theorem**.

Problem 12.3: Prove that for any positive integer n,

$$(x+y)^n = \binom{n}{0}x^n + \binom{n}{1}x^{n-1}y + \binom{n}{2}x^{n-2}y^2 + \cdots + \binom{n}{n}y^n.$$

Solution for Problem 12.3: In order to get a term of the form $x^{n-k}y^k$ in the product $(x+y)^n$, we must choose k copies of $(x+y)$ to contribute a y term, and then the remaining $n-k$ copies of $(x+y)$ will contribute an x term. Since there are $\binom{n}{k}$ ways to choose k of the n $(x+y)$ terms to contribute a y, we know that there are $\binom{n}{k}$ terms of the form $x^{n-k}y^k$ in the expansion of product $(x+y)^n$. Therefore,

$$(x+y)^n = \sum_{k=0}^{n} \binom{n}{k}x^{n-k}y^k.$$

Alternatively, we can think of the coefficient of $x^{n-k}y^k$ as the number of ways to form an n-letter word using $n-k$ x's and k y's. Of course, we know that this number is $\binom{n}{k}$, since we have to choose k of the n positions in which to insert a y. $\square$

Sidenote: The Binomial Theorem is also valid for exponents that are not integers. That is, for all real numbers x, y, and r with $|x| > |y|$,

$$(x+y)^r = x^r + rx^{r-1}y + \frac{r(r-1)}{2!}x^{r-2}y^2 + \frac{r(r-1)(r-2)}{3!}x^{r-3}y^3 + \cdots$$

(The proof of this requires calculus, so you'll have to take our word for it.)

Continued on next page...

> **Sidenote:** ... *Continued from previous page*
>
> ♪   For example, if $x > 1$, then
>
> $$\sqrt{x+1} = (x+1)^{\frac{1}{2}}$$
>
> $$= x^{\frac{1}{2}} + \frac{1}{2}x^{-\frac{1}{2}} + \frac{\frac{1}{2}(-\frac{1}{2})}{2!}x^{-\frac{3}{2}} + \frac{\frac{1}{2}(-\frac{1}{2})(-\frac{3}{2})}{3!}x^{-\frac{5}{2}} + \cdots$$
>
> $$= x^{\frac{1}{2}} + \frac{1}{2x^{\frac{1}{2}}} - \frac{1}{8x^{\frac{3}{2}}} + \frac{1}{16x^{\frac{5}{2}}} - \frac{5}{128x^{\frac{7}{2}}} + \cdots$$
>
> Here are a couple of examples for practice:
>
> (a)   Use this expansion to estimate $\sqrt{5}$ to 2 decimal places.
>
> (b)   Compute the first 4 terms in the expansion of $(x+1)^{1/3}$. Use this to approximate $\sqrt[3]{9}$.

We'll see a lot more of the Binomial Theorem, although in a very different context, in Chapter 14. For now, we can use the Binomial Theorem to prove other interesting identities.

Problem 12.4: Prove that

$$\binom{n}{0} + \binom{n}{1} + \binom{n}{2} + \cdots + \binom{n}{n} = 2^n$$

using

(a)   algebra.

(b)   a committee-forming argument.

(c)   a block-walking argument.

Solution for Problem 12.4:   Note that using Σ-notation, we can rewrite our identity as

$$\sum_{k=0}^{n} \binom{n}{k} = 2^n.$$

(a)   The algebra solution is the simplest, but maybe not the most insightful. We see a sum of binomial coefficient on the LHS, all with the same "top" number, namely n. This looks a lot like the Binomial Theorem from Problem 12.3, except without all the x's and y's. How can we make the x's and y's from the Binomial Theorem go away, so that we are left with just the sum of the coefficients? Simple: we just plug in $x = y = 1$, we get

$$\sum_{k=0}^{n} \binom{n}{k} 1^{n-k} 1^k = (1+1)^n.$$

Then making the obvious simplifications $1^{n-k} = 1$, $1^k = 1$, and $1 + 1 = 2$, we get our result:

$$\sum_{k=0}^{n} \binom{n}{k} = 2^n.$$

(b) The LHS of the identity counts the number of ways, starting with a club with n members, to form a 0-person committee, or a 1-person committee, or a 2-person committee, etc., up to an n-person committee. In other words, the LHS counts the number of ways to form *any* committee from a club with n members (where an empty committee and a committee of the entire club are both possibilities). On the other hand, we can form an arbitrary committee by deciding, for each person in the club, whether he is on the committee or not. Since each club member has 2 choices (on the committee or not on the committee) and there are n members, there are a total of 2^n ways to make all the choices and form the committee. Thus, the RHS of the identity also counts the number of ways to form any committee, and this establishes the identity.

(c) If we start at the top of Pascal's Triangle and take n steps, each either down-left or down-right, then we end up at some point in Row n. Since there are two choices for each step, there are 2^n such paths. However, every path ends up at one point of Row n, and the number of paths to points on Row n is $\binom{n}{0}$ (to the left-most point) $+\binom{n}{1}$ (to the next point) $+ \cdots + \binom{n}{n}$ (to the right-most point). Thus the LHS and the RHS both count all possible paths to Row n of Pascal's Triangle, and this establishes the identity.

$\square$

As we saw in solution (a) to the previous problem, the Binomial Theorem is a powerful tool for evaluating sums of binomial coefficients. Let's see another example.

Problem 12.5: For any positive integer n, compute the sum

$$\binom{n}{0} + \binom{n}{2} + \binom{n}{4} + \cdots.$$

Solution for Problem 12.5:   We see that this is the sum of all of the binomial coefficients of the form $\binom{n}{k}$ where k is even. If we were not missing the terms with odd k, then we could use the result from Problem 12.4 to compute the sum:

$$\binom{n}{0} + \binom{n}{1} + \binom{n}{2} + \binom{n}{3} + \cdots = 2^n. \tag{$*$}$$

How do we get rid of the odd terms?

We can cleverly use the Binomial Theorem again to make the odd terms go away. Specifically, watch what happens when we let $x = 1$ and $y = -1$ in the Binomial Theorem:

$$(1 - 1)^n = \binom{n}{0}1^n + \binom{n}{1}1^{n-1}(-1)^1 + \binom{n}{2}1^{n-2}(-1)^2 + \binom{n}{3}1^{n-3}(-1)^3 + \cdots$$

$$= \binom{n}{0} - \binom{n}{1} + \binom{n}{2} - \binom{n}{3} + \cdots.$$

But, for any positive integer n, the left side of the above expression is just $(1-1)^n = 0^n = 0$. So we simply have

$$\binom{n}{0} - \binom{n}{1} + \binom{n}{2} - \binom{n}{3} + \cdots = 0. \tag{$**$}$$

When we add equations $(*)$ and $(**)$, all of the $\binom{n}{k}$ terms with odd k cancel, and we're left with 2 times the sum of the all of the $\binom{n}{k}$ terms with even k:

$$2\left(\binom{n}{0} + \binom{n}{2} + \binom{n}{4} + \cdots\right) = 2^n.$$

We divide by 2 to get our answer:

$$\binom{n}{0} + \binom{n}{2} + \binom{n}{4} + \cdots = 2^{n-1}.$$

□

> **Concept:** Often we can prove identities involving binomial coefficients by plugging appropriate values into the Binomial Theorem.

You'll see some more difficult examples of this in the Exercises and Challenge Problems later in the chapter.

We'll finish this section with the **Hockey Stick identity**.

Problem 12.6: Prove that

$$\binom{k}{k} + \binom{k+1}{k} + \binom{k+2}{k} + \cdots + \binom{n}{k} = \binom{n+1}{k+1}$$

using

(a) a block-walking argument.

(b) algebra, specifically Pascal's identity and induction.

(c) a committee-forming argument.

In summation notation, we would write this as

$$\sum_{i=k}^{n} \binom{i}{k} = \binom{n+1}{k+1}.$$

Solution for Problem 12.6:

(a) The RHS counts the number of paths to the point in Row $(n + 1)$ of Pascal's Triangle that is $k + 1$ spaces over from the left edge. Call that point A. All such paths must at some point cross the diagonal above A consisting of all points that are k spaces over from the left edge. For example, in Figure 12.1 below on the next page, all paths from the top of the triangle to A must pass through one of the circled dots (a sample path is shown by the dashed line):

> **Extra!** *I don't like my hockey sticks touching other sticks, and I don't like them crossing one another, and I kind of have them hidden in the corner... I think it's essentially a matter of taking care of what takes care of you.* – Wayne Gretzky

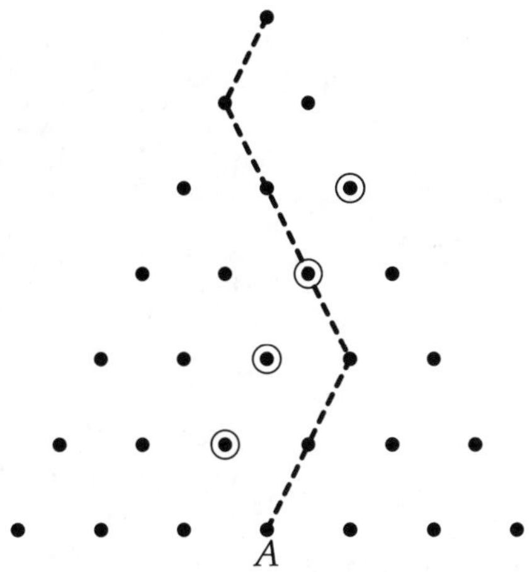

Figure 12.1: Paths to A must cross the circled diagonal

However, some paths to A may pass through more than one of the marked points, as shown by the path in Figure 12.2 below.

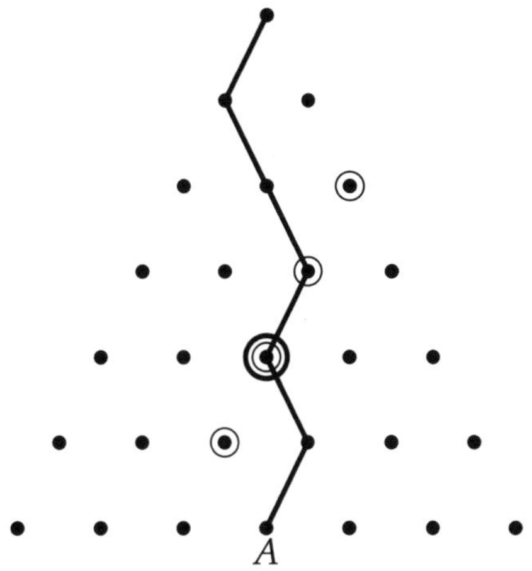

Figure 12.2: Path to A through more than 1 marked point

After each path leaves the diagonal of circled points, there is only one way to get to A: take down-left steps until reaching A. So, for each path to a circled point, there's exactly one way to continue that path to A without passing through another circled point. Therefore, each path to A corresponds to a path to exactly one of the marked diagonal points. Since the sum of the numbers of paths to the various diagonal points is exactly the LHS of the identity, this proves the result.

(b) To prove the Hockey Stick identity using algebra, we repeatedly apply Pascal's identity from Problem 12.2. In order to apply Pascal's identity, we need to change the first term on the LHS from $\binom{k}{k}$ to $\binom{k+1}{k+1}$. (Since these are both equal to 1, there's no problem making this change.) Now we just repeatedly apply Pascal's identity:

Extra! *You always admire what you really don't understand.* – Blaise Pascal

$$\binom{k}{k} + \binom{k+1}{k} + \binom{k+2}{k} + \cdots + \binom{n}{k} = \binom{k+1}{k+1} + \binom{k+1}{k} + \binom{k+2}{k} + \cdots + \binom{n}{k}$$

$$= \binom{k+2}{k+1} + \binom{k+2}{k} + \cdots + \binom{n}{k}$$

$$= \binom{k+3}{k+1} + \cdots + \binom{n}{k}$$

$$= \quad \vdots$$

$$= \binom{n}{k+1} + \binom{n}{k} = \binom{n+1}{k+1}.$$

Each step of the above computation is a result of applying Pascal's identity on the first two terms.

The above argument is somewhat imprecise; those vertical dots in the line before the end leave us with a lack of rigor. To make the proof rigorous, we should use mathematical induction. So let us prove, for any fixed nonnegative integer k, and any integer $n \geq k$, that

$$\sum_{i=k}^{n} \binom{i}{k} = \binom{n+1}{k+1}.$$

Base case: Note that the base case is $n = k$. The identity in the base case is just

$$\binom{k}{k} = \binom{k+1}{k+1},$$

which of course is true since both terms are equal to 1.

Inductive step: Assume that the result is true for some positive integer $n \geq k$, and consider the identity for $n + 1$. We will pull the $n + 1$ term out of the summation on the LHS, so that we can apply the inductive hypothesis:

$$\sum_{i=k}^{n+1} \binom{i}{k} = \sum_{i=k}^{n} \binom{i}{k} + \binom{n+1}{k}$$

$$= \binom{n+1}{k+1} + \binom{n+1}{k}$$

$$= \binom{n+2}{k+1},$$

where in the last step we are using Pascal's identity. Therefore, the result is true for $n + 1$.

So, by induction, the result holds for all nonnegative integers k and all integers $n \geq k$.

(c) The RHS is the number of ways to choose a $(k + 1)$-person committee from a club with $(n + 1)$ members. We need to find a way for the LHS to also count those committees.

Suppose that the club members are numbered from #1, #2, up to #$(n + 1)$, and let's suppose that the highest-numbered club member on the committee is named committee president. We can then count the committees using casework based on which member becomes president.

No member lower than #$(k + 1)$ can be president, and if person #j is president, then we have to choose the remaining k members of the committee from the $j - 1$ people ranked below the president. Therefore the number of committees is

$$\sum_{j=k+1}^{n+1} \binom{j-1}{k}.$$

We will now do a standard manipulation with $\sum$-expressions: changing the index of summation. If we let $i = j - 1$, we get

$$\sum_{j=k+1}^{n+1} \binom{j-1}{k} = \sum_{i=k}^{n} \binom{i}{k}.$$

Note how the lower and upper bound of the summation changed when we changed the summation index. This expression is exactly the LHS of the identity, so we are done.

$\square$

Exercises

12.2.1 Show that
$$\binom{n}{0} - \binom{n}{1} + \binom{n}{2} - \binom{n}{3} + \cdots + (-1)^n \binom{n}{n} = 0.$$

12.2.2 Prove that
$$k\binom{n}{k} = n\binom{n-1}{k-1}$$

(a) using algebra.

(b) using a committee-forming argument.

12.2.3 Prove that
$$\binom{n}{0} + 2\binom{n}{1} + 4\binom{n}{2} + \cdots + 2^n\binom{n}{n} = 3^n$$

(a) using algebra.

(b)★ using a committee-forming argument. **Hints:** 239

12.2.4 Prove that for all positive integers $0 \le k \le r \le n$,
$$\binom{n}{r}\binom{r}{k} = \binom{n}{k}\binom{n-k}{r-k}$$

(a) using a committee-forming argument.

(b) using algebra.

12.2.5★ For all integers $n \ge 2$, find a closed-form expression for $\binom{n}{1} + 3\binom{n}{3} + 5\binom{n}{5} + 7\binom{n}{7} + \cdots$. **Hints:** 163

12.3   More Identities

Problems

Problem 12.7: For any nonnegative integers m, n, and r, we wish to find a closed form for the sum

$$\binom{m}{0}\binom{n}{r} + \binom{m}{1}\binom{n}{r-1} + \binom{m}{2}\binom{n}{r-2} + \cdots + \binom{m}{r}\binom{n}{0}.$$

(a)   Experiment with small values of m, n, and r.

(b)   Compute the sum for the values $m = 5$, $n = 4$, $r = 3$. Try to write the final answer in terms of one or more binomial coefficients. Does this lead to a conjecture for the closed-form formula of the general sum?

(c)   Look for a counting argument that proves your conjecture from part (b).

Problem 12.8: For any nonnegative integers m, n, and r, find a closed form for the sum

$$\binom{m}{0}\binom{n}{r} + \binom{m}{1}\binom{n}{r+1} + \cdots + \binom{m}{m}\binom{n}{r+m}.$$

Problem 12.9: Evaluate

$$1\binom{11}{1} + 2\binom{11}{2} + 3\binom{11}{3} + \cdots + 11\binom{11}{11}.$$

Problem 12.10: Evaluate

$$\frac{19}{99} + \frac{(19)(18)}{(99)(98)} + \frac{(19)(18)(17)}{(99)(98)(97)} + \cdots + \frac{19!}{(99)(98)\cdots(81)}.$$

In this section we move on to identity problems that are a bit more difficult. For many of these problems, we'll see how $\sum$-notation will make the identities more manageable.

Part of the challenge in the problems in this section is that we're only providing one side of the identity. You'll be provided a messy expression, and the goal is to find a nicer form for it.

Problem 12.7: For any nonnegative integers m, n, and r, find a closed form for the sum

$$\binom{m}{0}\binom{n}{r} + \binom{m}{1}\binom{n}{r-1} + \binom{m}{2}\binom{n}{r-2} + \cdots + \binom{m}{r}\binom{n}{0}.$$

Solution for Problem 12.7: We might try an example first, to see if we can figure out what it is supposed to be equal to. For instance, if we plug in $m = 2$, $n = 3$, and $r = 2$, we get:

$$\binom{2}{0}\binom{3}{2} + \binom{2}{1}\binom{3}{1} + \binom{2}{2}\binom{3}{0} = 3 + 6 + 1 = 10.$$

No obvious pattern there. Let's try some bigger numbers, such as $m = 5, n = 4, r = 3$:

$$\binom{5}{0}\binom{4}{3} + \binom{5}{1}\binom{4}{2} + \binom{5}{2}\binom{4}{1} + \binom{5}{3}\binom{4}{0} = 4 + 30 + 40 + 10 = 84.$$

Since we know that we're dealing with binomial coefficients, we should try to write 84 as a binomial coefficient, and we note that $84 = \binom{9}{3}$. This just happens to be $\binom{m+n}{r}$ for our example. And now, if we look back at our first example, we also see that with $m = 2, n = 3, r = 2$, our answer was $\binom{m+n}{r} = \binom{5}{2} = 10$. So we conjecture that the closed form that we are seeking is $\binom{m+n}{r}$.

> **Concept:** Experiment with small values to try to see a pattern. Look to write numbers in terms of binomial coefficients.

In summation notation, the identity that we're attempting to prove is

$$\sum_{k=0}^{r} \binom{m}{k}\binom{n}{r-k} = \binom{m+n}{r}.$$

The RHS suggests to think of a club with $m + n$ people, from which we are choosing an r-person committee. The $m + n$ suggests that the club is broken up into two types of people, so we let the club have m men and n women. To achieve the expression on the LHS, we can count the committees using casework based on the number of men on the committee. If there are k men (and thus necessarily $r - k$ women), there are $\binom{m}{k}$ ways to choose the men and $\binom{n}{r-k}$ ways to choose the women, and hence $\binom{m}{k}\binom{n}{r-k}$ ways to choose the committee (since the choices of men and women are independent). When we sum the cases over all values of k, we have our identity. □

The identity

$$\sum_{k=0}^{r} \binom{m}{k}\binom{n}{r-k} = \binom{m+n}{r}$$

from Problem 12.7 is often known as **Vandermonde's identity**.

There is one detail that we might have overlooked, though, and that is the case where r is larger than either m or n. For example, if $m = 6, n = 8, r = 10$, then the first term of the LHS of Vandermonde's identity is $\binom{6}{0}\binom{8}{10}$, and perhaps we should be concerned about coefficients like $\binom{8}{10}$ appearing in our identity. We will leave you to think about this as an Exercise. We will also see a bit of this in our next problem.

> **Problem 12.8:** For any nonnegative integers m, n, and r, find a closed form for the sum
>
> $$\binom{m}{0}\binom{n}{r} + \binom{m}{1}\binom{n}{r+1} + \cdots + \binom{m}{m}\binom{n}{r+m}.$$

Solution for Problem 12.8: This expression is

$$\sum_{k=0}^{m} \binom{m}{k}\binom{n}{r+k}.$$

This looks very similar to Vandermonde's identity from the last problem. But it's not quite the same format. We'd like to use Vandermonde's identity to evaluate this identity, if we can.

> **Concept:** Try to use identities that you already know to prove new identities.

We can make our current expression look more like Vandermonde's identity by applying $\binom{m}{k} = \binom{m}{m-k}$:

$$\sum_{k=0}^{m} \binom{m}{k}\binom{n}{r+k} = \sum_{k=0}^{m} \binom{m}{m-k}\binom{n}{r+k}.$$

Now the "lower" terms of the binomial coefficient sum to $m + r$, a constant, just like in Vandermonde's identity. If only the sum were from $k = 0$ to $m + r$, and not m, then we could just apply Vandermonde's.

But we still can! All of the terms with $k > m$ are 0, since $\binom{m}{m-k} = 0$ if $k > m$. Therefore,

$$\sum_{k=0}^{m} \binom{m}{m-k}\binom{n}{r+k} = \sum_{k=0}^{m+r} \binom{m}{m-k}\binom{n}{r+k},$$

and we can apply Vandermonde's identity to this last sum to get $\binom{m+n}{m+r}$. $\square$

> **Problem 12.9:** Evaluate
> $$1\binom{11}{1} + 2\binom{11}{2} + 3\binom{11}{3} + \cdots + 11\binom{11}{11}.$$

Solution for Problem 12.9: We present two methods for simplifying this expression.

Method 1: We notice that all of the terms of the sum are of the form $k\binom{11}{k}$ for some value of k. Perhaps there is a way to manipulate this term into something a little easier to deal with. For this, we perform some algebra. (If you did Exercise 12.2.2, then you will already have done this manipulation.)

$$k\binom{11}{k} = k\frac{11!}{k!(11-k)!}$$
$$= \frac{11!}{(k-1)!(11-k)!} \qquad \text{(since } k \text{ cancels with the largest term in } k!\text{)}$$
$$= \frac{11!}{(k-1)!(10-(k-1))!}$$
$$= 11\frac{10!}{(k-1)!(10-(k-1))!}$$
$$= 11\binom{10}{k-1}.$$

Note that the key step was rewriting $(11 - k)!$ as $(10 - (k - 1))!$. We thought to do this so that we could make the denominator of our fraction look more like the denominator of a binomial coefficient, which always has the form $r!(n-r)!$ for some numbers n and r. Since we already had a $(k-1)!$ term in the denominator, we wanted to make the other term look like $(n-(k-1))!$ for some value of n.

So now our sum is

$$1\binom{11}{1} + 2\binom{11}{2} + 3\binom{11}{3} + \cdots + 11\binom{11}{11} = 11\binom{10}{0} + 11\binom{10}{1} + \cdots + 11\binom{10}{10}.$$

Now we know what to do! Rewrite the right side as

$$11\left(\binom{10}{0} + \binom{10}{1} + \cdots + \binom{10}{10}\right) = 11(2^{10}) = 11264,$$

where at the end we use the result from Problem 12.4.

Method 2: Denote the sum by S:

$$S = 1\binom{11}{1} + 2\binom{11}{2} + \cdots + 11\binom{11}{11}.$$

Using the identity $\binom{n}{k} = \binom{n}{n-k}$, we can rewrite S as:

$$S = 1\binom{11}{10} + 2\binom{11}{9} + \cdots + 11\binom{11}{0}.$$

When we add these two different expressions for S together, we see that every binomial coefficient in the sum will have coefficient 11. Therefore:

$$2S = 11\binom{11}{0} + 11\binom{11}{1} + \cdots + 11\binom{11}{11}$$
$$= 11\left(\binom{11}{0} + \binom{11}{1} + \cdots + \binom{11}{11}\right)$$
$$= 11(2^{11}).$$

So, dividing by 2, we get that $S = 11(2^{10}) = 11264$. $\square$

Our final problem in this chapter will present a more detailed example of how to explore an unfamiliar expression.

Problem 12.10: Evaluate

$$\frac{19}{99} + \frac{(19)(18)}{(99)(98)} + \frac{(19)(18)(17)}{(99)(98)(97)} + \cdots + \frac{19!}{(99)(98)\cdots(81)}.$$

Solution for Problem 12.10: We could try to evaluate it by hand . . .

No, that sounds like a bad idea. Numbers like 99 and 19 are pretty big and ugly. And we suspect that the numbers themselves are not important. So let's try to write the expression in terms of variables.

> **Concept:** In expressions that include big numbers, often it is easier to consider a more general form of the expression, in which we replace the big numbers with variables. Then, once we've simplified the general expression, we can simply plug in our big numbers in place of the variables to get our answer.

So suppose that our first term is $\frac{a}{b}$, where a and b are positive integers with $a \leq b$. (In our problem, we have $a = 19$ and $b = 99$.) Then we see that our sum is

$$\frac{a}{b} + \frac{a(a-1)}{b(b-1)} + \frac{a(a-1)(a-2)}{b(b-1)(b-2)} + \cdots .$$

Our sum stops when the numerator is $a!$, meaning that it has a terms. So the final summand also has a terms in the denominator, and the sum is

$$\frac{a}{b} + \frac{a(a-1)}{b(b-1)} + \cdots + \frac{a!}{b(b-1)\cdots(b-(a-1))}.$$

Make sure that you stop the denominator at the factor of $(b-(a-1))$ and not at $(b-a)$, which would be one term too many.

Maybe we still have no idea what to do, so let's try evaluating a similar expression, but with smaller numbers. For instance, let's make the first term $\frac{3}{5}$ and work out the expression:

$$\frac{3}{5} + \frac{(3)(2)}{(5)(4)} + \frac{3!}{(5)(4)(3)} = \frac{3}{5} + \frac{3}{10} + \frac{1}{10} = 1.$$

Hmmm. Let's try $\frac{4}{7}$:

$$\frac{4}{7} + \frac{(4)(3)}{(7)(6)} + \frac{(4)(3)(2)}{(7)(6)(5)} + \frac{4!}{(7)(6)(5)(4)} = \frac{4}{7} + \frac{2}{7} + \frac{4}{35} + \frac{1}{35} = 1.$$

Double hmmm. This is probably not a coincidence. We might be pretty strongly tempted to conjecture that

$$\frac{a}{b} + \frac{a(a-1)}{b(b-1)} + \cdots + \frac{a!}{b(b-1)\cdots(b-(a-1))} = 1.$$

Let's try a more extreme case, like $a = 1$ and $b = 10$:

$$\frac{1}{10}.$$

Oops. There's only one term. In fact, whenever $a = 1$, the sum is just $\frac{1}{b}$.

> **Concept:** Don't jump to conclusions. Check lots of different types of examples until you're fairly sure about your conjectures, then try to prove them. There are few things more frustrating than expending a lot of energy trying to prove things that aren't true.

Let's look at the other extreme case, where $a = b$. Then the sum is

$$\frac{a}{a} + \frac{a(a-1)}{a(a-1)} + \cdots + \frac{a!}{a!} = a.$$

We get a terms, each equal to 1, so the sum is a.

We should try to pick our examples more systematically, and we should try varying only one quantity at a time. (The fewer moving parts there are, the simpler it is to figure out what's going on.) So let's fix $a = 3$ and try various values of b:

$$\frac{3}{3} + \frac{(3)(2)}{(3)(2)} + \frac{(3)(2)(1)}{(3)(2)(1)} = 1 + 1 + 1 = 3,$$

$$\frac{3}{4} + \frac{(3)(2)}{(4)(3)} + \frac{(3)(2)(1)}{(4)(3)(2)} = \frac{3}{4} + \frac{1}{2} + \frac{1}{4} = \frac{3}{2},$$

$$\frac{3}{5} + \frac{(3)(2)}{(5)(4)} + \frac{(3)(2)(1)}{(5)(4)(3)} = \frac{3}{5} + \frac{3}{10} + \frac{1}{10} = 1,$$

$$\frac{3}{6} + \frac{(3)(2)}{(6)(5)} + \frac{(3)(2)(1)}{(6)(5)(4)} = \frac{1}{2} + \frac{1}{5} + \frac{1}{20} = \frac{3}{4}.$$

We see that two of our answers have 3 in the numerator. Moreover, if we write the rest as improper fractions with 3 in the numerator, we see that our answers are $\frac{3}{1}$, $\frac{3}{2}$, $\frac{3}{3}$, and $\frac{3}{4}$. It looks like if the sum starts with $a = 3$ and any $b \geq 3$, then the sum will equal $\frac{3}{b-2}$.

Let's try the same thing for $a = 4$. (We'll leave some of the computational details out.)

$$\frac{4}{4} + \frac{12}{12} + \frac{24}{24} + \frac{24}{24} = 1 + 1 + 1 + 1 = 4 = \frac{4}{1},$$

$$\frac{4}{5} + \frac{12}{20} + \frac{24}{60} + \frac{24}{120} = \frac{4}{5} + \frac{3}{5} + \frac{2}{5} + \frac{1}{5} = 2 = \frac{4}{2},$$

$$\frac{4}{6} + \frac{12}{30} + \frac{24}{120} + \frac{24}{360} = \frac{2}{3} + \frac{2}{5} + \frac{1}{5} + \frac{1}{15} = \frac{4}{3},$$

$$\frac{4}{7} + \frac{12}{42} + \frac{24}{210} + \frac{24}{840} = \frac{4}{7} + \frac{2}{7} + \frac{4}{35} + \frac{1}{35} = 1 = \frac{4}{4}.$$

Once again, we see that if $a = 4$ and $b \geq 4$, the sum equals $\frac{4}{b-3}$.

So now we feel confident conjecturing that

$$\frac{a}{b} + \frac{a(a-1)}{b(b-1)} + \cdots + \frac{a!}{b(b-1)\cdots(b-(a-1))} = \frac{a}{b-a+1}.$$

How do we go about proving it?

Since we're trying to prove a formula for positive integers, it is natural to think about using mathematical induction. But we have two variables in our formula: which one do we induct on? Let's play with the equation a little bit before deciding.

Remember, often the goal in induction proofs is to reduce the equation in terms of n down to the equation in terms of $n - 1$. So we'll look for an algebraic manipulation that makes the expression "smaller" in some sense. We notice that $\frac{a}{b}$ is a factor of every term on the LHS, so let's factor it out of the LHS:

$$\frac{a}{b}\left(1 + \frac{(a-1)}{(b-1)} + \cdots + \frac{(a-1)!}{(b-1)\cdots(b-a+1)}\right).$$

Aha—if we can use an inductive hypothesis on everything in the parenthesis except for the initial 1, we see that this would be equal to:

$$\frac{a}{b}\left(1 + \frac{a-1}{(b-1)-(a-1)+1}\right) = \frac{a}{b}\left(1 + \frac{a-1}{b-a+1}\right)$$
$$= \frac{a}{b}\left(\frac{b}{b-a+1}\right)$$
$$= \frac{a}{b-a+1}.$$

Now we see how to proceed. We can prove the identity by induction on b. The base case is when $b = 1$, in which case we must have $a = 1$ and the identity is simply $\frac{1}{1} = \frac{1}{1-1+1} = 1$, which is true. And we have proven the inductive step above: if the identity is true for $b - 1$, then it is true for b as well. Thus the identity holds for all b.

> **WARNING!!** Don't stop here! Don't forget to answer the original question!
> ☢

Finally, we can answer the question as originally posed: we have $a = 19$ and $b = 99$, so by the identity that we have just proven, the answer is $\frac{19}{99-19+1} = \frac{19}{81}$. □

> **Concept:** The type of exploration that we did in Problem 12.10 is very much how
> ⊙═⇛ an actual mathematician goes about problem solving. We started with an ugly expression, we tried some simple examples, and this led to a nice conjecture. But further exploration—in particular, exploration of the extreme cases—quickly shot this conjecture down. More systematic exploration led us to a more accurate conjecture that fit all of the cases that we had examined. Then, with a goal in sight, we were able to fashion a proof.

Exercises

12.3.1 Prove the following:

$$\binom{\binom{n}{2}}{2} = 3\binom{n+1}{4}$$

(a) Algebraically.

(b)$\star$ With a counting argument. **Hints:** 334, 134

12.3.2 Find a counting argument to prove that $1 \times 1! + 2 \times 2! + 3 \times 3! + \cdots + n \times n! = (n+1)! - 1$. **Hints:** 63

12.3.3 Evaluate

$$17\binom{20}{0} + 16\binom{20}{1} + 15\binom{20}{2} + \cdots + 0\binom{20}{17} + (-1)\binom{20}{18} + (-2)\binom{20}{19} + (-3)\binom{20}{20}.$$

12.3.4 Explain why Vandermonde's identity is still valid even if $r > m$ or $r > n$.

12.3.5★ During our exploration of the example where $a = 4$ in our solution to Problem 12.10, we saw the following sums (for $b = 6$ and $b = 7$):

$$\frac{4}{6} + \frac{12}{30} + \frac{24}{120} + \frac{24}{360} = \frac{2}{3} + \frac{2}{5} + \frac{1}{5} + \frac{1}{15} = \frac{4}{3},$$
$$\frac{4}{7} + \frac{12}{42} + \frac{24}{210} + \frac{24}{840} = \frac{4}{7} + \frac{2}{7} + \frac{4}{35} + \frac{1}{35} = 1 = \frac{4}{4}.$$

If we had written these fractions over a common denominator, we would have seen the following:

$$\frac{4}{6} + \frac{12}{30} + \frac{24}{120} + \frac{24}{360} = \frac{2}{3} + \frac{2}{5} + \frac{1}{5} + \frac{1}{15} = \frac{10 + 6 + 3 + 1}{15},$$
$$\frac{4}{7} + \frac{12}{42} + \frac{24}{210} + \frac{24}{840} = \frac{4}{7} + \frac{2}{7} + \frac{4}{35} + \frac{1}{35} = \frac{20 + 10 + 4 + 1}{35}.$$

(a) Look in Pascal's Triangle for the terms in the numerators and the (common) denominators of the last fractions in each row above. Use a combinatorial identity to simplify the sums in the numerators. **Hints:** 189

(b) Generalize this for arbitrary a and b, and use this to present another solution to the problem.

12.4 Summary

➤ There are three general methods for proving combinatorial identities:

1. Algebra
2. A block-walking argument
3. A committee-forming argument

One or more of these methods (especially algebra) may also involve mathematical induction.

➤ When trying to prove identities by committee-forming arguments:

- Usually try to find a counting explanation for the simpler side first.
- A single coefficient of the form $\binom{n}{r}$ means that a good place to start is choosing r items from a set of n items.
- A sum of coefficients may mean trying a casework argument.

➤ The identities from Section 12.2 in Problems 12.1–12.6 are the most common, and you should be familiar with all of them.

➤ Use identities that you already know to prove new identities.

Also remember these important problem-solving techniques:

> **Concept:** Experiment with small values to try to see a pattern. Look to write numbers in terms of binomial coefficients.

> **Concept:** In expressions that include big numbers, often it is easier to consider a more general form of the expression, in which we replace the big numbers with variables. Then, once we've simplified the general expression, we can simply plug in our big numbers in place of the variables to get our answer.

> **Concept:** Don't jump to conclusions. Check lots of different types of examples until you're fairly sure about your conjectures, then try to prove them. There are few things more frustrating than expending a lot of energy trying to prove things that aren't true.

REVIEW PROBLEMS

12.11 Find a closed-form expression for

$$\sum_{k=0}^{n} \binom{n}{k}\binom{m}{k}.$$

12.12 Show that

$$\sum_{k=1}^{n} \binom{n}{k}\binom{n}{k-1} = \binom{2n}{n-1}.$$

12.13 Prove that

$$(n-r)\binom{n}{r} = n\binom{n-1}{r}$$

(a) using a counting argument.

(b) using algebra.

12.14 Prove that

$$\binom{n}{1} + 6\binom{n}{2} + 6\binom{n}{3} = n^3$$

(a) using algebra.

(b) using a counting argument.

12.15 Prove that for any positive integers p and q,

$$\binom{q+2}{2}\binom{p}{0} + \binom{q+1}{1}\binom{p+1}{1} + \binom{q}{0}\binom{p+2}{2} = \binom{p+q+3}{2}.$$

(Source: Mandelbrot)

12.16 Find a closed-form expression for

$$\sum_{i=0}^{\lfloor \frac{n}{2} \rfloor} \binom{n}{i},$$

where $\lfloor x \rfloor$ is the greatest integer less than or equal to x.

Challenge Problems

12.17 The *geometric mean* of a set of n positive numbers is the positive n^{th} root of the product of the n numbers. Consider each non-empty subset of the set $\{a_1, a_2, \ldots, a_n\}$, where all the a_i are positive. Suppose we take the geometric mean of each of these subsets. Then we take the geometric mean of these geometric means. Show that this geometric mean equals the geometric mean of the numbers $\{a_1, a_2, \ldots, a_n\}$. *(Source: Canada)* **Hints:** 132, 69

12.18 Find a closed form for even n:

$$\sum_{k=0}^{n/2} \left[2^{2k} \cdot 2k \binom{n}{2k} \right].$$

Hints: 160

12.19 Find

$$\sum_{k=0}^{49} (-1)^k \binom{99}{2k}.$$

(Source: AMC) **Hints:** 212, 93

12.20★ Prove that

$$\sum_{i=0}^{n} 2^i \binom{n}{i} \binom{n-i}{\lfloor (n-i)/2 \rfloor} = \binom{2n+1}{n},$$

where $\lfloor x \rfloor$ is the greatest integer less than or equal to x. **Hints:** 114, 280

12.21★ For any positive integer n, compute the sum

$$\binom{n}{0} + \binom{n}{3} + \binom{n}{6} + \cdots.$$

Hints: 286, 97

12.22★ For any positive integers $0 < k \le n$, define the number $\left\{{n \atop k}\right\}$ to be the number of ways to partition the set $\{1, 2, \ldots, n\}$ into k non-empty disjoint subsets. These numbers are called the **Stirling numbers** of the second kind (there are also Stirling numbers of the first kind, but we're not going to discuss them here). For example, $\left\{{4 \atop 3}\right\} = 6$, because we have the following 6 partitions of $\{1, 2, 3, 4\}$ into 3 disjoint non-empty subsets:

$$\{1\}, \{2\}, \{3, 4\} \qquad \{1\}, \{3\}, \{2, 4\} \qquad \{1\}, \{4\}, \{2, 3\}$$
$$\{2\}, \{3\}, \{1, 4\} \qquad \{2\}, \{4\}, \{1, 3\} \qquad \{3\}, \{4\}, \{1, 2\}$$

(a) Find simple formulas for $\left\{{n \atop 1}\right\}$ and $\left\{{n \atop n}\right\}$.

(b) Find simple formulas for $\left\{{n \atop 2}\right\}$ and $\left\{{n \atop n-1}\right\}$.

(c) Prove the Pascal-like identity

$$\left\{{n \atop k-1}\right\} + k\left\{{n \atop k}\right\} = \left\{{n+1 \atop k}\right\},$$

for all $1 < k \le n$.

(d)★ Prove that

$$\left\{{n \atop k}\right\} = \frac{1}{k!} \sum_{j=0}^{k} (-1)^j \binom{k}{j} (k-j)^n,$$

via the use of a suitable counting argument. **Hints:** 259, 42, 296

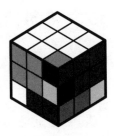

L'état, c'est moi (I am the state) – Louis XIV

CHAPTER **13**

_____**Events With States**

13.1 Introduction

Many "events" are actually sequences of smaller events. For example, baseball's World Series produces a winner only after a series of up to 7 games, each of which has its own winner. The games themselves consist of individual innings, and an individual game might last an arbitrarily long time (if the game is tied after 9 innings and extra innings need to be played).

A **state** is a description of an intermediate stage of an event. For example, in the World Series, a state might be "the teams are tied at 2 games apiece." States allow us to break up complex events into more manageable simple events. It's a bit hard to precisely describe exactly what is a state, but the problems in this chapter should give you a pretty good idea of what states are and how we use them.

We will also look at some different types of problems that involve states. One such type of problem deals with **random walks**, which are processes in which a person or thing is moving around some universe (which might be a line, a plane, the surface of some polyhedron, etc.) and in which the direction of movement of any particular step of the walk is randomly chosen.

Another type of problem that we often use states to solve is problems about 2-player strategy games. We can think of intermediate positions of such a game as states, some of which are winning positions (from which the player that makes the next move can win) and the rest are losing positions. (We're only going to consider games that do not have ties, although games with ties can be studied using similar techniques.) Thinking of a game in terms of its intermediate states is often a useful way to analyze the game's strategies.

Extra!   *We are by nature observers, and thereby learners. That is our permanent state.*
⮕⮕⮕⮕   – Ralph Waldo Emerson

13.2   State Diagrams and Random Walks

Problem 13.1: Many sports playoffs use a "best-of-n" system, in which the first player or team to win some number of games wins the event. For example, baseball's World Series is a "best-of-7" event, in which the first team to win 4 games wins the series. It's called "best-of-7" because at most 7 games need to be played (after 6 games, the teams might be tied at 3 games apiece).

In parts (a)–(c), assume that the two teams are evenly matched, so that each team has a 50% chance of winning any individual game.

(a)   What is the expected number of games played in a best-of-3 series (where the first team to win 2 games wins the series)?

(b)   What is the expected number of games played in a best-of-5 series (where the first team to win 3 games wins the series)?

(c)   What is the expected number of games played in a best-of-7 series (where the first team to win 4 games wins the series)?

(d)   Suppose that one team has a $\frac{2}{3}$ probability of winning each game. Now what is the expected number of games played in a best-of-7 series?

Problem 13.2:

(a)   A bug is sitting on vertex A of a regular tetrahedron. At the start of each minute, he randomly chooses one of the edges at the vertex he is currently sitting on and crawls along that edge to the adjacent vertex. It takes him one minute to crawl to the next vertex, at which point he chooses another edge (at random) and starts crawling again. What is the probability that, after 6 minutes, he is back at vertex A?

(b)   Do the same problem as in part (a), but this time the bug is crawling along the edges of a cube.

Problem 13.3: A **random walk** is a process in which an actor (a person, a bug, a space alien, or some other creature) moves from point to point, and where the direction of movement at each step is randomly chosen. For example, Problem 13.2(a) describes a random walk along the edges of a tetrahedron. Suppose that a person conducts a random walk on a line: she starts at 0 and each minute randomly moves either 1 unit in the positive direction or 1 unit in the negative direction (with equal probability).

(a)   What is the probability that she is back at the origin after n moves?

(b)   What is her expected distance from the origin after n moves?

(c)   Suppose we place "barriers" at some positive integer a and some negative integer b: when our walker reaches one the barriers, the walk ends. What is the probability that the walk ends at the positive barrier?

One use of states is to keep track of intermediate information in a problem. The following is a typical example of how we use states to solve a problem.

Problem 13.1: Many sports playoffs use a "best-of-n" system, in which the first player or team to win some number of games wins the event. For example, baseball's World Series is a "best-of-7" event, in which the first team to win 4 games wins the series. It's called "best-of-7" because at most 7 games need to be played (after 6 games, the teams might be tied at 3 games apiece).

In parts (a)–(c), assume that the two teams are evenly matched, so that each team has a 50% chance of winning any individual game.

(a) What is the expected number of games played in a best-of-3 series (where the first team to win 2 games wins the series)?

(b) What is the expected number of games played in a best-of-5 series (where the first team to win 3 games wins the series)?

(c) What is the expected number of games played in a best-of-7 series (where the first team to win 4 games wins the series)?

(d) Suppose that one team has a $\frac{2}{3}$ probability of winning each game. Now what is the expected number of games played in a best-of-7 series?

Solution for Problem 13.1:

(a) Let's call one team A and the other team B. We can easily list the different possible outcomes and their probabilities:

Outcome	# of games	Probability
AA	2	$\frac{1}{4}$
BB	2	$\frac{1}{4}$
ABA	3	$\frac{1}{8}$
ABB	3	$\frac{1}{8}$
BAA	3	$\frac{1}{8}$
BAB	3	$\frac{1}{8}$

So there is a $\frac{1}{2}$ probability of the series lasting 2 games, and a $\frac{1}{2}$ probability of the series lasting 3 games, for an expected length of $2(\frac{1}{2}) + 3(\frac{1}{2}) = 2.5$.

There's an even quicker, "think about it" solution. Suppose A wins the first game. Then there's a $\frac{1}{2}$ probability that A wins the second game (and the series lasts 2 games) and a $\frac{1}{2}$ probability that B wins the second game (and the series lasts 3 games). The same argument holds if B wins the first game.

But this kind of reasoning is not going to work very well as the length of the series increases. As the series gets longer, we need to be a bit more systematic.

One way to think of the best-of-3 series is as a walk on a 2×2 grid, as in Figure 13.1 below. Every time A wins a game we take a step up, and every time B wins a game we take a step to the right. If we hit the top of the grid, then A has won 2 games, and also the series; similarly, if we hit the right side of the grid, then *B* has won 2 games, and also the series.

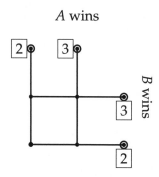

Figure 13.1: State diagram for best-of-3 series

Each point of Figure 13.1 represents a **state**: an intermediate point of the event. The circled states (along the top and right edges) represent the possible final outcomes, and the boxed numbers next to those circles indicate the total number of games in a series with that outcome.

We can start at the bottom-left of the grid and label each state with the probability of reaching it. Each state's probability, except for the beginning and end states, is

$$(1/2)(\text{probability of the state below}) + (1/2)(\text{probability of the state to the left}).$$

The probability of reaching any of the end states is $\frac{1}{2}$ times the probability of the state just before the end state. Figure 13.2 below shows the process of filling in the probabilities of the states. Note that we start with probability 1 at the beginning state.

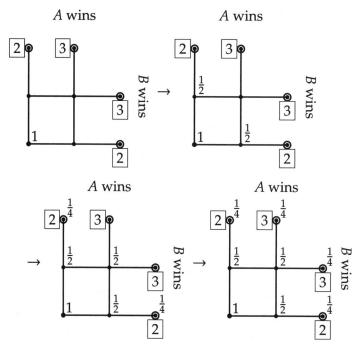

Figure 13.2: Best-of-3 series diagram with probabilities

Now we can just read off the relevant probabilities from the grid. There is a $\frac{1}{4} + \frac{1}{4} = \frac{1}{2}$ probability that the series will end in 2 games, and a $\frac{1}{4} + \frac{1}{4} = \frac{1}{2}$ probability that the series will end in 3 games, giving an expected length of $2(\frac{1}{2}) + 3(\frac{1}{2}) = 2.5$.

(b) We can look at a 3-by-3 state diagram, as shown at right. Note that the sum of the probabilities of the ending states is

$$\frac{1}{8} + \frac{3}{16} + \frac{3}{16} + \frac{3}{16} + \frac{3}{16} + \frac{1}{8} = 1,$$

which is a quick check that we haven't made a mistake.

We see that the expected length of the best-of-5 series is

$$3\left(\frac{1}{8} + \frac{1}{8}\right) + 4\left(\frac{3}{16} + \frac{3}{16}\right) + 5\left(\frac{3}{16} + \frac{3}{16}\right) = \frac{33}{8} = 4.125.$$

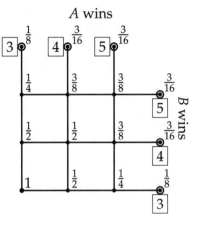

(c) We can look at a 4-by-4 state diagram, shown below:

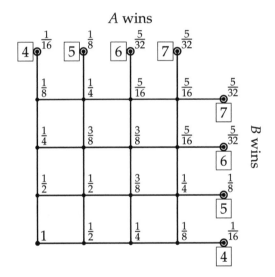

We see that the expected length of the best-of-7 series is

$$4\left(\frac{1}{16} + \frac{1}{16}\right) + 5\left(\frac{1}{8} + \frac{1}{8}\right) + 6\left(\frac{5}{32} + \frac{5}{32}\right) + 7\left(\frac{5}{32} + \frac{5}{32}\right) = \frac{93}{16} = 5.8125.$$

(d) The real power of state diagrams can be seen when team A has a $\frac{2}{3}$ chance of winning each game and team B has a $\frac{1}{3}$ chance. We can still use the same diagram, but now the probability at each state is

(2/3)(probability of the state below) + (1/3)(probability of the state to the left).

The completed diagram is shown in Figure 13.3 on the next page:

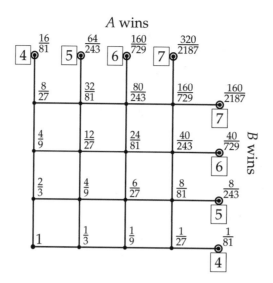

Figure 13.3: Team A wins any game with probability 2/3

For example, looking at the state corresponding to the point $(1, 3)$, where A has won 3 games and B has won 1 game, we see that the probability of reaching this state is

$$\frac{2}{3}\left(\frac{12}{27}\right) + \frac{1}{3}\left(\frac{8}{27}\right) = \frac{32}{81}.$$

Now we can make a chart of the probabilities of the various series lengths:

Length	Probability
4	$\frac{16}{81} + \frac{1}{81} = \frac{17}{81}$
5	$\frac{64}{243} + \frac{8}{243} = \frac{72}{243}$
6	$\frac{160}{729} + \frac{40}{729} = \frac{200}{729}$
7	$\frac{320}{2187} + \frac{160}{2187} = \frac{480}{2187}$

As a check, note that

$$\frac{17}{81} + \frac{72}{243} + \frac{200}{729} + \frac{480}{2187} = 1.$$

Then our expected series length is

$$4\left(\frac{17}{81}\right) + 5\left(\frac{72}{243}\right) + 6\left(\frac{200}{729}\right) + 7\left(\frac{480}{2187}\right) = \frac{12036}{2187} = 5\frac{1101}{2187} \approx 5.50.$$

$\square$

Note that we can read off additional data from the diagram in Figure 13.3. For example, the probability of team A winning the series is

$$\frac{16}{81} + \frac{64}{243} + \frac{160}{729} + \frac{320}{2187} = \frac{1808}{2187} \approx 82.67\%.$$

> **Concept:** State diagrams give us a convenient way to organize complicated data about an event with many intermediate steps.

Problem 13.2:

(a) A bug is sitting on vertex A of a regular tetrahedron. At the start of each minute, he randomly chooses one of the edges at the vertex he is currently sitting on and crawls along that edge to the adjacent vertex. It takes him one minute to crawl to the next vertex, at which point he chooses another edge (at random) and starts crawling again. What is the probability that, after 6 minutes, he is back at vertex A?

(b) Do the same problem as in part (a), but this time the bug is crawling along the edges of a cube.

Solution for Problem 13.2:

(a) Let's call the four points of the tetrahedron A, B, C, and D. During each minute, the bug is equally likely to crawl from his current vertex to any of the other three vertices. We can represent this information as a state diagram with four states, as shown below:

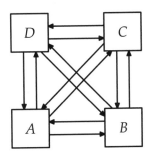

As you can see, this is a pretty messy diagram, and we haven't even included any probability information in the diagram. However, there is a major simplification to our diagram that we can make. We really don't need 4 different states, as there are essentially only two states that we care about: "on A" and "not on A."

> **Concept:** Try to make your states as simple as possible. Often, in a problem involving states, identifying the simplest possible set of states is a major step towards solving the problem.

If the bug is "on A", he moves to "not on A" with probability 1.

If he is "not on A", he moves to "on A" with probability $\frac{1}{3}$ and stays at "not on A" (though on a different vertex) with probability $\frac{2}{3}$.

We can represent this with the simpler state diagram shown below:

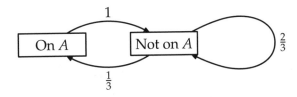

This allows us to set up a recursion. Let p_n be the probability that the bug is on vertex A after n minutes and q_n be the probability that the bug is not on vertex A after n minutes. Of course, $p_n + q_n = 1$ for all n, since the bug must always be either at A or not at A at the end of any minute.

We see that if the bug is at A after $n-1$ minutes, then he will be at "not A" after n minutes with probability 1. We also see that if the bug is at "not A" after $n-1$ minutes, then with probability $\frac{1}{3}$ he will move to A, and with probability $\frac{2}{3}$ he will stay at "not A."

Therefore, we can write the system of equations

$$p_n = \frac{1}{3}q_{n-1},$$

$$q_n = p_{n-1} + \frac{2}{3}q_{n-1}.$$

Now solving the problem—that is, finding p_6—is a simple matter of plugging in the initial condition $p_0 = 1$, $q_0 = 0$ and computing the probabilities:

n	0	1	2	3	4	5	6
p_n	1	0	$\frac{1}{3}$	$\frac{2}{9}$	$\frac{7}{27}$	$\frac{20}{81}$	$\frac{61}{243}$
q_n	0	1	$\frac{2}{3}$	$\frac{7}{9}$	$\frac{20}{27}$	$\frac{61}{81}$	$\frac{182}{243}$

So the probability that the bug is back on A after 6 minutes is $\frac{61}{243}$.

Note that we could solve the recursion explicitly, using the techniques from Chapter 10. Rewrite $p_n = \frac{1}{3}q_{n-1}$ as $p_n = \frac{1}{3}(1 - p_{n-1})$. Then write this as

$$p_n + \frac{1}{3}p_{n-1} = \frac{1}{3}.$$

The general form of the solution will be

$$p_n = c\left(-\frac{1}{3}\right)^n + \frac{1}{4},$$

and the initial condition $p_0 = 1$ gives $c = \frac{3}{4}$. Therefore

$$p_n = \frac{3}{4}\left(-\frac{1}{3}\right)^n + \frac{1}{4}.$$

Plugging in $n = 6$ gives

$$p_6 = \frac{3}{4} \cdot \frac{1}{729} + \frac{1}{4} = \frac{61}{243}.$$

Also note that as n grows very large, the probability p_n gets very close to $\frac{1}{4}$ (since $\frac{1}{3}$ raised to a high power is a very small number). This is not surprising: after the bug has been crawling around for a very long time, it should be roughly equally likely that it is at any of the 4 vertices.

(b) On a cube, the states are a bit more complicated. The bug can be on A, one move away from A (on an adjacent vertex), two moves away from A (on the other end of a diagonal of a cube face), or three moves away from A (on the opposite vertex from A).

The state diagram, with the associated probabilities for moving between states, is shown below:

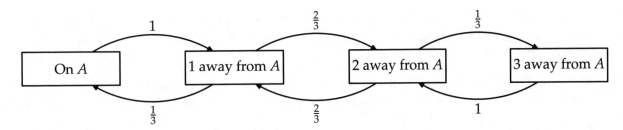

Let a_n, b_n, c_n, and d_n be the probabilities that the bug at time n is on A, 1 move away, 2 moves away, and 3 moves away, respectively. Then we can write the system of recursions:

$$a_n = \frac{1}{3}b_{n-1},$$

$$b_n = a_{n-1} + \frac{2}{3}c_{n-1},$$

$$c_n = \frac{2}{3}b_{n-1} + d_{n-1},$$

$$d_n = \frac{1}{3}c_{n-1}.$$

Once again, since we only care about $n = 6$, it is simple enough to make a table and list the values:

n	0	1	2	3	4	5	6
a_n	1	0	$\frac{1}{3}$	0	$\frac{7}{27}$	0	$\frac{61}{243}$
b_n	0	1	0	$\frac{7}{9}$	0	$\frac{61}{81}$	0
c_n	0	0	$\frac{2}{3}$	0	$\frac{20}{27}$	0	$\frac{182}{243}$
d_n	0	0	0	$\frac{2}{9}$	0	$\frac{20}{81}$	0

So the answer is $\frac{61}{243}$. Hmmm... that's the same answer as part (a). Coincidence? I don't think so! (Try to explain to yourself why the numbers in parts (a) and (b) are the same.)

□

Problem 13.3: A **random walk** is a process in which an actor (a person, a bug, a space alien, or some other creature) moves from point to point, and where the direction of movement at each step is randomly chosen. For example, Problem 13.2(a) describes a random walk along the edges of a tetrahedron. Suppose that a person conducts a random walk on a line: she starts at 0 and each minute randomly moves either 1 unit in the positive direction or 1 unit in the negative direction (with equal probability).

(a) What is the probability that she is back at the origin after n moves?

(b) What is her expected distance from the origin after n moves?

(c) Suppose we place "barriers" at some positive integer a and some negative integer b: when our walker reaches one the barriers, the walk ends. What is the probability that the walk ends at the positive barrier?

Solution for Problem 13.3: Although there are states involved in this problem (as in most random walk problems), for this particular problem it's best to think "holistically" and view the entire walk at once, rather than considering individual states.

(a) The person ends up back at 0 after n moves if she has made an equal number of left and right moves. This is impossible if n is odd. If n is even, then she must make $n/2$ left moves and $n/2$ right moves to end up back at 0. These moves can be arranged in $\binom{n}{n/2}$ ways, and each arrangement is equally likely. Since there are 2^n equally likely possible sequences of left and right moves, and $\binom{n}{n/2}$ of them result in returning to the origin after n moves, the desired probability is

$$\frac{\binom{n}{n/2}}{2^n}$$

if n is even. As mentioned above, if n is odd, then the probability is 0.

(b) Let's start by examining the cases for small n. If $n = 0$, then she must be 0 units from the origin, and if $n = 1$, then she must be 1 unit from the origin.

   If $n = 2$, then she is at the origin with probability $\frac{1}{2}$ and 2 units away from the origin with probability $\frac{1}{2}$. So the expected distance is 1.

   If $n = 3$, then she is 1 away from the origin with probability $\frac{3}{4}$ and 3 units away from origin with probability $\frac{1}{4}$. So the expected distance is $\frac{3}{2}$.

   More specifically, in the $n = 3$ case, we see that she is at the following positions with the following probabilities:

Note the numerators of the probabilities: they're just the entries of Row 3 of Pascal's Triangle. This makes sense, since the number of paths to any given point is the same as the number of paths to the corresponding entry of Pascal's Triangle. Further, the denominator of each probability is $8 = 2^3$, since that is the total number of paths.

So more generally, we can make a chart of the possible n-step walks and their resulting position:

Steps left	Steps right	Number of paths	Distance from 0
n	0	$\binom{n}{0}$	n
$n-1$	1	$\binom{n}{1}$	$n-2$
$n-2$	2	$\binom{n}{2}$	$n-4$
		$\vdots$	
0	n	$\binom{n}{n}$	n

In general, if we take i steps left and $n - i$ steps right, then there are $\binom{n}{i}$ ways to arrange the steps, and we will end up at position $(-i) + (n - i) = n - 2i$, so we will be at a distance of $|n - 2i|$ from the origin.

Therefore, since there are 2^n possible walks of length n, the expected distance from the origin after n moves is:

$$\sum_{i=0}^{n} \frac{\binom{n}{i}}{2^n} |n - 2i|.$$

We can simplify this a little bit by moving the common 2^n term outside the summation. We can also take advantage of the symmetry present to get rid of the absolute value: walks that end to the left of the origin are symmetric to walks that end to the right of the origin. These simplifications give us:

$$\sum_{i=0}^{n} \frac{\binom{n}{i}}{2^n} |n - 2i| = \frac{1}{2^n} \sum_{i=0}^{n} \binom{n}{i} |n - 2i|$$

$$= \frac{1}{2^{n-1}} \sum_{i=0}^{\lfloor \frac{n}{2} \rfloor} \binom{n}{i} (n - 2i).$$

We can split this into two separate sums:

$$\frac{1}{2^{n-1}} \sum_{i=0}^{\lfloor \frac{n}{2} \rfloor} \binom{n}{i} (n - 2i) = \frac{n}{2^{n-1}} \sum_{i=0}^{\lfloor \frac{n}{2} \rfloor} \binom{n}{i} - \frac{1}{2^{n-2}} \sum_{i=0}^{\lfloor \frac{n}{2} \rfloor} i \binom{n}{i}.$$

The identity $i\binom{n}{i} = n\binom{n-1}{i-1}$ allows us to rewrite the second sum, so that we have:

$$\frac{n}{2^{n-1}} \sum_{i=0}^{\lfloor \frac{n}{2} \rfloor} \binom{n}{i} - \frac{1}{2^{n-2}} \sum_{i=0}^{\lfloor \frac{n}{2} \rfloor} i \binom{n}{i} = \frac{n}{2^{n-1}} \sum_{i=0}^{\lfloor \frac{n}{2} \rfloor} \binom{n}{i} - \frac{n}{2^{n-2}} \sum_{i=1}^{\lfloor \frac{n}{2} \rfloor} \binom{n-1}{i-1}.$$

Because of the $\lfloor \frac{n}{2} \rfloor$ upper limit of the sums, it is a bit easier to compute them if we consider the even and odd cases separately. If n is odd, then the first sum is 2^{n-1} and the second sum is $2^{n-2} - \frac{1}{2}\binom{n-1}{\frac{n-1}{2}}$, so the whole expression is equal to

$$\frac{n}{2^{n-1}} \binom{n-1}{\frac{n-1}{2}}.$$

If n is even, then the first sum is $2^{n-1} + \frac{1}{2}\binom{n}{\frac{n}{2}}$ and the second sum is 2^{n-2}, so the whole expression is equal to

$$\frac{n}{2^n} \binom{n}{\frac{n}{2}}.$$

We check these expressions with the cases of $n = 0, 1, 2, 3$ that we have already computed, and we see that they match.

(c) Let p_n be the probability that if the person is at position n, then she will end at the positive barrier at a. Note that $p_a = 1$ and $p_b = 0$, and that for any integer $b < n < a$, we have

$$p_n = \frac{1}{2} p_{n-1} + \frac{1}{2} p_{n+1} = \frac{p_{n-1} + p_{n+1}}{2},$$

since the person is equally likely to move left or right. Therefore, the sequence

$$p_b, \; p_{b+1}, \; p_{b+2}, \; \ldots, \; p_{a-2}, \; p_{a-1}, \; p_a$$

is an arithmetic sequence, since each term is the average of its two adjacent terms. There are $a - b$ terms from $p_b = 0$ to $p_a = 1$, so the common difference between adjacent terms is $\frac{1}{b-a}$. Therefore, for any $b \le n \le a$, we have $p_n = \dfrac{n - b}{a - b}$. To solve the problem, we compute $p_0 = \dfrac{-b}{a - b}$.

$\square$

> **Sidenote:** There are many more questions that we can ask about random walks on the number line. Two that seem natural to ask are:
>
> - What is the probability that the walker will eventually return to the origin?
>
> - What is the expected time before the walker returns to the origin for the first time?
>
> It turns out that the walker will return to the origin with probability 1, although this is difficult to prove. This is even true for a random walk on the 2-dimensional plane (in which the walker moves north, south, east, or west, each with probability $\frac{1}{4}$). However, in three dimensions, this is not true! A random walker in 3-D space, who starts at the origin, only has probability of about 0.34 of ever returning to the origin, even though the walk may continue for an infinite amount of time.
>
> Although on the number line the probability that the walker will return to the origin is 1, the expected time until this occurs is not finite. So we are guaranteed to come back home, but it may take an arbitrarily long time.

Exercises

13.2.1   A moth starts at vertex A of a certain cube and is trying to get to vertex B, which is opposite A, in five or fewer steps, where a step consists in traveling along an edge from one vertex to another. The moth will stop as soon as it reaches B. In how many ways can the moth achieve its objective? *(Source: HMMT)*

13.2.2   A bug starts at one vertex of a cube and moves along the edges of the cube. At each vertex the bug will choose to travel along one of the three edges emanating from that vertex, where each edge has equal probability of being chosen, and all choices are independent. What is the probability that after seven moves the bug will have visited every vertex exactly once? *(Source: AMC)*

13.2.3   The Red Sox play the Yankees in a best-of-seven series that ends as soon as one team wins four games. Suppose that the probability that the Red Sox win Game n is $\frac{n-1}{6}$. What is the probability that the Red Sox will win the series? *(Source: HMMT)*

13.2.4★   Six ants simultaneously stand on the six vertices of a regular octahedron, with each ant at a different vertex. Simultaneously and independently, each ant moves from its vertex to one of the four adjacent vertices, each with equal probability. What is the probability that no two ants arrive at the same vertex? *(Source: AMC)* **Hints:** 201

13.3 Events With Infinite States

> **Problems**

Problem 13.4: In a tennis game, if the players are tied at "deuce," then the game continues until one player has won two more points that the other player. If Homer and Marge are playing tennis and reach deuce, and Homer has a $\frac{3}{5}$ probability of winning any given point, then what is the probability that Homer wins the game?

Problem 13.5: What is the expected number of rolls of a fair 6-sided die until a ⚁ is rolled?

Problem 13.6: Anna writes a sequence of integers starting with the number 12. Each subsequent integer she writes is chosen randomly with equal chance from among the positive divisors of the previous integer (including the possibility of the integer itself). She keeps writing integers until she writes the integer 1 for the first time, and then she stops. For example, one possible sequence is

$$12, 6, 6, 3, 3, 3, 1.$$

What is the expected value of the number of terms in Anna's sequence? *(Source: USAMTS)*

Problem 13.7: We have two bins, A and B. Initially there are 3 balls in bin A and no balls in bin B. We proceed with a series of moves as follows: on each move, one of the three balls is randomly chosen, and it is moved from its bin to the other bin. Find the expected number of moves until the first time that all of the balls are simultaneously in bin B.

Problem 13.8: The game of craps is played by one player with two dice as follows. The dice are rolled and:

- If the sum is 7 or 11, the player wins immediately.

- If the sum is 2, 3, or 12, the player loses immediately.

- If the sum is any other number, that number becomes the *point*. The dice are then re-rolled until either the point or a 7 is rolled. If the point is rolled before a 7, the player wins; if a 7 is rolled before the point, the player loses.

What is the player's probability of winning?

Often the events that we are analyzing can (theoretically) last an infinitely long time. This sometimes requires us to use a bit of creativity in analyzing them. For example:

Problem 13.4: In a tennis game, if the players are tied at "deuce," then the game continues until one player has won two more points that the other player. If Homer and Marge are playing tennis and reach deuce, and Homer has a $\frac{3}{5}$ probability of winning any given point, then what is the probability that Homer wins the game?

Solution for Problem 13.4: We show the state diagram in Figure 13.4. Note that "Ad Homer" and "Ad Marge" mean that Homer and Marge, respectively, lead by 1 point.

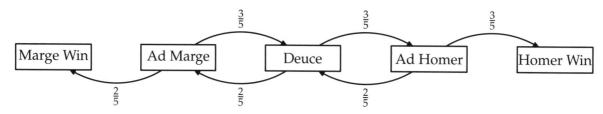

Figure 13.4: Homer & Marge play tennis

The diagram helps us to visualize what's going on. Starting at deuce, there are three possibilities for the way that the next two points happen.

- Homer wins them both, and wins the game. This happens with probability $\frac{3}{5} \cdot \frac{3}{5} = \frac{9}{25}$.

- Marge wins them both, and wins the game. This happens with probability $\frac{2}{5} \cdot \frac{2}{5} = \frac{4}{25}$.

- Homer and Marge each win a point, and they go back to deuce. This happens with probability $\frac{3}{5} \cdot \frac{2}{5} + \frac{2}{5} \cdot \frac{3}{5} = \frac{12}{25}$.

As a quick check, we note that the probabilities sum to $\frac{9}{25} + \frac{4}{25} + \frac{12}{25} = \frac{25}{25} = 1$.

> **Concept:** If it's easy to do, a quick check of your work in the middle of a problem can prevent errors from propagating through a long solution.

Now we take an algebraic approach. Let p be the probability that we're looking for: the probability that Homer wins the game, given that it is currently at deuce. Then we can write the equation

$$p = \frac{9}{25} + \frac{12}{25}p,$$

noting that Homer can win right away (with probability 9/25), or the game returns to deuce (with probability 12/25) giving Homer another chance to win with probability p.

The equation simplifies to $\frac{13}{25}p = \frac{9}{25}$, so $p = \frac{9}{13}$. $\square$

The solution to Problem 13.4 illustrates an important principle:

> **Concept:** In problems with state information, it often helps to introduce variable(s) that store information about the intermediate state(s).

For example, in Problem 13.4, we introduced the variable p to represent the probability of success at an intermediate state. Quite often, in state problems, we will be able to write equations using these one or more intermediate variables.

Here's another example, this time with expected value.

Problem 13.5: What is the expected number of rolls of a fair 6-sided die until a ⚁ is rolled?

Solution for Problem 13.5: Let μ represent the expected number of rolls. (Recall that the Greek letter μ, called "mu", is often used to represent expected value; we'll be using it a lot in this section.)

With probability $\frac{1}{6}$, we roll ⚁ on the very first roll. However, with probability $\frac{5}{6}$, we don't roll ⚁, in which case we're right back where we started, except that we've already used one roll and we still expect to have μ rolls to go.

Therefore, we can write the following equation for μ:

$$\mu = \frac{1}{6}(1) + \frac{5}{6}(1 + \mu) = 1 + \frac{5}{6}\mu.$$

Another way to think about this equation is that we're going to use one roll no matter what, and if we don't roll ⚁, then we're still going to expect to need μ additional rolls to get our ⚁.

Solving the above equation gives $\mu = 6$, so we expect to need 6 rolls. (This hopefully makes some intuitive sense as well.) $\square$

We can check our work by calculating the expected value more explicitly. There is a $\frac{1}{6}$ chance that we'll need 1 roll, a $\frac{5}{6} \cdot \frac{1}{6}$ chance we'll need 2 rolls (if we don't get ⚁ on the first roll but get it on the second), a $\frac{5}{6} \cdot \frac{5}{6} \cdot \frac{1}{6}$ chance we'll need 3 rolls (no ⚁ on the first two rolls, but ⚁ on the third), and so on. So the expected value is

$$\left(\frac{1}{6}\right)(1) + \left(\frac{5}{6}\right)\left(\frac{1}{6}\right)(2) + \left(\frac{5}{6}\right)^2 \left(\frac{1}{6}\right)(3) + \cdots = \sum_{k=1}^{\infty} \left(\frac{5}{6}\right)^{k-1} \left(\frac{1}{6}\right)k.$$

We can simplify the above expression by pulling $\frac{1}{6}$ outside the sum, and moving a factor of $\frac{5}{6}$ from outside the sum to inside the sum:

$$\sum_{k=1}^{\infty} \left(\frac{5}{6}\right)^{k-1} \left(\frac{1}{6}\right)k = \frac{1}{6} \sum_{k=1}^{\infty} \left(\frac{5}{6}\right)^{k-1} k = \frac{1}{5} \sum_{k=1}^{\infty} \left(\frac{5}{6}\right)^{k} k.$$

How do we sum $\sum_{k=1}^{\infty} kr^k$ for some constant r with $|r| < 1$? Let $S = \sum_{k=1}^{\infty} kr^k$ be the sum. Then $rS = \sum_{k=1}^{\infty} kr^{k+1}$, and subtracting gives

$$S - rS = \sum_{k=1}^{\infty} r^k = \frac{1}{1-r} - 1.$$

In our example, plugging in $r = \frac{5}{6}$ gives $\frac{S}{6} = 5$, so $S = 30$, and $\mu = \frac{1}{5}(30) = 6$, as before.

Problem 13.6: Anna writes a sequence of integers starting with the number 12. Each subsequent integer she writes is chosen randomly with equal chance from among the positive divisors of the previous integer (including the possibility of the integer itself). She keeps writing integers until she writes the integer 1 for the first time, and then she stops. One such sequence is: 12, 6, 6, 3, 3, 3, 1. What is the expected value of the number of terms in Anna's sequence? *(Source: USAMTS)*

Solution for Problem 13.6: We could start by drawing the state diagram. The "state" is the current integer that Anna is writing, which could be any divisor of 12, namely 1, 2, 3, 4, 6, or 12. We can go from one state to another if the new number is a divisor of the old number (possibly equal). So the state diagram looks like:

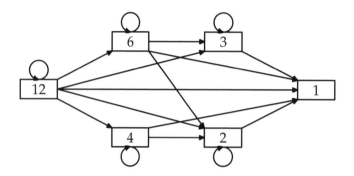

Note that the loops in the state diagram correspond to the possibility of writing the same number multiple times. Unfortunately, this diagram is a bit of a mess. We're probably better off just writing equations.

Let $\mu_1, \mu_2, \mu_3, \mu_4, \mu_6, \mu_{12}$ be the expected lengths of the sequence if it starts with 1, 2, 3, 4, 6, 12 (respectively). We can write equations for each of these in terms of the others. For example, starting with the first number "12", the remainder of the sequence is equally likely to start with a 1, 2, 3, 4, 6, or 12. Therefore, the equation is

$$\mu_{12} = 1 + \frac{1}{6}(\mu_1 + \mu_2 + \mu_3 + \mu_4 + \mu_6 + \mu_{12}).$$

We can write similar equations for the other states, and we get the following system of equations:

$$\mu_{12} = 1 + \frac{1}{6}(\mu_1 + \mu_2 + \mu_3 + \mu_4 + \mu_6 + \mu_{12}),$$

$$\mu_6 = 1 + \frac{1}{4}(\mu_1 + \mu_2 + \mu_3 + \mu_6),$$

$$\mu_4 = 1 + \frac{1}{3}(\mu_1 + \mu_2 + \mu_4),$$

$$\mu_3 = 1 + \frac{1}{2}(\mu_1 + \mu_3),$$

$$\mu_2 = 1 + \frac{1}{2}(\mu_1 + \mu_2),$$

$$\mu_1 = 1.$$

Note the last equation: $\mu_1 = 1$. Once we write a 1, we're done.

We can now start at the bottom equation and work our way up. We start with:

$$\mu_2 = 1 + \frac{1}{2}(1 + \mu_2) \quad \Rightarrow \quad \mu_2 = 3$$

and

$$\mu_3 = 1 + \frac{1}{2}(1 + \mu_3) \quad \Rightarrow \quad \mu_3 = 3.$$

Continuing, we see that

$$\mu_4 = 1 + \frac{1}{3}(4 + \mu_4) \quad \Rightarrow \quad \mu_4 = \frac{7}{2}$$

and

$$\mu_6 = 1 + \frac{1}{4}(7 + \mu_6) \quad \Rightarrow \quad \mu_6 = \frac{11}{3}.$$

Finally, we plug these all back into the top equation to get

$$\mu_{12} = 1 + \frac{1}{6}\left(1 + 3 + 3 + \frac{7}{2} + \frac{11}{3} + \mu_{12}\right) = \frac{121}{36} + \frac{\mu_{12}}{6},$$

so $\mu_{12} = \dfrac{121}{30}$. $\square$

The next problem has a similar flavor:

Problem 13.7: We have two bins, A and B. Initially there are 3 balls in bin A and no balls in bin B. We proceed with a series of moves as follows: on each move, one of the three balls is randomly chosen, and it is moved from its bin to the other bin. Find the expected number of moves until the first time that all of the balls are simultaneously in bin B.

Solution for Problem 13.7: There are 4 states: there are either 3, 2, 1, or 0 balls in A (and the rest of the balls are in B). We can draw a state diagram with the transition probabilities labeled:

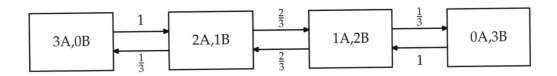

Note that the probabilities shown in the state diagram result from the fact that each ball is equally likely to be chosen.

Let μ_3, μ_2, μ_1 denote the expected number of moves until we finish, starting with $3, 2, 1$ balls in bin A, respectively. We can then write equations:

$$\mu_3 = 1 + \mu_2,$$

$$\mu_2 = 1 + \frac{1}{3}\mu_3 + \frac{2}{3}\mu_1,$$

$$\mu_1 = 1 + \frac{2}{3}\mu_2.$$

This system is not quite as simple to solve as the system from Problem 13.6, because in this problem, the connections between states are "two-way": we can go back and forth between states. (In Problem 13.6, the state diagram was essentially "one-way": once we left a state, we couldn't go back.) Nonetheless, this system is still pretty easy to solve.

We can plug μ_1 into the equation for μ_2 to get

$$\mu_2 = 1 + \frac{1}{3}\mu_3 + \frac{2}{3}\left(1 + \frac{2}{3}\mu_2\right),$$

which simplifies to give

$$\frac{5}{9}\mu_2 = \frac{5}{3} + \frac{1}{3}\mu_3.$$

Plug this into the top equation to get

$$\mu_3 = 1 + \frac{9}{5}\left(\frac{5}{3} + \frac{1}{3}\mu_3\right),$$

which gives $\mu_3 = 10$. So the expected number of moves is 10. $\square$

We finish this section with a classic casino game. (After calculating the probability of winning, you may not want to play!)

Problem 13.8: The game of craps is played by one player with two dice as follows. The dice are rolled and:

- If the sum is 7 or 11, the player wins immediately.

- If the sum is 2, 3, or 12, the player loses immediately.

- If the sum is any other number, that number becomes the *point*. The dice are then re-rolled until either the point or a 7 is rolled. If the point is rolled before a 7, the player wins; if a 7 is rolled before the point, the player loses.

What is the player's probability of winning?

Solution for Problem 13.8: It's probably easiest to start by making a chart:

Initial Roll	Probability	Probability of winning	Initial Roll	Probability	Probability of winning
2	$\frac{1}{36}$	0	8	$\frac{5}{36}$	$P(8$ rolled before $7)$
3	$\frac{2}{36}$	0	9	$\frac{4}{36}$	$P(9$ rolled before $7)$
4	$\frac{3}{36}$	$P(4$ rolled before $7)$	10	$\frac{3}{36}$	$P(10$ rolled before $7)$
5	$\frac{4}{36}$	$P(5$ rolled before $7)$	11	$\frac{2}{36}$	1
6	$\frac{5}{36}$	$P(6$ rolled before $7)$	12	$\frac{1}{36}$	0
7	$\frac{6}{36}$	1			

We still have the point probabilities to calculate. Let's just look at the probability that a 4 is rolled before a 7 (the others are similar). There are essentially two ways to think about it. One way is as in Problem 13.4. Let p denote the probability of rolling a 4 before a 7. On any given roll, there is a $\frac{3}{36}$ chance that we win right away (if we roll a 4) and a $\frac{27}{36}$ chance that we get another roll (if we don't roll a 4 or a 7). So the probability is

$$p = \frac{3}{36} + \frac{27}{36}p,$$

which solves to give $p = \frac{1}{3}$.

Alternatively, note that there are 3 rolls that win, 6 rolls that lose, and 27 rolls that are irrelevant. So we have 3 winning rolls out of 9 "outcome" rolls, for a probability of $\frac{3}{9} = \frac{1}{3}$.

We can do this calculation for all the point numbers and fill in our table:

Initial Roll	Probability	Probability of winning	Initial Roll	Probability	Probability of winning
2	$\frac{1}{36}$	0	8	$\frac{5}{36}$	$\frac{5}{11}$
3	$\frac{2}{36}$	0	9	$\frac{4}{36}$	$\frac{2}{5}$
4	$\frac{3}{36}$	$\frac{1}{3}$	10	$\frac{3}{36}$	$\frac{1}{3}$
5	$\frac{4}{36}$	$\frac{2}{5}$	11	$\frac{2}{36}$	1
6	$\frac{5}{36}$	$\frac{5}{11}$	12	$\frac{1}{36}$	0
7	$\frac{6}{36}$	1			

So the probability of winning at craps is

$$\frac{3}{36} \cdot \frac{1}{3} + \frac{4}{36} \cdot \frac{2}{5} + \frac{5}{36} \cdot \frac{5}{11} + \frac{6}{36} + \frac{5}{36} \cdot \frac{5}{11} + \frac{4}{36} \cdot \frac{2}{5} + \frac{3}{36} \cdot \frac{1}{3} + \frac{2}{36},$$

which simplifies to $\frac{244}{495} \approx 49.3\%$. $\square$

Since casinos typically pay even-money on a craps game (meaning that you have to wager \$1 to win \$1), if you bet \$1 you will win \$1 with (approximate) probability 0.493 and lose \$1 with probability 0.507, so the house will make an expected profit of \$0.507 − \$0.493 = \$0.014 for every dollar that you bet. (This is sometimes stated as a "house edge" of 1.4%.)

Exercises

13.3.1 Alice, Bob and Carol repeatedly take turns tossing a die. Alice begins; Bob always follows Alice; Carol always follows Bob; and Alice always follows Carol. Find the probability that Carol will be the first one to toss a ⚁. (The probability of obtaining a ⚁ on any toss is $\frac{1}{6}$, independent of the outcome of any other toss.) *(Source: AMC)*

13.3.2 Doug and Ryan are competing in the 2005 Wiffle Ball Home Run Derby. Each player takes a series of swings, and each swing results in either a home run or an out. Each player continues until he makes an out, which ends his series of swings. When Doug swings, the probability that he will hit a home run is $\frac{1}{3}$. When Ryan swings, the probability that he will hit a home run is $\frac{1}{2}$.

(a) What is the probability that Doug will hit more home runs than Ryan hits? **Hints:** 313

(b) What is the probability that they tie?

(Source: HMMT)

13.3.3 Going back to Homer and Marge's tennis game from Problem 13.4, what is the expected number of points that Homer and Marge will have to play, starting at deuce, until one of them wins the game?

13.3.4★ Find the probability that, in the process of repeatedly flipping a fair coin, one will encounter a run of 5 heads before one encounters a run of 2 tails. *(Source: AIME)* **Hints:** 214, 83

13.3.5★ Alfred and Bonnie play a game in which they take turns tossing a fair coin. The winner of a game is the first person to obtain a head. Alfred and Bonnie play this game several times with the stipulation that the loser of a game goes first in the next game. Suppose that Alfred goes first in the first game. Find the probability that he wins the sixth game. *(Source: AIME)* **Hints:** 27, 288

13.4 Two-player Strategy Games

Problem 13.9:
(a) Tina and Valentin play a game in which they start with 17 marbles in a pile. They take turns removing 1, 2, or 3 marbles from the pile. The player who removes the final marble wins the game. If Tina goes first and both play intelligently, who should win and why?

(b) Generalize to the situation where there are n marbles in the pile and each player may remove up to k marbles on his or her turn.

Problem 13.10: In the game **Heaps**, we start with one or more piles, each of which has a positive number of chips. (There may be different amounts of chips in each pile.) Each of the two players, in turn, can remove 1 chip from any pile or can remove 2 chips from a pile that contains exactly 2 chips. The player who removes the final chip wins.

(a) Suppose there is 1 pile. Determine which player will win and why.

(b) Suppose there are 2 piles. Determine which player will win and why.

Problem 13.11: The game of **Nim** is exactly the same as Heaps (from Problem 13.10) except that on each turn, the player can remove any number of chips from one pile. As with Heaps, the player who removes the final chip is the winner.

(a) Suppose there are 2 piles. Determine which player will win and why.

(b) Suppose there are 3 piles of 1, n, and $n + 1$ chips, for some positive integer n. Determine which player will win and why.

Problem 13.12: Recall the game of Chomp from Problem 4.16: We start with a 5×7 array of cookies, as in the picture below. The players alternate turns, and on each turn, the player may choose any cookie left on the board and remove (or "chomp") that cookie along with all the cookies above and/or to the right of the selected cookie. For example, a possible first move would be:

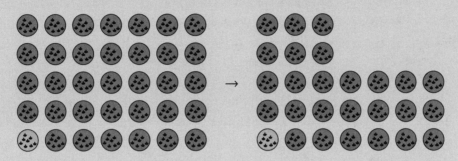

The cookie in the lower-left corner of the board is poison: the player who is forced to chomp it loses. Which player has a winning strategy?

In this section, we explore the theory of two-person games. Now of course we don't mean *any* two-person game, but only those that satisfy a fairly rigid set of conditions. All of the games that we are going to consider must have the following properties:

1. The game must be played by two players, who alternate moves. The "first player" is the player who makes the first move, and the other player is the "second player."

2. There must be no chance involved: no dice or coin flips during the game.

3. Both players must have complete information about all aspects of the game at all times. In other words, nothing is done in secret and neither player can possess any information that the other player does not have.

4. On each move, the player whose turn it is must have a finite number of possible actions.

5. The game must end, after a finite number of turns, with one player winning. No draws allowed.

These games can be analyzed through the use of states, as we will see in the first problem.

Problem 13.9:
(a) Tina and Valentin play a game in which they start with 17 marbles in a pile. They take turns removing 1, 2, or 3 marbles from the pile. The player who removes the final marble wins the game. If Tina goes first and both play intelligently, who should win and why?

(b) Generalize to the situation where there are n marbles in the pile and each player may remove up to k marbles on his or her turn.

Extra! *Talent wins games, but teamwork and intelligence wins championships.* – Michael Jordan
▐▌▶ ▐▌▶ ▐▌▶ ▐▌▶

Solution for Problem 13.9:

(a) We can study this game by looking at a smaller number of marbles in the pile. This is our usual problem-solving strategy of looking at smaller versions of a problem in order to gain insight into the larger version of the problem.

Obviously, if there are 1, 2, or 3 marbles in the pile, then the first player will win right away.

If there are 4 marbles in the pile, then no matter how many marbles the first player takes, she will have to leave 1, 2, or 3 marbles remaining, and then the second player will be able to win right away. So the first player will lose if there are 4 marbles.

If there are 5 marbles in the pile, then the first player can remove 1 marble, leaving 4 marbles remaining. Then, as we have just discussed above, the second player must lose: he is now playing the role of the first player in the 4-marble game.

Before continuing, let's introduce some terminology. A **winning position** is a position from which the player whose turn it is can win, no matter what his opponent does. A **losing position** is a position from which the player whose turn it is cannot win, assuming that his opponent plays intelligently. For our game, as we have discussed above, having 1, 2, 3, or 5 marbles is a winning position, and having 4 marbles is a losing position.

Let's show a state diagram with the positions that we have already discussed. We will show the winning positions in white and the losing positions in grey.

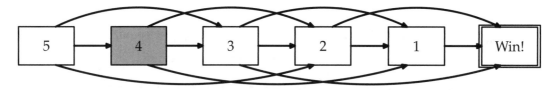

We notice that all moves from a losing position end up in a winning position, and every winning position either wins immediately or has some move to a losing position. In fact, this is how we can recursively define winning and losing positions:

- A winning position is one that either has some move that leads to an immediate win or has some move that leads to a losing position.

- A losing position is one that has every move leading to a winning position.

Think about the logic of this description. The winning player gets to use strategy: he or she can choose any next move, so as to win right away (as in the example where there are 1, 2, or 3 marbles left) or to force the opponent into a losing position (as in the example where there are 5 marbles). The losing player, on the other hand, is doomed to have the other player win no matter what he or she does, and therefore *all* of his or her moves must lead to winning positions (for the other player).

So we can continue to develop our state diagram backwards. If there are 6 or 7 marbles, then the player whose turn it is has a move to leave 4 marbles, so 6 and 7 marbles are winning positions. If there are 8 marbles, then the player whose turn it is must make a move that leaves 5, 6, or 7 marbles, all of which are winning positions. So 8 marbles is a losing position.

By now, you can probably see the pattern.

> **Concept:** This is the general way that we attack many problems in which we need to analyze a game. We look at smaller versions of the game, and try to notice a pattern. Then we try to prove that our pattern works.

We can continue and draw the diagram for all 17 possible positions:

So the player who goes first (Tina in our problem) should win, and she can win by removing 1 marble from the original position.

(b) In part (a), we saw a clear pattern: if the number of marbles in the pile is a multiple of 4, then it is a losing position, otherwise it is a winning position.

More generally, we can conjecture that if the possible moves are removing $1, 2, 3$, up to k marbles, then a position is a losing position if the number of marbles is a multiple of $k + 1$, and otherwise it is a winning position. To prove this, we need to show two steps:

Step 1: From every winning position, there is a move that either immediately wins or places the game into a losing position. If the number of marbles m is not a multiple of $(k + 1)$, then $m = q(k + 1) + r$ where q is a nonnegative integer and $0 < r < k + 1$. Removing r marbles is a valid move and leads either to an immediate win (if $q = 0$) or to the losing position with $q(k + 1)$ marbles (if $q > 0$).

Step 2: All moves from a losing position lead to a winning position. If the number of marbles m is a multiple of $(k + 1)$, then $m = q(k + 1)$ for some positive integer q. Removing r marbles, where $1 \le r \le k$, will leave $q(k + 1) - r$ marbles. We have the strict inequalities

$$q(k + 1) > q(k + 1) - r > q(k + 1) - (k + 1) = (q - 1)(k + 1),$$

so the number of marbles remaining after any move must be strictly between two multiples of $k + 1$, and hence is not a multiple of $k + 1$.

Thus any positive multiple of $k + 1$ is a losing position. To answer the original question, the first player will win unless n is a multiple of $k + 1$; in that case, the second player will win.

$\square$

> **Important:** All intermediate states of a two-player game (of the type that was described above) can be partitioned into two types: **winning positions** and **losing positions**.
>
> - A winning position either has some move that leads to an immediate win or has some move that leads to a losing position.
>
> - A losing position has every move leading to a winning position.
>
> Assuming that both players play intelligently, the player whose move it is will win if the current position is a winning position and will lose if it is a losing position.

Problem 13.10: In the game **Heaps**, we start with one or more piles, each of which has at least 2 chips. (There may be different amounts of chips in each pile.) Each of the two players, in turn, can remove 1 chip from any pile or can remove 2 chips from a pile that contains exactly 2 chips. The player who removes the final chip wins.

(a) Suppose there is 1 pile. Determine which player will win and why.

(b) Suppose there are 2 piles. Determine which player will win and why.

Solution for Problem 13.10:

(a) If you think about it a little bit, you'll quickly see that the 1-pile version of Heaps is pretty boring. The players don't have any choices: they are simply forced to take 1 chip off the pile, until they get down to 2 chips, at which point the player whose move it is (unless he or she is really stupid) will take the 2 chips and win the game. So 2 chips is a winning position, 3 chips is a losing position, 4 chips is a winning position, and so on. We can pretty easily see that a pile with an even number of chips is a winning position, and a pile with an odd number of chips is a losing position (ignoring the case of 1 chip, which should never happen).

(b) With 2 piles, now the players have a choice of moves. Let's look at some small versions of the game to see if we can make a guess at which positions are winning and which are losing.

Suppose there are 2 chips in each pile. This is a losing position, since whatever the first player does, the second player can just do the same thing in the other pile. We also see more generally that if there are an equal number of chips in each pile, then this is a losing position, since the second player can just copy the first player's moves in the other pile.

Let's make a chart of what we know so far:

Winning positions	Losing positions
Single pile with even number of chips	Single pile with odd number of at least 3 chips
Single pile with 1 chip	Two piles with the same number of chips

Now let's consider positions with a 1-chip pile together with a pile with at least 2 chips. (We already know that two 1-chip piles is losing.) If the larger pile has a odd number of chips, then the position is winning: the player takes the 1-chip pile and leaves a single pile with an odd number (at least 3) of chips, which is losing. If the larger pile has 2 chips, then the position is winning: the player takes 1 chip from the 2-chip pile and leaves the losing position of 2 piles with 1 chip each. On the other hand, if the larger pile has an even number of at least 4 chips, then removing either the 1-chip pile or a chip from the larger pile leaves a winning position, thus this position is losing.

Let's add these to our chart. To make the chart a bit more concise, we'll denote by (a, b) the position where one pile has a chips and the other pile has b chips. (Note that (a, b) and (b, a) represent the same position.) In all cases m represents a positive integer:

Winning positions	Losing positions
$(0, m)$ where m is even	$(0, m)$ where $m \geq 3$ is odd
$(0, 1)$	(m, m)
$(1, m)$ where $m \geq 3$ is odd	$(1, m)$ where $m \geq 4$ is even
$(1, 2)$	

If a player has a 2-chip pile available on his move, then he will win if the other pile has an odd number of chips, since he can take the 2 chips and leave a losing odd pile for his opponent. (The exception is the case where the other pile has only 1 chip, but this is still a winning position: take 1 chip from the 2-chip pile and leave the opponent with the losing position of 2 piles of 1 chip each.) On the other hand, if a player has a 2-chip pile available on his move and the other pile has an even number of more than 2 chips, then he will win by taking 1 chip from the 2-chip pile, leaving a 1-chip pile together with a larger even pile, which is losing.

Let's add these to our chart.

Winning positions	Losing positions
$(0, m)$ where m is even	$(0, m)$ where $m \geq 3$ is odd
$(0, 1)$	(m, m)
$(1, m)$ where $m \geq 3$ is odd	$(1, m)$ where $m \geq 4$ is even
$(2, m)$ where $m \neq 2$	

In particular, any position with a 2-chip pile and a larger pile is a winning position. So any player who can (eventually) force his opponent to leave him a 2-chip pile on his move can win.

We also note that if both piles have more than 2 chips, then either move will reduce the total number of chips by 1. In particular, any move will reverse the parity of the total number of chips, changing it from odd to even, or vice versa. The only time that a player can make a move that preserves the parity of the total number of chips is when he is able to remove a 2-chip pile.

Thus, we can conclude that if both piles have more than 2 chips, then the winning positions are those in which the total number of chips is odd. The winning strategy is simple: never take a chip from a 3-chip pile. If it's our turn, and both piles have at least 3 chips, and the total number of chips is odd, then we can never have two piles of 3 chips on our turn. Eventually, our opponent will have to leave us a 2-chip pile, and then we can win as described above.

Here is the final chart for the 2-pile version of Heaps:

Winning positions	Losing positions
$(0, m)$ where m is even	$(0, m)$ where $m \geq 3$ is odd
$(0, 1)$	(m, m)
$(1, m)$ where $m \geq 3$ is odd	$(1, m)$ where $m \geq 4$ is even
$(2, m)$ where $m \neq 2$	
(a, b) where $a, b \geq 3$, $a + b$ odd	(a, b) where $a, b \geq 3$, $a + b$ even

□

Once again, note the difference in the analysis of winning and losing positions. For each winning position, there must be *some* move that either wins the game immediately or leaves a losing position. On the other hand, for each losing position, *all* moves must lead to winning positions.

Also note the reversal between parts (a) and (b). In the 1-pile game, the winning positions (of more than 1 chip) are those with an even number of chips, but in the 2-pile game, the winning positions (of piles with at least 2 chips each) are those with an odd total number of chips.

It is possible to analyze this game for an arbitrary number of piles, but the analysis is more difficult. See Problem 13.25 for the strategy to the general game of Heaps—you'll be asked to prove that it works.

We'll continue with what is arguably the most famous and most studied game of the "taking chips from piles" family.

> **Problem 13.11:** The game of **Nim** is exactly the same as Heaps (from Problem 13.10) except that on each turn, the player can remove any number of chips from one pile. As with Heaps, the player who removes the final chip is the winner.
>
> (a) Suppose there are 2 piles. Determine which player will win and why.
>
> (b) Suppose there are 3 piles of 1, n, and $n + 1$ chips, for some positive integer n. Determine which player will win and why.

Solution for Problem 13.11:

(a) Just as in the 2-pile version of Heaps, we see that if the two piles are equal, then it is a losing position, because the second player can always just copy the first player's moves, but in the other pile.

So the first player's strategy is clear. If the piles are not the same size, then take enough chips from the larger pile in order to make them equal. This forces a losing position on the second player, so the first player will win. If the piles are equal to start with, though, then the first player is stuck with the initial losing position, and the second player will win.

Here is the chart for the 2-pile version of Nim:

Winning positions	Losing positions
(a, b) where $a \neq b$	(a, a)

(b) We'll use the abbreviation (a, b, c) to denote the game where there are three piles that have a, b, and c chips in them. (Of course, the order of the piles does not matter.) We start with the chart from the 2-pile game:

Winning positions	Losing positions
$(0, a, b)$ where $a \neq b$	$(0, a, a)$

We can start examining 3-pile positions with small values of n. If the game is $(1, 1, 2)$, then the first player can win by removing both chips from the pile with 2, leaving the losing 2-pile position $(1, 1, 0)$. So $(1, 1, 2)$ is a winning position.

Winning positions	Losing positions
$(0, a, b)$ where $a \neq b$	$(0, a, a)$
$(1, 1, 2)$	

In the process of analyzing $(1, 1, 2)$, we saw a general concept: any position in which there are 2 equal piles and any third pile is a winning position, because the player whose turn it is can remove the third pile, leaving a 2-equal-piles losing position. Let's add this to the chart:

Winning positions	Losing positions
$(0, a, b)$ where $a \neq b$	$(0, a, a)$
$(1, 1, 2)$	
(a, a, b)	

If the position is $(1,2,3)$, then removing any pile completely would leave a winning 2-unequal-piles position for the other player. The other possible moves and their resulting positions are:

$$1 \text{ chip from 2-chip pile: } (1,1,1)$$
$$1 \text{ chip from 3-chip pile: } (1,2,2)$$
$$2 \text{ chips from 3-chip pile: } (1,2,1)$$

All of the resulting positions are winning. Thus, all moves from $(1,2,3)$ lead to a win for the other player, and hence $(1,2,3)$ is a losing position.

Winning positions	Losing positions
$(0,a,b)$ where $a \neq b$	$(0,a,a)$
$(1,1,2)$	$(1,2,3)$
(a,a,b)	

Moving on to $n = 3$, we see immediately that $(1,3,4)$ is winning, because we can remove 2 chips from the 4-chip pile and leave the losing position $(1,3,2)$.

Winning positions	Losing positions
$(0,a,b)$ where $a \neq b$	$(0,a,a)$
$(1,1,2)$	$(1,2,3)$
(a,a,b)	
$(1,3,4)$	

Next is $n = 4$, giving the position $(1,4,5)$. Removing any entire pile leaves the 2-unequal-piles winning position. All of the other possible moves lead to winning positions, as the other player can make a move that leads back to a losing position. All of the possibilities are shown in the chart below:

$$1 \text{ chip from 4-chip pile: } (1,3,5) \rightarrow (1,3,2)$$
$$2 \text{ chips from 4-chip pile: } (1,2,5) \rightarrow (1,2,3)$$
$$3 \text{ chips from 4-chip pile: } (1,1,5) \rightarrow (1,1,0)$$
$$1 \text{ chip from 5-chip pile: } (1,4,4) \rightarrow (0,4,4)$$
$$2 \text{ chips from 5-chip pile: } (1,4,3) \rightarrow (1,3,2)$$
$$3 \text{ chips from 5-chip pile: } (1,4,2) \rightarrow (1,3,2)$$
$$4 \text{ chips from 5-chip pile: } (1,4,1) \rightarrow (1,0,1)$$

All of the positions in the right column of the above table are losing. Thus, all moves from $(1,4,5)$ lead to a win for the other player, and hence $(1,4,5)$ is a losing position.

Winning positions	Losing positions
$(0,a,b)$ where $a \neq b$	$(0,a,a)$
$(1,1,2)$	$(1,2,3)$
(a,a,b)	
$(1,3,4)$	$(1,4,5)$

Based on our experimentation, we are ready to conjecture that $(1,n,n+1)$ is a winning position if n is odd and is a losing position if n is even. We will prove this by induction. We have already done the base cases.

If we have the position $(1, 2k + 1, 2k + 2)$ for some positive integer k, then we can remove 2 chips from the $(2k + 2)$-chip pile and leave the position $(1, 2k + 1, 2k)$, which by the inductive hypothesis is a losing position. Thus $(1, 2k + 1, 2k + 2)$ is a winning position.

If we have the position $(1, 2k, 2k + 1)$ for some positive integer k, then as we discussed above, removing an entire pile leaves a 2-unequal-piles winning position. The other possibilities all allow the other player to make a move that returns to a losing position (by the inductive hypothesis) of the form $(1, 2m, 2m + 1)$ with $0 \leq m < k$, as follows:

$$2c \text{ chips from } (2k)\text{-chip pile, } 1 \leq c \leq k - 1: \quad (1, 2k - 2c, 2k + 1) \to (1, 2k - 2c, 2k - 2c + 1)$$
$$2c + 1 \text{ chips from } (2k)\text{-chip pile, } 0 \leq c \leq k - 1: (1, 2k - 2c - 1, 2k + 1) \to (1, 2k - 2c - 1, 2k - 2c - 2)$$
$$2c \text{ chips from } (2k + 1)\text{-chip pile, } 1 \leq c \leq k: \quad (1, 2k, 2k - 2c + 1) \to (1, 2k - 2c, 2k - 2c + 1)$$
$$2c + 1 \text{ chips from } (2k + 1)\text{-chip pile, } 1 \leq c \leq k - 1: \quad (1, 2k, 2k - 2c) \to (1, 2k - 2c + 1, 2k - 2c)$$
$$1 \text{ chip from } (2k + 1)\text{-chip pile:} \quad (1, 2k, 2k) \to (0, 2k, 2k)$$

All moves lead to winning positions, so $(1, 2k, 2k + 1)$ is a losing position.

$\square$

In general, Nim is a very difficult game to analyze; even the relatively small examples that we've considered had some subtle points, and it's not at all easy to see how to generalize our arguments to a larger number of piles. You can think about some more special cases of Nim if you like; we will tell you the general strategy in Problem 13.26, and let you prove that it works.

> **Sidenote:** There are many websites that contain free applets for you to practice play-
> ♪ ing Nim. We have links to some of our favorites on this book's website,
> listed on page vi. You may want to try playing Nim on these websites
> before reading Problem 13.26.

Problem 13.12: Recall the game of Chomp from Problem 4.16: We start with a 5×7 array of cookies, as in the picture below. The players alternate turns, and on each turn, the player may choose any cookie left on the board and remove (or "chomp") that cookie along with all the cookies above and/or to the right of the selected cookie. For example, a possible first move would be:

The cookie in the lower-left corner of the board is poison: the player who is forced to chomp it loses. Which player has a winning strategy?

Solution for Problem 13.12: As we saw in Problem 4.16, this game has $\binom{12}{5} = 792$ possible positions, and trying to analyze them all seems like a nightmare.

We can analyze some special cases. Clearly a $1 \times n$ board is a winning position if $n > 1$: just chomp all but the poison cookie. By symmetry, an $n \times 1$ board is winning too.

A $2 \times n$ board is a winning position as well: just chomp a single cookie in the corner, so that you're left with a top row of $n - 1$ cookies and a bottom row of n cookies. If your opponent chomps k cookies from the top row, then chomp k cookies from the bottom row to leave a position with the same shape ($n - k - 1$ cookies in the top row and $n - k$ cookies in the bottom row). If your opponent chomps from the bottom row, then we're back to a $2 \times m$ board (for some $m < n$), and we repeat the process by chomping just a single cookie. Eventually the board will get chomped down to the single poison cookie, which your opponent will be forced to eat.

Extending this reasoning to larger, more complicated-shaped boards is, however, very difficult. But we don't need to! All the question asked us is *which player* had the winning strategy, not to determine what that strategy is. We can show via a very clever argument that the initial board position—in fact, any rectangular position (except for the single cookie at the end)—is a winning position for the player whose move it is.

Suppose the initial position was a losing position. That means that any move leads to a winning position. In particular, it means that the chomp board with the single upper-right cookie missing is a winning position. That means that there must be some winning move from this position. However, any move that can be made from the board with one cookie missing can also be made from the original board! This means that the original board has a winning move, which is a contradiction of the original assumption that the original board is losing. Therefore, the initial position is a winning position. □

What's extremely interesting about Chomp is that even though it's relatively straightforward to show, as we did, that the initial position is a winning position for the player who goes first, it is unknown what the specific winning strategy is. This is an open research question.

Concept: Many problems have "nonconstructive" solutions like this one, in which we can show the existence of some object (in this case, a winning strategy), but we can't exhibit precisely what that object is.

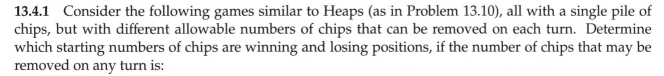

Exercises

13.4.1 Consider the following games similar to Heaps (as in Problem 13.10), all with a single pile of chips, but with different allowable numbers of chips that can be removed on each turn. Determine which starting numbers of chips are winning and losing positions, if the number of chips that may be removed on any turn is:

(a) 1 or 2.

(b) 1, 2, 3, 4, 5, or 6.

(c) 1 or 4.

(d) any power of 2.

(e) any number of chips that is a divisor of the current stack size, except that the whole stack may be taken only if there are 1 or 2 chips remaining.

13.4.2 Sean and Larry are playing a game on a blackboard. Sean starts by writing the number 1, and then in alternating turns (starting with Larry), each player multiplies the current number by any number from 2 through 9 (inclusive) and writes the new number on the board. The first player to write a number larger than 1000 wins. A sample game might be (each number is preceded by "S" or "L" to indicate who wrote it):

$$S1, L3, S12, L60, S420, L1260; \text{ Larry wins.}$$

Which player should win and why? **Hints:** 249

13.4.3 The game of **Split** is played with two piles of chips. On each turn, a player removes one pile and splits the remaining pile into two new piles in any manner he chooses. (For example, the game might start with piles of 9 and 12 chips; the first player could remove the 9-chip pile and split the other pile into piles of 8 and 4 chips.) A player loses if he or she is unable to move. What are the winning starting positions? **Hints:** 344

13.4.4 If a game of Nim starts with 3 piles of 2, 4, and 6 chips, which player should win?

13.4.5 Two players take turns drawing diagonals on a 100-sided regular polygon, subject to the condition that no two drawn diagonals can cross. A player who cannot move loses. Which player has a winning strategy? **Hints:** 5, 200

13.4.6★ Wythoff's game is similar to Nim: the game starts with two piles of chips, and on each move a player may either remove any number of chips from one pile, or an equal number of chips from both piles. The player who takes the last chip is the winner.

(a) Suppose that the game begins with piles of 5 and 6 chips. Which player should win and why?

(b)★ Determine the winning and losing positions. **Hints:** 303, 16, 120

13.5 Summary

➤ A state is a description of an intermediate stage of an event. States allow us to break up complex events into more manageable simple events.

➤ For a problem with multiple states, it often helps to draw a diagram. Draw a box for each state and arrows for the transitions between states. Label probabilities or expected values as appropriate.

➤ Try to make your states as simple as possible. Often, in a problem involving states, identifying the simplest possible set of states is a major step towards solving the problem.

➤ It often helps to introduce variable(s) that store information about the intermediate state(s). This is especially true in events that can last for an arbitrarily long sequence of states.

➤ A particular type of state problem is a 2-player game. We try to identify winning positions and losing positions.

➤ When analyzing a game, it often helps to look at smaller versions of the game, and try to notice a pattern. Then we try to prove that our pattern works.

We also saw the following general problem-solving concepts:

> **Concept:** If it's easy to do, a quick check of your work in the middle of a problem can prevent errors from propagating through a long solution.

> **Concept:** Many problems have "nonconstructive" solutions, in which we can show the existence of some object (for example, a winning strategy for a 2-player game), but we can't exhibit precisely what that object is.

REVIEW PROBLEMS

13.13 Andrea flips a fair coin repeatedly, continuing until she either flips two heads in a row (the sequence HH) or flips tails followed by heads (the sequence TH). What is the probability that she will stop after flipping HH? *(Source: HMMT)*

13.14 A $3 \times 3 \times 3$ cube is composed of 27 small $1 \times 1 \times 1$ cubes, and little passageways allow for movement between adjacent cubes. (Cubes are adjacent if they share a face; cubes sharing only an edge are not adjacent.) A mouse is placed in a corner cube (the shaded cube in the picture at right), and a piece of cheese is placed in the center $1 \times 1 \times 1$ cube. Each minute, the mouse moves randomly to an adjacent cube. Find the expected number of minutes before the mouse reaches the cheese.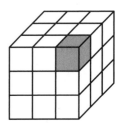

13.15 A bag contains two red beads and two green beads. You reach into the bag and pull out a bead, replacing it with a red bead regardless of the color you pulled out. What is the expected number of replacements needed so that all beads in the bag are red? *(Source: AMC)*

13.16 Recall that in a tennis game, if the players are tied at "deuce," then the game continues until one player has won two more points than the other player. Homer and Marge are again playing tennis and have reached deuce. Unfortunately for them, Bart is now the umpire of the game, and during any point, he will call "Let" with probability $\frac{1}{3}$, meaning that Homer and Marge will have to play that point again. If Homer has a $\frac{3}{5}$ probability of winning any given point in which Bart does not call "Let," then what is the expected number of points necessary (including points in which Bart calls "Let") before the game is completed?

13.17 A space alien is doing a random walk starting at 0 on the nonnegative number line. From 0 the alien must take one step to the right (to 1), but from any positive position on the line, the alien will move one step left or right with equal probability. What is the expected number of times that the alien will revisit 0 before the first time that he visits 5?

13.18 The figure at right is a map of part of a city: the small rectangles are city blocks and the lines are streets. Each morning a student walks from intersection A to intersection B, always walking along streets shown, always going east or south. For variety, at each intersection where he has a choice, he chooses with equal probability whether to go east or south. Find the probability that, on any given morning, he walks through intersection C. *(Source: AMC)*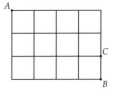

13.19   We play a game. The pot starts at $0. On every turn, you flip a fair coin. If you flip heads, I add $100 to the pot. If you flip tails, I take all of the money out of the pot, and you are assessed a "strike." You can stop the game before any flip and collect the contents of the pot, but if you get 3 strikes, the game is over and you win nothing. Find, with proof, the expected value of your winnings if you follow an optimal strategy. *(Source: USAMTS)*

13.20   A particle moves in the Cartesian plane from one lattice point to another according to the following rules:

1. From any lattice point (a, b), the particle may move only to $(a + 1, b)$, $(a, b + 1)$, or $(a + 1, b + 1)$.

2. There are no right angle turns in the particle's path. That is, the sequence of points visited contains neither a subsequence of the form (a, b), $(a + 1, b)$, $(a + 1, b + 1)$ nor a subsequence of the form (a, b), $(a, b + 1)$, $(a + 1, b + 1)$.

How many different paths can the particle take from $(0, 0)$ to $(5, 5)$? *(Source: AIME)*

Challenge Problems

13.21   In a game similar to three card monte, the dealer places three cards on the table: the queen of spades and two red cards. The cards are placed in a row, with the queen in the center; the card configuration is thus RQR. The dealer proceeds with a series of moves. On each move, the dealer randomly switches the center card with one of the two edge cards (so the configuration after the first move is either RRQ or QRR). What is the probability that, after 2004 moves, the center card is the queen? *(Source: HMMT)*

13.22   A calculator has a display, which shows a nonnegative integer N, and a button, which replaces N by a random integer chosen from the set $\{0, 1, \ldots, N - 1\}$, provided that $N > 0$. Initially, the display holds the number $N = 2003$. If the button is pressed repeatedly until $N = 0$, what is the probability that the numbers 1, 10, 100, and 1000 will all show up on the display at some point? *(Source: HMMT)* **Hints:** 116, 46

13.23   There are 6 peas in a glass, 4 floating on the top and 2 sitting on the bottom. At each five second interval, a random number of peas from 0 to 2 sink from the top to the bottom and a random number from 0 to 2 rise from the bottom to the top. (If there is only 1 pea left, it moves or stays put with equal probability.) What is the probability that all six peas are on the top before all six are on the bottom? *(Source: Mandelbrot)* **Hints:** 206

13.24   You and I each have $14. I flip a fair coin repeatedly. If it comes up heads, I give you a dollar, but if it comes up tails, you give me a dollar. What is the expected number of flips until one of us runs out of money? *(Source: Mandelbrot)* **Hints:** 91, 317

13.25★   Here is the general analysis for Heaps with an arbitrary number of piles.

Let A be the number of 2-chip piles, and let B be the number of piles with either an even number (greater than 2) of chips or a single chip. The winning positions are those in which either A or B is odd.

Prove that it is correct, and describe the winning strategy. **Hints:** 1, 56

13.26★ Here is the general analysis for Nim with an arbitrary number of piles.

Represent the number of chips in each pile by a binary number (for example, 6 chips is the binary number 110). The losing positions are those in which there are an even number of 1's in each digit-position, counted across all the piles. (For example, piles of 7, 5, 3, and 1 chips is a losing position, because the binary representations 111, 101, 11, and 1 have an even number of 1's in the ones, twos, and fours places.)

Prove that it is correct, and describe the winning strategy. **Hints:** 245, 287

You must try to generate happiness within yourself. – Ernie Banks

CHAPTER **14**

Generating Functions

14.1 Introduction

In this chapter, we will explore a really clever algebraic device called a **generating function**. Generating functions allow us to use powerful algebra techniques to approach difficult counting problems.

Suppose that $a_0, a_1, a_2, \ldots$ is a sequence of numbers. The corresponding generating function for this sequence is

$$f(x) = a_0 + a_1 x + a_2 x^2 + \cdots.$$

That is, the coefficient of x^k is the kth term of the sequence. If the original sequence is finite, then the generating function is just a polynomial, but if the sequence is infinite, then the generating function is an "infinite polynomial," more properly called a **power series**.

Here are some examples:

- If our sequence is $1, 5, 10, 10, 5, 1$—Row 5 of Pascal's Triangle–then the generating function is

$$1 + 5x + 10x^2 + 10x^3 + 5x^4 + x^5.$$

  Notice that this is equal to $(1 + x)^5$ by the Binomial Theorem. We will explore this relationship further in Section 14.3.

- If our sequence is $1, 3, 6, 10, 15, 21, \ldots$—the triangular numbers, or all the numbers of the form $\binom{k}{2}$ for positive integers $k \geq 2$—then the generating function is

$$1 + 3x + 6x^2 + 10x^3 + 15x^4 + 21x^5 + \cdots.$$

  In Section 14.4, we will see some special properties of this generating function.

- If our sequence is $0, 1, 1, 2, 3, 5, 8, 13, \ldots$—the Fibonacci numbers—then the generating function is

$$x + x^2 + 2x^3 + 3x^4 + 5x^5 + 8x^6 + 13x^7 + \cdots .$$

Although this seems rather unwieldy, it does have some nice properties, as we will see in Section 14.6★.

14.2 Basic examples of Generating Functions

 **Problems**

All of these problems are easy. But try to answer them using generating functions.

Problem 14.1: There are three urns on a table. One has 2 red balls, one has 2 green balls, and one has 3 blue balls. Balls of the same color are indistinguishable. In how many ways can I choose 4 balls from the urns?

Problem 14.2: As in Problem 14.1, there are three urns on a table: one has 2 red balls, one has 2 green balls, and one has 3 blue balls. But now, all balls of the same color are distinguishable. In how many ways can I choose 4 balls from the urns?

Problem 14.3: In how many ways can we roll three dice to get the sum 9?

Problem 14.4: Suppose that I have two weird dice. One die has ⚀ on three faces, ⚂ on two faces, and ⚄ on the sixth face. The other die has ⚀ on one face, ⚃ on four faces, and ⚅ on the sixth face. If I roll both dice, what is the probability that the sum is greater than 6 but less than 10?

In simple problems in which we are counting the number of possibilities across multiple events, generating functions provide a useful tool for keeping track of the number of possibilities. This is a bit confusing to describe in words, but hopefully some simple examples will illustrate what we mean.

All of the problems in this section are very easy, and could easily be solved by counting methods that you already know. We present them here for illustrative purposes, as their transparency hopefully makes it easy to see what's going on when we introduce generating functions.

Problem 14.1: There are three urns on a table. One has 2 red balls, one has 2 green balls, and one has 3 blue balls. Balls of the same color are indistinguishable. In how many ways can I choose 4 balls from the urns?

Solution for Problem 14.1: Of course, this is an easy problem, and we could just list the possibilities in a table, as shown on the next page:

Extra! *Not without a shudder may the human hand reach into the mysterious urn of destiny.*
– Friedrich Schiller

Red	Green	Blue
0	1	3
0	2	2
1	0	3
1	1	2
1	2	1
2	0	2
2	1	1
2	2	0

So there are 8 ways to choose the balls.

Now let's see how we can use generating functions to count the number of possibilities.

From the red urn, because the balls are indistinguishable, our only choice is how many of the balls to choose. We can choose 0 red balls, 1 red ball, or 2 red balls. In particular, since the two balls are indistinguishable, if we choose 1 ball, we don't care which one it is, so there's really only one way to choose 1 ball. Therefore, we have 1 way to choose 0 balls, 1 way to choose 1 ball, and 1 way to choose 2 balls. So, the generating function for the number of ways to choose balls from the urn with the red balls is

$$1 + x + x^2 = 1 \cdot x^0 + 1 \cdot x^1 + 1 \cdot x^2.$$

The coefficient "1" of x^0 represents the number of ways to choose 0 red balls (there is only 1 way), the coefficient "1" of x^1 represents the number of ways to choose 1 red ball, and the coefficient "1" of x^2 represents the number of ways to choose 2 red balls.

If we wished to be really precise, we should say that $1 + x + x^2$ is "the generating function for the sequence of the numbers of ways to choose k balls from the urn with the red balls." However, it is a bit verbose for us to have to use this much detail every time we talk about a generating function. Instead, we will simply say that this function is "the generating function for the urn with the red balls" or even more simply, "the generating function for the red balls." As long as the context of the problem or solution makes clear what it is that the coefficients of the generating function represent, it's all right to be somewhat imprecise about the words that we use to describe the function.

By exactly the same process as that for the urn with the red balls, we find that the generating function for the urn with the green balls is also

$$1 + x + x^2 = 1 \cdot x^0 + 1 \cdot x^1 + 1 \cdot x^2,$$

since there is 1 way to choose 0 green balls, 1 way to choose 1 green ball, and 1 way to choose 2 green balls.

Finally, the generating function for the urn with the blue balls is

$$1 + x + x^2 + x^3 = 1 \cdot x^0 + 1 \cdot x^1 + 1 \cdot x^2 + 1 \cdot x^3,$$

since there is 1 way to choose each of 0, 1, 2, or 3 blue balls.

Now we get to the step where we see how generating functions can help us. To get the generating function for the ways to pick balls from multiple urns, it turns out that all we need to do is multiply the three generating functions together:

$$(1 + x + x^2)(1 + x + x^2)(1 + x + x^2 + x^3) = (1 + 2x + 3x^2 + 2x^3 + x^4)(1 + x + x^2 + x^3)$$
$$= 1 + 3x + 6x^2 + 8x^3 + 8x^4 + 6x^5 + 3x^6 + x^7.$$

We can now read the coefficient of x^4 and see that there are 8 ways to choose 4 balls.

Why does this work? Let's write out the individual products in the above equation that make up the x^4 term:

$$(1 \cdot x \cdot x^3) + (1 \cdot x^2 \cdot x^2) + (x \cdot 1 \cdot x^3) + (x \cdot x \cdot x^2) + (x \cdot x^2 \cdot x) + (x^2 \cdot 1 \cdot x^2) + (x^2 \cdot x \cdot x) + (x^2 \cdot x^2 \cdot 1) = 8x^4.$$

Notice how these 8 products exactly correspond with the 8 entries from our chart earlier:

Red	Green	Blue	Monomial
0	1	3	$1 \cdot x \cdot x^3$
0	2	2	$1 \cdot x^2 \cdot x^2$
1	0	3	$x \cdot 1 \cdot x^3$
1	1	2	$x \cdot x \cdot x^2$
1	2	1	$x \cdot x^2 \cdot x$
2	0	2	$x^2 \cdot 1 \cdot x^2$
2	1	1	$x^2 \cdot x \cdot x$
2	2	0	$x^2 \cdot x^2 \cdot 1$

We see that each product corresponds to a way to choose the 4 balls. That's because each product is composed of three powers of x: the first power of x corresponds to the number of red balls, the second power of x corresponds to the number of green balls, and the third power of x corresponds to the number of blue balls. (Note that we consider $1 = x^0$ as a power of x corresponding to selecting 0 balls.) Therefore, adding up the products is the same as counting the number of ways to choose 4 balls. $\square$

Problem 14.2: As in Problem 14.1, there are three urns on a table: one has 2 red balls, one has 2 green balls, and one has 3 blue balls. But now, all balls of the same color are distinguishable. In how many ways can I choose 4 balls from the urns?

Solution for Problem 14.2: Once again, the answer to this is easy. There are 7 distinguishable balls in total, and we are choosing 4 of them, so this can be done in $\binom{7}{4} = 35$ ways.

We can also look at this problem using generating functions. The generating function for the red balls is
$$1 + 2x + x^2 = 1 \cdot x^0 + 2 \cdot x^1 + 1 \cdot x^2,$$

because there is 1 way to choose 0 red balls, 2 ways to choose 1 red ball (since the balls are distinguishable), and 1 way to choose 2 red balls. (Again, if we wanted to be precise, we should say "the generating function for the number of ways to choose red balls," but since the context is clear, we'll simply say "the generating function for the red balls.")

Similarly, the generating function for the green balls is
$$1 + 2x + x^2 = 1 \cdot x^0 + 2 \cdot x^1 + 1 \cdot x^2,$$

and the generating function for the blue balls is

$$1 + 3x + 3x^2 + x^3 = 1 \cdot x^0 + 3 \cdot x^1 + 3 \cdot x^2 + 1 \cdot x^3.$$

Therefore, the generating function for choosing balls from all of the urns is the product of the individual generating functions:

$$(1 + 2x + x^2)(1 + 2x + x^2)(1 + 3x + 3x^2 + x^3).$$

Multiplying this out gives

$$(1 + 4x + 6x^2 + 4x^3 + x^4)(1 + 3x + 3x^2 + x^3) = 1 + 7x + 21x^2 + 35x^3 + 35x^4 + 21x^5 + 7x^6 + x^7,$$

so by reading off the coefficient of the x^4 term, we see that there are 35 ways to choose 4 balls.

Again, let's see in detail why this works. We'll write out the individual products from the above equation that make up the x^4 term:

$$(1 \cdot 2x \cdot x^3) + (1 \cdot x^2 \cdot 3x^2) + (2x \cdot 1 \cdot x^3) + (2x \cdot 2x \cdot 3x^2) + (2x \cdot x^2 \cdot 3x) + (x^2 \cdot 1 \cdot 3x^2) + (x^2 \cdot 2x \cdot 3x) + (x^2 \cdot x^2 \cdot 1) = 35x^4.$$

As in Problem 14.1, these 8 products exactly correspond with the ways to choose particular numbers of balls from each urn:

Number of balls			# of ways	
Red	Green	Blue	to choose	Monomial
0	1	3	2	$1 \cdot 2x \cdot x^3$
0	2	2	3	$1 \cdot x^2 \cdot 3x^2$
1	0	3	2	$2x \cdot 1 \cdot x^3$
1	1	2	12	$2x \cdot 2x \cdot 3x^2$
1	2	1	6	$2x \cdot x^2 \cdot 3x$
2	0	2	3	$x^2 \cdot 1 \cdot 3x^2$
2	1	1	6	$x^2 \cdot 2x \cdot 3x$
2	2	0	1	$x^2 \cdot x^2 \cdot 1$

As in the corresponding table from Problem 14.1, we see that each product corresponds to choosing a particular number of balls from each urn. The difference between Problem 14.1 and the current problem is that now each product has a coefficient attached to it, which counts the number of ways of choosing the corresponding numbers of balls from each urn.

For example, look at the 4th line of the table. There are 2 ways to choose 1 red ball from the red urn, 2 ways to choose 1 green ball from the green urn, and 3 ways to choose 2 blue balls from the blue urn. Therefore, there are $2 \cdot 2 \cdot 3 = 12$ ways to choose this particular combination of balls. This is reflected in the coefficient of the product: $2x \cdot 2x \cdot 3x^2 = 12x^4$. □

Note that in the solution to Problem 14.2, we could have simplified the algebra quite a bit. We see that $1 + 2x + x^2 = (1 + x)^2$ and $1 + 3x + 3x^2 + x^3 = (1 + x)^3$, so the generating function for the entire set of balls is

$$(1 + x)^2(1 + x)^2(1 + x)^3 = (1 + x)^7,$$

and then naturally the coefficient of x^4 is $\binom{7}{4} = 35$, by the Binomial Theorem.

We'll see more of this in Section 14.3.

Problem 14.3: In how many ways can we roll three dice to get the sum 9?

Solution for Problem 14.3: When rolling a die, there is 1 way to get a ⚀, 1 way to get a ⚁, and so on. Therefore, the generating function for one die is

$$x + x^2 + x^3 + x^4 + x^5 + x^6.$$

If we sum three dice, the generating function is

$$(x + x^2 + x^3 + x^4 + x^5 + x^6)^3,$$

and we are looking for the coefficient of the x^9 term in this product.

First, let's factor x^3 out of our generating function:

$$x^3(1 + x + x^2 + x^3 + x^4 + x^5)^3.$$

To find the coefficient of the x^9 term of the above function, we can simply find the coefficient of the x^6 term of $(1 + x + x^2 + x^3 + x^4 + x^5)^3$. At this point, we don't have a good algebraic tool for quickly finding this coefficient, so we're forced to multiply this expression out:

$$
\begin{aligned}
&(1 + x + x^2 + x^3 + x^4 + x^5)^3 \\
&= (1 + x + x^2 + x^3 + x^4 + x^5)^2(1 + x + x^2 + x^3 + x^4 + x^5) \\
&= (1 + 2x + 3x^2 + 4x^3 + 5x^4 + 6x^5 + 5x^6 + 4x^7 + 3x^8 + 2x^9 + x^{10})(1 + x + x^2 + x^3 + x^4 + x^5) \\
&= 1 + 3x + 6x^2 + 10x^3 + 15x^4 + 21x^5 + 25x^6 + 27x^7 + 27x^8 + 25x^9 + 21x^{10} + 15x^{11} + 10x^{12} + 6x^{13} + 3x^{14} + x^{15}.
\end{aligned}
$$

The coefficient of the x^6 term is 25, so there are 25 ways to roll a sum of 9. □

The last step in the above solution was pretty messy. We didn't actually need to multiply the whole polynomial out, since all we really cared about was the x^6 term. So we could have isolated the parts of the expression that produce an x^6 term:

$$(\cdots + 2x + 3x^2 + 4x^3 + 5x^4 + 6x^5 + 5x^6 + \cdots)(1 + x + x^2 + x^3 + x^4 + x^5).$$

We just compute the products that will produce x^6:

$$(2x)(x^5) + (3x^2)(x^4) + (4x^3)(x^3) + (5x^4)(x^2) + (6x^5)(x) + (5x^6)(1) = (2 + 3 + 4 + 5 + 6 + 5)x^6 = 25x^6,$$

and we arrive at the answer 25 much more quickly.

Also, you probably noticed some patterns in the above computation. We'll see a better way to do this problem in Section 14.4.

Problem 14.4: Suppose that I have two weird dice. One die has ⚁ on three faces, ⚂ on two faces, and ⚃ on the sixth face. The other die has ⚀ on one face, ⚃ on four faces, and ⚅ on the sixth face. If I roll both dice, what is the probability that the sum is greater than 6 but less than 10?

Solution for Problem 14.4: Certainly we could do this by casework, but let's see how generating functions eliminate the need to consider separate cases.

The generating function for the first die is $3x^2 + 2x^3 + x^5$ and the generating function for the second die is $x + 4x^4 + x^6$. When we take the sum of the dice, we multiply the generating functions to get

$$(3x^2 + 2x^3 + x^5)(x + 4x^4 + x^6) = 3x^3 + 2x^4 + 13x^6 + 8x^7 + 3x^8 + 6x^9 + x^{11}.$$

Note that the coefficients sum to 36, encompassing all 36 possible rolls. There are $8 + 3 + 6 = 17$ ways to roll a sum of 7, 8, or 9 (we simply read the coefficients of the x^7, x^8, and x^9 terms), so the probability is $\frac{17}{36}$. $\square$

As noted, we could have done Problem 14.4 by casework, but the beauty of the generating function solution is that it concisely encapsulates all possible outcomes of the rolls. We can now read this information from the generating function quite easily. For example, suppose we wanted to compute the expected value of a roll of the dice. We can read the necessary information directly from the generating function:

$$E(\text{roll}) = \frac{3(3) + 2(4) + 13(6) + 8(7) + 3(8) + 6(9) + 1(11)}{36} = \frac{240}{36} = \frac{20}{3}.$$

Concept: One of the big advantages of generating functions over casework is that generating functions typically encapsulate all of the relevant cases at once.

Exercises

14.2.1 After dinner, Marilyn, John, and Bill all pitch in for the tip. Marilyn and John will each leave $1, $2, or $3, whereas Bill will leave $1 or $2. In how many ways can they leave a $6 tip?

14.2.2 Use generating functions to find the number of integer solutions to $a + b + c + d = 7$, where b must be even, c must be odd, and $0 \le a, b, c, d \le 4$.

14.2.3 Suppose 4 friends go to a carnival, and each plays a game in which they can win $2, win $1, break even (win or lose nothing), lose $1, or lose $2. In how many different ways can the friends collectively break even? Solve using generating functions and then, for practice, solve via a counting argument. **Hints:** 178, 281

14.2.4★ A "fair die" is a die that rolls 1, 2, 3, 4, 5, 6, each with probability $\frac{1}{6}$. A "fair pair of dice" is a pair of dice that rolls a sum of 2 with probability $\frac{1}{36}$, a sum of 3 with probability $\frac{2}{36}$, and so on. In other words, a fair pair of dice will roll sums with the same probabilities as a pair of fair dice. Assume that a die must have a positive integer on each face. Is a fair pair of dice necessarily a pair of fair dice? **Hints:** 140, 186, 136

Extra! *I shall never believe that God plays dice with the world.* – Albert Einstein

Not only does God play dice, but. . . He sometimes throws them where they cannot be seen.
– Stephen Hawking

14.3 The Binomial Theorem (as a Generating Function)

Problems

Problem 14.5: Joe is ordering 3 hot dogs at Alpine City Beef. He can choose among five toppings for each hot dog: ketchup, mustard, relish, onions, and sauerkraut. In how many different ways can he choose 6 toppings total for his three hot dogs?

Problem 14.6: We have 25 people at our party. We all agree to pitch in $1 or $2 each for snacks, except for Curtis, who will pay $3, $5, or $9. In how many different ways can we come up with exactly $39 for snacks?

As we saw briefly in Problem 14.2, we may think of the polynomial produced by the Binomial Theorem as a special case of a generating function. For example, if we write

$$(x + 1)^n = \binom{n}{0} + \binom{n}{1}x + \binom{n}{2}x^2 + \cdots + \binom{n}{n}x^n,$$

then we see that the coefficient of x^k is $\binom{n}{k}$ for any $0 \le k \le n$. We know that $\binom{n}{k}$ counts the number of ways to choose a subset of k items from a set of n items, and thus we can view $(x + 1)^n$ as the generating function for the number of ways to choose a subset from a set of n items. As with our previous examples of generating functions, the coefficient of x^k counts something that depends on k.

Usually, the generating function arising from the Binomial Theorem doesn't help much in solving problems, since we can often solve such problems by other means. But at times, it clarifies the solution a bit, or gives us a better understanding of *why* the solution works—which you should always be striving for!

Problem 14.5: Joe is ordering 3 hot dogs at Alpine City Beef. He can choose among five toppings for each hot dog: ketchup, mustard, relish, onions, and sauerkraut. In how many different ways can he choose 6 toppings total for his three hot dogs?

Solution for Problem 14.5: On each individual hot dog, there is $\binom{5}{0} = 1$ way for Joe to have 0 toppings, $\binom{5}{1} = 5$ ways for Joe to have 1 topping, $\binom{5}{2} = 10$ ways for Joe to have 2 toppings, and so on. Therefore, the generating function for the number of ways to choose toppings on each individual hot dog is

$$1 + 5x + 10x^2 + 10x^3 + 5x^4 + x^5 = (1 + x)^5.$$

Since there are 3 hot dogs, the generating function for the number of ways to choose toppings for all 3 hot dogs is

$$((1 + x)^5)^3 = (1 + x)^{15}.$$

Since Joe wants to pick 6 toppings total, we want the coefficient of x^6 in this generating function. By the Binomial Theorem, this coefficient is $\binom{15}{6} = 5005$, so Joe has 5005 different choices. $\square$

There is a pretty simple counting explanation for why $\binom{15}{6}$ is the answer—can you determine it?

Problem 14.6: We have 25 people at our party. We all agree to pitch in $1 or $2 each for snacks, except for Curtis, who will pay $3, $5, or $9. In how many different ways can we come up with exactly $39 for snacks?

Solution for Problem 14.6: Each of the 24 non-Curtis people at the party has 1 way to contribute $1 and 1 way to contribute $2. Therefore, the generating function for each of these people is $(x + x^2)$. On the other hand, Curtis's generating function is $(x^3 + x^5 + x^9)$. When we combine these functions to get the overall generating function for the possible snack contributions, we get

$$(x + x^2)^{24}(x^3 + x^5 + x^9).$$

Since we want to count the ways that the contributions sum to $39, we are looking for the coefficient of the x^{39} term of the above. The issue now is how best to calculate that coefficient. We want to simplify the algebra as much as possible.

Concept: Often the key step in using generating functions is finding the most efficient algebraic method to compute the coefficient(s) that you need.

We can simplify our expression a bit by factoring x out of each of the first 24 factors and x^3 out of the last factor:

$$x^{27}(1 + x)^{24}(1 + x^2 + x^6).$$

So the coefficient of x^{39} in the above expression is the same as the coefficient of x^{12} in

$$(1 + x)^{24}(1 + x^2 + x^6).$$

How do we find the x^{12} coefficient of the above polynomial? If we break it up into three parts:

$$(1 + x)^{24} + x^2(1 + x)^{24} + x^6(1 + x)^{24},$$

we see that the x^{12} coefficient is equal to the x^{12} coefficient of $(1 + x)^{24}$, plus the x^{10} coefficient of $(1 + x)^{24}$, plus the x^6 coefficient of $(1 + x)^{24}$. Now we can use the Binomial Theorem, which tells us that the sum of these coefficients is

$$\binom{24}{12} + \binom{24}{10} + \binom{24}{6} = 4{,}800{,}008.$$

So there are 4,800,008 ways that we can all come up with $39. $\square$

Exercises

14.3.1 What is the simpler counting explanation for the answer to Problem 14.5?

14.3.2 6 boys and 8 girls go on a nature hike. Each boy will pick either 0 or 2 flowers, and each girl will pick either 0 or 3 flowers. In how many ways can the group collectively pick 20 flowers?

14.3.3 Find the number of ways to collect $22 from 30 people, if the first 29 people each contribute nothing or $1, and the 30th person contributes nothing, $2, or $5.

14.3.4 The Supreme Court of Aopslandia has 13 judges, consisting of 1 Chief Justice and 12 Associate Judges. The Chief Justice's vote counts for 3, so that, for example, if the Chief Justice and 7 Associate Judges vote "yes" and the other 5 Associate Judges vote "no," the vote total is 10-5. In how many ways can the Court vote 8-7? (Solve using generating functions.)

14.3.5 Use the coefficient of x^r in $(1 + x)^{m+n}$ to show that

$$\sum_{k=0}^{r} \binom{m}{k}\binom{n}{r-k} = \binom{m+n}{r}.$$

14.4 Distributions (as Generating Functions)

> **Problems**

> **Problem 14.7:** Find the number of solutions in nonnegative integers to the equation $a + b + c = 30$.

> **Problem 14.8:** Find an expression for the coefficient of x^k in the infinite polynomial
> $$(1 + x + x^2 + x^3 + \cdots)^n.$$

> **Problem 14.9:** Find the number of ways to distribute \$15 to five people, if each person gets a whole number of dollars between \$0 and \$4 (inclusive).

> **Problem 14.10:** I'm doling out 100 identical pieces of candy to 5 kids. The two youngest kids want at most 1 piece. The middle kid will take any number of pieces. The two oldest kids each demand an odd number of pieces. In how many ways can I distribute the candy?

> **Problem 14.11:** Chelsea, Barbara, and Jenna go to dinner, and have to pay a \$50 bill. Chelsea insists on contributing a number of dollars that is a multiple of 3, whereas Barbara and Jenna each insist on contributing an odd number of dollars. In how many ways can they pay the bill?

So far, the generating function problems that we've done have really been "toy" problems, in that we could have solved them pretty easily with less sophisticated techniques. But when it comes to distribution problems, generating functions begin to flex their muscles a bit, and give us better insights into what's really going on.

Along the way, we'll see some pretty interesting algebra!

> **Problem 14.7:** Find the number of solutions in nonnegative integers to the equation $a + b + c = 30$.

Solution for Problem 14.7: This is a basic distribution problem like those that we studied in Chapter 7. We think of arranging a row of thirty 1's and two dividers. The 1's to the left of the first divider will give the value of a, the 1's between the two dividers will give the value of b, and the 1's to the right of the second divider will give the value of c. Since there are 32 items that we have to arrange, we need to simply choose 2 of the 32 positions to be dividers, and the rest will be 1's. Therefore, the number of ways to arrange the 1's and dividers (and thus the number of solutions to the equation) is $\binom{32}{2}$. (If this explanation didn't make sense to you, go back and reread Chapter 7.)

Now let's see another approach using generating functions.

The integer a can be any nonnegative integer, so in particular it can be 0, 1, 2, 3, ... More precisely, there is 1 way in which a can be 0, there is 1 way in which a can be 1, there is 1 way in which a can be 2, and so on for every nonnegative integer. Therefore, the generating function for a is

$$1 + x + x^2 + x^3 + \cdots,$$

where the expression continues on towards infinity. (Such a "function" is called a **power series**.) Note that we don't stop the generating function at the x^{30} term, even though we know that a can't be more than 30. The reason is that the problem statement itself doesn't constrain a to be at most 30; the problem only states that a must be a nonnegative integer. As we'll see in just a bit, the algebra is actually much simpler if we have an (infinite) power series for the generating function of a, rather than a regular (finite) polynomial.

Similarly, the generating function for each of b and c is also

$$1 + x + x^2 + x^3 + \cdots.$$

Therefore, the generating function of the sum $a + b + c$ is

$$(1 + x + x^2 + x^3 + \cdots)^3,$$

and we are looking for the coefficient of x^{30} in this expression. But how do we cube an infinite polynomial?

Let's try squaring it first:

$$(1 + x + x^2 + x^3 + \cdots)(1 + x + x^2 + x^3 + \cdots).$$

We can then just collect term by term:

$$
\begin{aligned}
1 \cdot 1 &= 1 \\
1 \cdot x + x \cdot 1 &= 2x \\
1 \cdot x^2 + x \cdot x + x^2 \cdot 1 &= 3x^2 \\
1 \cdot x^3 + x \cdot x^2 + x^2 \cdot x + x^3 \cdot 1 &= 4x^3
\end{aligned}
$$

In general, we see that the x^k term in the product is:

$$1 \cdot x^k + x \cdot x^{k-1} + x^2 \cdot x^{k-2} + \cdots + x^k \cdot 1 = (k+1)x^k.$$

Therefore,

$$(1 + x + x^2 + x^3 + \cdots)^2 = 1 + 2x + 3x^2 + 4x^3 + \cdots.$$

Now we can multiply this by the third infinite polynomial:

$$(1 + 2x + 3x^2 + 4x^3 + \cdots)(1 + x + x^2 + x^3 + \cdots).$$

As above, we collect term by term:

$$
\begin{aligned}
1 \cdot 1 &= 1 \\
1 \cdot x + 2x \cdot 1 &= (1+2)x = 3x \\
1 \cdot x^2 + 2x \cdot x + 3x^2 \cdot 1 &= (1+2+3)x^2 = 6x^2 \\
1 \cdot x^3 + 2x \cdot x^2 + 3x^2 \cdot x + 4x^3 \cdot 1 &= (1+2+3+4)x^3 = 10x^3
\end{aligned}
$$

In general, we see that the x^k term in the product is:

$$
1 \cdot x^k + 2x \cdot x^{k-1} + 3x^2 \cdot x^{k-2} + \cdots + (k+1)x^k \cdot 1 \quad = \quad (1+2+3+\cdots+(k+1))x^k = \frac{(k+1)(k+2)}{2}x^k.
$$

Therefore, the coefficient of x^k in $(1 + x + x^2 + x^3 + \cdots)^3$ is $\dfrac{(k+1)(k+2)}{2}$, and thus the answer to our original problem is $\frac{(31)(32)}{2} = 496$. $\square$

Of course, we knew from our previous work that we could find the distributions by thinking about arranging 30 integers and 2 dividers in a row, which can be done in $\binom{32}{2} = 496$ ways. But our work in the solution to Problem 14.7 shows an algebraic way to arrive at this same number, where we note that $\dfrac{(k+1)(k+2)}{2} = \dbinom{k+2}{2}$.

This nice pattern should get us thinking about whether a similar pattern holds in general.

Problem 14.8: Find an expression for the coefficient of x^k in the infinite polynomial

$$
(1 + x + x^2 + x^3 + \cdots)^n.
$$

Solution for Problem 14.8: We've already seen a nice pattern for $n = 1, 2, 3$:

$$
\begin{aligned}
(1 + x + x^2 + x^3 + \cdots)^1 &= 1 + x + x^2 + x^3 + \cdots \\
(1 + x + x^2 + x^3 + \cdots)^2 &= 1 + 2x + 3x^2 + 4x^3 + \cdots \\
(1 + x + x^2 + x^3 + \cdots)^3 &= 1 + 3x + 6x^2 + 10x^3 + \cdots
\end{aligned}
$$

Do those coefficients on the right side of the above equations look familiar? Adept counters should recognize these as binomial coefficients. So let's rewrite our expressions in terms of binomial coefficients:

$$
(1 + x + x^2 + x^3 + \cdots)^1 = \binom{0}{0} + \binom{1}{0}x + \binom{2}{0}x^2 + \binom{3}{0}x^3 + \cdots
$$

$$
(1 + x + x^2 + x^3 + \cdots)^2 = \binom{1}{1} + \binom{2}{1}x + \binom{3}{1}x^2 + \binom{4}{1}x^3 + \cdots
$$

$$
(1 + x + x^2 + x^3 + \cdots)^3 = \binom{2}{2} + \binom{3}{2}x + \binom{4}{2}x^2 + \binom{5}{2}x^3 + \cdots
$$

The pattern seems clear! We conjecture that

$$
(1 + x + x^2 + x^3 + \cdots)^n = \binom{n-1}{n-1} + \binom{n}{n-1}x + \binom{n+1}{n-1}x^2 + \binom{n+2}{n-1}x^3 + \cdots,
$$

or, written in summation notation,

$$\left(\sum_{k=0}^{\infty} x^k \right)^n = \sum_{k=0}^{\infty} \binom{n-1+k}{n-1} x^k.$$

Note that the upper bound of each summation is ∞, indicating that the sum is infinite.

Now that we know the statement that we want to prove, we can prove it using mathematical induction. We've already done the base case $n = 1$ above, so let's go straight to the inductive step.

Assume that the result is true for some positive integer n. We examine the expression for $n + 1$ by pulling out the first n sums and using our inductive hypothesis:

$$\left(\sum_{k=0}^{\infty} x^k \right)^{n+1} = \left(\left(\sum_{k=0}^{\infty} x^k \right)^n \right) \left(\sum_{k=0}^{\infty} x^k \right)$$

$$= \left(\sum_{k=0}^{\infty} \binom{n-1+k}{n-1} x^k \right) \left(\sum_{k=0}^{\infty} x^k \right).$$

Let's examine the coefficient of x^m in this last expression, for some positive integer m. An x^m term in this product comes from multiplying a term from the first sum with a term from $(1 + x + x^2 + \cdots + x^m)$ in the second sum. Therefore, the x^m coefficient of the product is the sum of the first $m + 1$ coefficients of the first sum. In other words, the product is

$$\sum_{m=0}^{\infty} \left(\sum_{i=0}^{m} \binom{n-1+i}{n-1} \right) x^m.$$

But examine that sum inside the parentheses of the last expression. It is the sum

$$\binom{n-1}{n-1} + \binom{n}{n-1} + \cdots + \binom{n+m-1}{n-1},$$

which is exactly the type of expression to which we can apply the Hockey Stick identity (from Section 12.6). By the Hockey Stick identity, we know that this expression equals $\binom{n+m}{n}$, and thus our generating function is

$$\sum_{m=0}^{\infty} \binom{n+m}{n} x^m = \sum_{m=0}^{\infty} \binom{(n+1)-1+m}{(n+1)-1} x^m,$$

completing our induction proof. $\square$

We have one more bit of algebraic manipulation up our sleeve. What's a more concise way to write

$$1 + x + x^2 + x^3 + \cdots?$$

This is a geometric series where the ratio between successive terms is x. Therefore, we can sum it:

$$(1 + x + x^2 + x^3 + \cdots) = \sum_{k=0}^{\infty} x^k = \frac{1}{1-x}.$$

Of course, as you know, we can only sum the geometric series like this if $|x| < 1$. However, we're not going to worry about the value of x. We think of the variable x as essentially being a placeholder, meaning that we don't ever actually plug anything in for x; rather, we use x as a "dummy" variable that allows us to keep track of our algebraic information. As long as the algebraic manipulation works for *some* (nonzero) values of x, we're OK. When you take a calculus course that includes the study of power series, you will learn some of the tools that can make these arguments more mathematically rigorous, but for now it is fine for you to assume that making the substitution

$$1 + x + x^2 + x^3 + \cdots = \frac{1}{1-x}$$

is legal. So, our formula from Problem 14.8 becomes:

> **Important:**
>
> $$\frac{1}{(1-x)^n} = \left(\sum_{k=0}^{\infty} x^k\right)^n = \sum_{k=0}^{\infty} \binom{n-1+k}{n-1} x^k = \sum_{k=0}^{\infty} \binom{n-1+k}{k} x^k.$$

This formula makes lots of casework-intensive distribution problems easy, as we see in the next problem.

> **Problem 14.9:** Find the number of ways to distribute \$15 to five people, if each person gets a whole number of dollars between \$0 and \$4 (inclusive).

Solution for Problem 14.9: The generating function for the amount that each person receives is $1 + x + x^2 + x^3 + x^4$, so the generating function for the total distribution is $(1 + x + x^2 + x^3 + x^4)^5$. We could multiply it out and take the x^{15} coefficient, but we can be a bit more clever by making the algebraic manipulation

$$1 + x + x^2 + x^3 + x^4 = \frac{1 - x^5}{1 - x}.$$

Now when we take the 5th power, we have

$$(1 + x + x^2 + x^3 + x^4)^5 = \frac{(1 - x^5)^5}{(1 - x)^5}.$$

We can now expand the numerator, and we know how to deal with the denominator:

$$\frac{(1 - x^5)^5}{(1 - x)^5} = (1 - 5x^5 + 10x^{10} - 10x^{15} + 5x^{20} - x^{25})\left(\binom{4}{4} + \binom{5}{4}x + \binom{6}{4}x^2 + \binom{7}{4}x^3 + \cdots\right).$$

We're trying to find the coefficient of x^{15} in this expression, and we can see that the x^{15} terms come about when the x^0, x^5, x^{10}, and x^{15} terms of the first factor are multiplied by the x^{15}, x^{10}, x^5, and x^0 terms, respectively, of the second factor. Therefore, the coefficient of the x^{15} term is:

$$\binom{19}{4} - 5\binom{14}{4} + 10\binom{9}{4} - 10\binom{4}{4} = 3876 - 5005 + 1260 - 10 = 121.$$

Thus, there are 121 ways to distribute the money. $\square$

> **Problem 14.10:** I'm doling out 100 identical pieces of candy to 5 kids. The two youngest kids want at most 1 piece. The middle kid will take any number of pieces. The two oldest kids each demand an odd number of pieces. In how many ways can I distribute the candy?

Solution for Problem 14.10: The generating function for each of the two youngest kids is $(1 + x)$. The generating function for the middle kid is $(1 + x + x^2 + x^3 + \cdots)$. The generating function for the each of the oldest kids is $(x + x^3 + x^5 + x^7 + \cdots)$. So their combined generating function is

$$(1 + x)^2(1 + x + x^2 + x^3 + \cdots)(x + x^3 + x^5 + x^7 + \cdots)^2.$$

We need to manipulate this product algebraically so that it's easier to work with.

First, note that we can factor x out of the $(x + x^3 + x^5 + x^7 + \cdots)$ terms, to get

$$x^2(1 + x)^2(1 + x + x^2 + x^3 + \cdots)(1 + x^2 + x^4 + x^6 + \cdots)^2.$$

Now we can use the fact that the last two factors are geometric series to write the function as

$$\frac{x^2(1 + x)^2}{(1 - x)(1 - x^2)^2}.$$

Now we notice that

$$\frac{1 + x}{1 - x^2} = \frac{1 + x}{(1 + x)(1 - x)} = \frac{1}{1 - x},$$

so we can simplify our generating function to

$$\frac{x^2}{(1 - x)^3}.$$

We want the coefficient of the x^{100} term of this expression, which is the same as the coefficient of the x^{98} term of $\dfrac{1}{(1 - x)^3}$, which we know to be $\dbinom{100}{2} = 4{,}950.$ $\square$

The casework solution would have been really ugly. But using generating functions bundles up the casework, so that in some sense we can handle all the cases at once using algebra. And, as we saw, often the algebra simplifies immensely!

> **Concept:** For distribution problems that seem like they might require nasty casework, consider using generating functions.

Also, even though we went through some relatively complicated algebraic manipulation, what we ended up with, $\binom{100}{2}$, was a fairly simple answer. This suggests that there might be a simple counting explanation for this simple answer, and knowing what the answer is might help us find it.

Recall that $\binom{100}{2}$ counts the number of ways to distribute 101 items into 3 nonempty piles (we have 101 items and we have to insert 2 dividers into the 100 slots in between the items). So we'll try to reformulate the problem into a problem involving 101 pieces of candy, as follows. We add 1 extra piece of candy to make it 101 total pieces, and divide the candy into 3 piles, which we can do in $\binom{100}{2}$ ways.

From the first pile, we take back our "extra" piece of candy, and give the rest to the middle kid. From the second pile, if the number of candies is odd, we give them all to the first old kid; if the number is even, we give 1 to the first young kid, and the rest to the first old kid. We do a similar thing with the third pile and the other old and young kids.

We will leave it as an Exercise to show that this procedure gives a 1-1 correspondence

$$\left\{\begin{matrix}\text{distributions of 101 candies}\\ \text{into 3 nonempty piles}\end{matrix}\right\} \quad\leftrightarrow\quad \left\{\begin{matrix}\text{distributions of 100 candies as}\\ \text{in Problem 14.10}\end{matrix}\right\}.$$

Problem 14.11: Chelsea, Barbara, and Jenna go to dinner, and have to pay a $50 bill. Chelsea insists on contributing a number of dollars that is a multiple of 3, whereas Barbara and Jenna each insist on contributing an odd number of dollars. In how many ways can they pay the bill?

Solution for Problem 14.11: Chelsea's generating function is

$$1 + x^3 + x^6 + x^9 + \cdots = \frac{1}{1 - x^3},$$

and Barbara and Jenna each have a generating function of

$$x + x^3 + x^5 + x^7 + \cdots = \frac{x}{1 - x^2}.$$

Therefore, the combined generating function is

$$\frac{x^2}{(1 - x^3)(1 - x^2)^2},$$

and we are seeking the coefficient of the x^{50} term of this function, which is the x^{48} coefficient of

$$\frac{1}{(1 - x^3)(1 - x^2)^2}.$$

Unfortunately, there doesn't seem to be any further algebraic manipulation that we can use to give a quick answer. We have to resort to casework. We can see that Chelsea must in fact contribute a multiple of 6, because Barbara's and Jenna's contributions will sum to an even number. Therefore, the only terms of $\frac{1}{1-x^3}$ that we need be concerned with are $(1 + x^6 + x^{12} + x^{18} + \cdots)$. If we let c_n denote the coefficient of x^n in $\frac{1}{(1-x^2)^2}$, then we seek the x^{48} coefficient of

$$(1 + x^6 + x^{12} + x^{18} + \cdots)(c_0 + c_2 x^2 + c_4 x^4 + \cdots).$$

Thus our answer is

$$c_{48} + c_{42} + c_{36} + \cdots + c_6 + c_0.$$

But we know that

$$\frac{1}{(1 - x^2)^2} = 1 + 2x^2 + 3x^4 + 4x^6 + \cdots,$$

so that if n is even, then $c_n = \frac{n}{2} + 1$. Thus, our answer is

$$25 + 22 + 19 + \cdots + 4 + 1 = 117.$$

Since our answer of 117 doesn't seem to satisfy any nice pattern, and since there seems to be no good way to simplify the above sum, we can be relatively sure that we've found the simplest possible solution. □

14.4.1   Find an expression for the coefficient of x^k in the generating function

$$\frac{1}{(1 - x^m)^n}.$$

14.4.2   Change Problem 14.10 to: we have 100 pieces of candy, two kids who will take only an odd number of pieces of candy, one who will take any number, one who will take 0 or 1, and one who will take 0 or 5. Now how many ways are there for us to distribute the candy?

14.4.3   Prove the 1-1 correspondence shown on page 313 after the solution to Problem 14.10. **Hints:** 149

14.4.4   In how many different ways can I collect a total of 20 dollars from 4 different children and 3 different adults if each child can contribute up to 6 dollars and each adult can give up to 10 dollars (and we assume that each individual gives a nonnegative whole number of dollars)? **Hints:** 161

14.4.5   Tina randomly selects two distinct numbers from the set $\{1, 2, 3, 4, 5\}$ and Sergio randomly selects a number from the set $\{1, 2, \ldots, 10\}$. What is the probability that Sergio's number is larger than the sum of the two numbers chosen by Tina? *(Source: AMC)* **Hints:** 295

14.5   The Generating Function for Partitions

Recall the following definition from Chapter 4:

> **Definition:**   A **partition** of a positive integer n is a decomposition of n into a sum of positive integers, not all necessarily distinct, without regard to the order in which we list the integers in the sum.

For example, 3, $1 + 2$, and $1 + 1 + 1$ are the partitions of 3. We don't consider $1 + 2$ and $2 + 1$ to be different partitions, since we don't care in what order we list the integers that we're summing. And note that we're allowed to repeat integers (so that $1 + 1 + 1$ is indeed a valid partition of 3).

Problems

> **Problem 14.12:**
> (a)   List the partitions of 4.
>
> (b)   List the partitions of 5.
>
> (c)   List the partitions of 6.

Problem 14.13: Let p_n be the number of partitions of n. (For example, $p_3 = 3$ and $p_4 = 5$.) Determine the generating function

$$f(x) = p_0 + p_1 x + p_2 x^2 + \cdots$$

for counting the number of partitions of any positive integer. (By convention, we set $p_0 = 1$.)

The problems that we've done with generating functions so far were problems that we already knew how to do without generating functions. True, generating functions gave us a nice way to keep track of the information that we needed to solve the problems, but we could have solved the problems without generating functions, using good old-fashioned counting techniques.

Now, we'll see some more powerful uses of generating functions in problems that would be fairly difficult to solve via other means. In particular, generating functions give us nice solutions to hard problems involving partitions. Before trying to determine how to use generating functions with partitions, let's work through a few more small examples, so that we have some data to play with.

Problem 14.12:

(a) List the partitions of 4.

(b) List the partitions of 5.

(c) List the partitions of 6.

Solution for Problem 14.12:

(a) The partitions of 4 are:

$$4, \quad 3 + 1, \quad 2 + 2, \quad 2 + 1 + 1, \quad 1 + 1 + 1 + 1.$$

There are 5 of them.

(b) The partitions of 5 are:

$$5, \quad 4 + 1, \quad 3 + 2, \quad 3 + 1 + 1, \quad 2 + 2 + 1, \quad 2 + 1 + 1 + 1, \quad 1 + 1 + 1 + 1 + 1.$$

There are 7 of them.

(c) The partitions of 6 are:

$$6, \quad 5 + 1, \quad 4 + 2, \quad 4 + 1 + 1, \quad 3 + 3, \quad 3 + 2 + 1, \quad 3 + 1 + 1 + 1,$$
$$2 + 2 + 2, \quad 2 + 2 + 1 + 1, \quad 2 + 1 + 1 + 1 + 1, \quad 1 + 1 + 1 + 1 + 1 + 1.$$

There are 11 of them. □

Problem 14.13: Let p_n be the number of partitions of n. (For example, $p_3 = 3$ and $p_4 = 5$.) Determine the generating function

$$f(x) = p_0 + p_1 x + p_2 x^2 + \cdots$$

for counting the number of partitions of any positive integer. (By convention, we set $p_0 = 1$.)

Solution for Problem 14.13: As we saw in Problem 14.12, the first few terms of this generating function will be

$$1 + x + 2x^2 + 3x^3 + 5x^4 + 7x^5 + 11x^6 + \cdots.$$

There doesn't seem to be any nice pattern to these coefficients.

Let's think about how we might construct the various partitions in terms of polynomials. For example, let's again list the partitions of 5:

$$5, \quad 4+1, \quad 3+2, \quad 3+1+1, \quad 2+2+1, \quad 2+1+1+1, \quad 1+1+1+1+1.$$

We can write each of these as monomials that multiply to give x^5:

$$x^5, \quad x^4 \cdot x, \quad x^3 \cdot x^2, \quad x^3 \cdot x \cdot x, \quad x^2 \cdot x^2 \cdot x, \quad x^2 \cdot x \cdot x \cdot x, \quad x \cdot x \cdot x \cdot x \cdot x.$$

Now let's group equal monomials:

$$x^5, \quad x^4 \cdot x, \quad x^3 \cdot x^2, \quad x^3 \cdot (x)^2, \quad (x^2)^2 \cdot x, \quad x^2 \cdot (x)^3, \quad (x)^5.$$

More generally, we can think of a partition of n as a choice of exponents $a_1, a_2, a_3, \ldots$ such that

$$(x)^{a_1} \cdot (x^2)^{a_2} \cdot (x^3)^{a_3} \cdots = x^n.$$

This choice corresponds to the partition of n consisting of a_1 1's, a_2 2's, a_3 3's, and so on, since we have that

$$a_1 + 2a_2 + 3a_3 + \cdots = n.$$

We can very cleverly encapsulate all this information for all n at once. Consider the power series

$$(1 + x + x^2 + x^3 + \cdots)(1 + x^2 + x^4 + x^6 + \cdots)(1 + x^3 + x^6 + x^9 + \cdots) \cdots.$$

Every x^n term in the product will come from choosing an $(x)^{a_1}$ term from the first infinite polynomial, an $(x^2)^{a_2}$ term from the second infinite polynomial, and so on, such that

$$(x)^{a_1} \cdot (x^2)^{a_2} \cdot (x^3)^{a_3} \cdots = x^n.$$

The number of ways that we can do this is exactly the number of ways that n can be broken up into a sum of a_1 1's, a_2 2's, a_3 3's, and so on. But that's exactly the number of partitions of n.

So we have the generating function that we want:

$$f(x) = (1 + x + x^2 + x^3 + \cdots)(1 + x^2 + x^4 + x^6 + \cdots)(1 + x^3 + x^6 + x^9 + \cdots) \cdots.$$

We can simplify the way that we write $f(x)$ by using the fact that each term is a geometric series. Doing this, we see that

$$f(x) = \frac{1}{(1-x)(1-x^2)(1-x^3)\cdots} = \prod_{k=1}^{\infty} \frac{1}{(1-x^k)}.$$

The notation $\prod$ is a shorthand for infinite products; it functions in exactly the same way for products as $\sum$ does for sums. $\square$

Now that we have the generating function for partitions, try to solve the next few problems using the generating function.

Problems

Problem 14.14: Show that the number of partitions of n that have no number repeated equals the number of partitions of n that have only odd numbers. For example, if $n = 4$, then the partitions with no number repeated are 4 and $3 + 1$, and the partitions with only odd numbers are $3 + 1$ and $1 + 1 + 1 + 1$.

Problem 14.15: Show that the number of partitions of n in which no part appears exactly once is equal to the number of partitions in which no part is 1 greater or 1 less than a multiple of 6. For example, if $n = 7$, then the partitions with no part appearing exactly once are $1 + 1 + 1 + 2 + 2$ and $1 + 1 + 1 + 1 + 1 + 1 + 1$, and the partitions with no part 1 greater or 1 less than a multiple of 6 are $2 + 2 + 3$ and $3 + 4$.

Problem 14.16: For any partition Π of n, let

$$f(\Pi) = \text{the number of 1's in } \Pi$$

and

$$g(\Pi) = \text{the number of distinct numbers in } \Pi.$$

(a) List the partitions of 5.

(b) Compute $f(\Pi)$ and $g(\Pi)$ for each partition Π of 5.

(c) Add all the $f(\Pi)$'s and all the $g(\Pi)$'s from part (b). What do you observe?

(d) For any n, let $s(n)$ be the sum of the $f(\Pi)$ over all partitions Π of n, and let $t(n)$ be the sum of the $g(\Pi)$ over all partitions Π of n. Prove that $s(n) = t(n)$ for all n.
(Source: USAMO)

Problem 14.14: Show that the number of partitions of n that have no number repeated equals the number of partitions of n that have only odd numbers.

Solution for Problem 14.14: Since we are trying to prove that two sets are the same size, we might try to look for a 1-1 correspondence. However, after a little experimentation, you'll probably see that an obvious 1-1 correspondence proves elusive.

So instead we'll try another tack. Let's determine the generating functions for both sets and show that they're the same.

Concept: We can show that two sequences are equal by showing that their generating functions are the same.

Recall that the generating function for all partitions is

$$(1 + x + x^2 + x^3 + \cdots)(1 + x^2 + x^4 + x^6 + \cdots)(1 + x^3 + x^6 + x^9 + \cdots)\cdots,$$

where the monomial we choose from each term corresponds to the number of times that number appears in the partition. For example, choosing the monomial $x^6 = (x^2)^3$ from the second term means that the corresponding partition will contain three 2's.

For the first set in our problem—partitions with no number repeated—we're only allowed to choose 0 or 1 ones, 0 or 1 twos, 0 or 1 threes, etc., to make up a partition. Therefore, the generating function is

$$(1 + x)(1 + x^2)(1 + x^3)\cdots.$$

In other words, each term only lets us choose 0 or 1 of each number to be in our partition.

For the second set in our problem—partitions that have only odd numbers—we're not allowed to have any 2's, 4's, 6's, etc. So the generating function is

$$(1 + x + x^2 + x^3 + \cdots)(1 + x^3 + x^6 + x^9 + \cdots)(1 + x^5 + x^{10} + x^{15} + \cdots)\cdots.$$

We've eliminated the polynomials that correspond to even numbers.

In order to prove that the numbers of partitions in each set are the same, we must show that these generating functions are the same; that is, we must show that:

$$(1 + x)(1 + x^2)(1 + x^3)\cdots = (1 + x + x^2 + x^3 + \cdots)(1 + x^3 + x^6 + x^9 + \cdots)(1 + x^5 + x^{10} + x^{15} + \cdots)\cdots.$$

The right side of the above will look a lot nicer if we sum the geometric series, so that what we want to show is:

$$(1 + x)(1 + x^2)(1 + x^3)\cdots = \frac{1}{(1 - x)(1 - x^3)(1 - x^5)\cdots}.$$

Are these equal?

All of the terms on the RHS are of the form $1/(1 - x^k)$ for some k. So let's write the LHS in that form as well, by replacing $1 + x$ by $\frac{1-x^2}{1-x}$, and similarly for all the other terms on the LHS:

$$\left(\frac{1 - x^2}{1 - x}\right)\left(\frac{1 - x^4}{1 - x^2}\right)\left(\frac{1 - x^6}{1 - x^3}\right)\cdots.$$

Notice that all of the numerator terms are of the form $1 - x^k$ for even k. These will all cancel out with denominator terms, and leave us in the denominator with all of the terms of the form $1 - x^l$ for odd l. But that's what we have on the RHS!

Therefore, the generating functions are equal, proving the assertion. $\square$

WARNING!!
☢

We've played a bit fast and loose with some of the algebra here. We need to be a bit more careful when manipulating infinite polynomials like this. When you study calculus, you will learn more about the formal theory of power series (infinite polynomials), and you will see that the things that we are doing are in fact legal. But for now, you'll have to accept our word for it.

Problem 14.15: Show that the number of partitions of n in which no part appears exactly once is equal to the number of partitions in which no part is 1 greater or 1 less than a multiple of 6. For example, if $n = 7$, then the partitions with no part appearing exactly once are $1 + 1 + 1 + 2 + 2$ and $1 + 1 + 1 + 1 + 1 + 1 + 1$, and the partitions with no part 1 greater or 1 less than a multiple of 6 are $2 + 2 + 3$ and $3 + 4$.

Solution for Problem 14.15: Once again, we will try to write the generating functions for both sets of partitions, and show that they are equal.

For the first type of partition—those with no part appearing exactly once—we may have 0, 2, 3, 4, etc. of each number. Therefore, the generating function is

$$(1 + x^2 + x^3 + x^4 + \cdots)(1 + x^4 + x^6 + x^8 + \cdots)(1 + x^6 + x^9 + x^{12} + \cdots) \cdots .$$

Note how the x is missing from the first factor, the x^2 is missing from the second factor, and so on, because those are the terms that correspond to exactly one of that number appearing in the partition. We can simplify the geometric series in the above expression to

$$\left(1 + \frac{x^2}{1 - x}\right)\left(1 + \frac{x^4}{1 - x^2}\right)\left(1 + \frac{x^6}{1 - x^3}\right) \cdots . \tag{$*$}$$

For the second type of partition, we simply omit the factors of the generating function that correspond to numbers that are 1 more or less than a multiple of 6. So we only have the factors corresponding to 2, 3, 4, 6, 8, 9, ..., and the generating function is

$$\frac{1}{1 - x^2} \cdot \frac{1}{1 - x^3} \cdot \frac{1}{1 - x^4} \cdot \frac{1}{1 - x^6} \cdot \frac{1}{1 - x^8} \cdots . \tag{$**$}$$

We need to show that these two generating functions are the same.

The factors of $(*)$ can be simplified using some clever algebra. For example, consider the first factor of $(*)$:

$$1 + \frac{x^2}{1 - x} = \frac{1 - x + x^2}{1 - x}$$
$$= \frac{(1 - x + x^2)(1 + x)}{(1 - x)(1 + x)}$$
$$= \frac{1 + x^3}{1 - x^2}.$$

So the first generating function $(*)$ can be rewritten as

$$\left(\frac{1 + x^3}{1 - x^2}\right)\left(\frac{1 + x^6}{1 - x^4}\right)\left(\frac{1 + x^9}{1 - x^6}\right)\left(\frac{1 + x^{12}}{1 - x^8}\right) \cdots .$$

Notice that the $1 + x^3$ term in the numerator will cancel with a factor of the $1 - x^6$ term in the denominator, leaving a $1 - x^3$ term in the denominator. More generally, all of the terms in the numerator are of the form $1 + x^{3k}$ for some positive integer k, and they will cancel with a factor of the $1 - x^{6k}$ term in the denominator to leave a $1 - x^{3k}$ term in the denominator.

When we have done all the canceling, we will be left with 1 in the numerator, and with all terms in the denominator of the form $1 - x^l$ for l even and $1 - x^{3k}$ for k odd. These are exactly the terms of the form $1 - x^m$ where m is not 1 more or less than a multiple of 6 (since all numbers that are 1 more or less than a multiple of 6 are odd but not a multiple of 3). This is exactly the generating function in (**), and thus the two generating functions are the same. □

The next problem involves a fairly sophisticated generating functions argument.

Problem 14.16: For any partition Π of n, let

$$f(\Pi) = \text{the number of 1's in } \Pi$$

and

$$g(\Pi) = \text{the number of distinct numbers in } \Pi.$$

Further, let $s(n)$ be the sum of the $f(\Pi)$ over all partitions Π of n, and let $t(n)$ be the sum of the $g(\Pi)$ over all partitions Π of n. Prove that $s(n) = t(n)$ for all n. (Source: USAMO)

Solution for Problem 14.16: This one is a bit more complicated than previous problems that we've looked at, so let's work through an example so that we're sure what's going on.

Suppose that $n = 5$. We'll list all the partitions of 5 in a table, together with $f(\Pi)$ and $g(\Pi)$ for each partition Π. Note that $f(\Pi)$ is simply the number of 1's in Π, and that $g(\Pi)$ is the number of *distinct* numbers in Π (so that, for example, $g(3 + 1 + 1)$ is 2, not 3, since there are two distinct numbers: 1 and 3).

Π	$f(\Pi)$	$g(\Pi)$
5	0	1
$4 + 1$	1	2
$3 + 2$	0	2
$3 + 1 + 1$	2	2
$2 + 2 + 1$	1	2
$2 + 1 + 1 + 1$	3	2
$1 + 1 + 1 + 1 + 1$	5	1

Now we see that $s(5)$, which is the sum of all the $f(\Pi)$'s in the table, is $0 + 1 + 0 + 2 + 1 + 3 + 5 = 12$, and that $t(5)$, which is the sum of all the $g(\Pi)$'s in the table, is $1 + 2 + 2 + 2 + 2 + 1 = 12$. So, indeed, $s(5) = t(5) = 12$.

Looking at the table, there seems to be no easy way to construct a 1-1 correspondence; indeed, it's not even clear what the sets are that we would try to put into correspondence. So we'll instead try the same tactic that we've used in the previous two problems: create a generating function for $s(n)$, create a generating function for $t(n)$, and show that they're the same.

Let's start with $s(n)$. This is the sum of all the 1's over all the partitions of n. We need to find a good way to organize this sum. One logical idea would be to organize the partitions of n based on the number of 1's in them. For example, going back to our $n = 5$ case, we would group the partitions as shown below:

Number of ones	0	1	2	3	4	5
Partitions	5	4 + 1	3 + 1 + 1	2 + 1 + 1 + 1		1 + 1 + 1 + 1 + 1
	3 + 2	2 + 2 + 1				

So we can say that $s(n)$ is

$$1(\text{\# of partitions with 1 one})$$
$$+ \quad 2(\text{\# of partitions with 2 ones})$$
$$+ \quad 3(\text{\# of partitions with 3 ones})$$
$$+ \quad \vdots$$
$$+ \quad n(\text{\# of partitions with } n \text{ ones}).$$

But what happens to a partition of n with k ones if we delete the ones? We get a partition of $n - k$ that doesn't have *any* ones. More precisely, there is an obvious 1-1 correspondence:

$$\{\text{partitions of } n \text{ with } k \text{ ones}\} \quad \leftrightarrow \quad \{\text{partitions of } n - k \text{ with no ones}\}$$

Therefore, we can rewrite $s(n)$ as:

$$1(\text{\# of partitions of } n - 1 \text{ with no ones})$$
$$+ \quad 2(\text{\# of partitions of } n - 2 \text{ with no ones})$$
$$+ \quad 3(\text{\# of partitions of } n - 3 \text{ with no ones})$$
$$+ \quad \vdots$$
$$+ \quad n(\text{\# of partitions of zero with no ones}).$$

Recall that, by convention, we say that zero has one partition (namely the "empty" partition), so that the last term in the above sum makes sense.

Now we can try to come up with a generating function. If we want the generating functions for partitions without any ones, we simply eliminate the first factor $(1 + x + x^2 + x^3 + \cdots)$ from the partition generating function, since this is the term that counts the number of 1's in any given partition. Therefore, the generating function that counts partitions without any ones is

$$M(x) = (1 + x^2 + x^4 + x^6 + \cdots)(1 + x^3 + x^6 + x^9 + \cdots)\cdots = \frac{1}{(1 - x^2)(1 - x^3)\cdots} = \prod_{k=2}^{\infty} \frac{1}{1 - x^k}.$$

We've given the function the name $M(x)$ so that we can conveniently refer to it later. Just to clarify, remember that the coefficient of x^k in $M(x)$ is the number of partitions of k without any ones.

But $M(x)$ is not exactly what we want for $s(n)$. Recall that $s(n)$ is equal to

$$1(\text{\# of partitions of } n - 1 \text{ with no ones})$$
$$+ \quad 2(\text{\# of partitions of } n - 2 \text{ with no ones})$$
$$+ \quad 3(\text{\# of partitions of } n - 3 \text{ with no ones})$$
$$+ \quad \vdots$$
$$+ \quad n(\text{\# of partitions of zero with no ones}).$$

Using $M(x)$, we can rewrite this as

$$
\begin{aligned}
& 1(\text{coefficient of } x^{n-1} \text{ in } M(x)) \\
+\ & 2(\text{coefficient of } x^{n-2} \text{ in } M(x)) \\
+\ & 3(\text{coefficient of } x^{n-3} \text{ in } M(x)) \\
+\ & \vdots \\
+\ & n(\text{coefficient of } x^{0} \text{ in } M(x)).
\end{aligned}
$$

If we let c_k denote the coefficient of x^k in $M(x)$, then

$$s(n) = 1c_{n-1} + 2c_{n-2} + 3c_{n-3} + \cdots + nc_0.$$

How can we get this to be the coefficient of x^n in some function? To do this, we would need to multiply $M(x)$ by something that has coefficient 1 for x, so that the $c_{n-1}x^{n-1}$ term in $M(x)$ gets multiplied by $1x$ to give a $1c_{n-1}x^n$ term in our new function. Similarly, we want to multiply $M(x)$ by something with coefficient 2 for x^2, coefficient 3 for x^3, and so on. In particular, we want

$$S(x) = (x + 2x^2 + 3x^3 + 4x^4 + \cdots)M(x).$$

Now $s(n)$ is exactly the coefficient of x^n in $S(x)$. Finally, notice that

$$(x + 2x^2 + 3x^3 + 4x^4 + \cdots) = x(1 + 2x + 3x^2 + 4x^3 + \cdots) = \frac{x}{(1-x)^2}.$$

Therefore, our generating function for $s(n)$ is

$$S(x) = \frac{x}{(1-x)^2(1-x^2)(1-x^3)\cdots} = \frac{x}{1-x}\prod_{k=1}^{\infty}\frac{1}{1-x^k}.$$

So far, so good. Now let's turn our attention to $t(n)$.

As with $s(n)$, we'd like to organize the partitions of n in a way that lets us count up the distinct numbers in a more manageable fashion. Let's again go back to our example with $n = 5$:

Π	$g(\Pi)$
5	1
4 + 1	2
3 + 2	2
3 + 1 + 1	2
2 + 2 + 1	2
2 + 1 + 1 + 1	2
1 + 1 + 1 + 1 + 1	1

Recall that $g(\Pi)$ is the number of distinct numbers in Π, and that $t(n)$ is the sum of the $g(\Pi)$'s over all partitions Π of n. Rather than just counting the distinct numbers in each partition, let's add a column to our table, listing the distinct numbers that appear in each partition:

Π	$g(\Pi)$	Numbers
5	1	5
$4+1$	2	1,4
$3+2$	2	2,3
$3+1+1$	2	1,3
$2+2+1$	2	1,2
$2+1+1+1$	2	1,2
$1+1+1+1+1$	1	1

Does this give us any ideas?

Instead of counting how many distinct numbers are in each partition, let's do the reverse, and count how many partitions contain each number. In other words, let's rearrange the table:

Number	Partitions containing number	Number of partitions
1	$4+1, 3+1+1, 2+2+1, 2+1+1+1, 1+1+1+1+1$	5
2	$3+2, 2+2+1, 2+1+1+1$	3
3	$3+2, 3+1+1$	2
4	$4+1$	1
5	5	1

Note that $t(n) = 5 + 3 + 2 + 1 + 1 = 12$ using the above table. Make sure that you see why this way of counting also counts $t(n)$: any given partition with k distinct numbers will appear in k rows of the above table, and thus will be counted k times.

So we can write $t(n)$ as

$$\begin{aligned}
&(\text{\# of partitions of } n \text{ containing a } 1) \\
+\ &(\text{\# of partitions of } n \text{ containing a } 2) \\
+\ &(\text{\# of partitions of } n \text{ containing a } 3) \\
+\ &\ \vdots \\
+\ &(\text{\# of partitions of } n \text{ containing an } n).
\end{aligned}$$

But if a partition of n contains a k, and we remove the k, what are we left with? We're left with a partition of $n - k$. Therefore, there is a 1-1 correspondence:

$$\{\text{partitions of } n \text{ containing a } k\} \quad \leftrightarrow \quad \{\text{partitions of } n - k\}$$

So we can rewrite $t(n)$ as

$$\begin{aligned}
&(\text{\# of partitions of } n - 1) \\
+\ &(\text{\# of partitions of } n - 2) \\
+\ &(\text{\# of partitions of } n - 3) \\
+\ &\ \vdots \\
+\ &(\text{\# of partitions of } 0).
\end{aligned}$$

As we know, the generating function for the number of partitions is

$$P(x) = \frac{1}{(1-x)(1-x^2)(1-x^3)\cdots} = \prod_{k=1}^{\infty} \frac{1}{1-x^k}.$$

So the coefficient of x^n in the generating function for $t(n)$ is the sum of the coefficients of $1, x, x^2, \ldots, x^{n-1}$ in $P(x)$. We can get a generating function with these coefficients by multiplying $P(x)$ by $(x + x^2 + x^3 + x^4 + \cdots)$, since if we write

$$P(x) = p_0 + p_1 x + p_2 x^2 + p_3 x^3 + \cdots,$$

then we have

$$(x + x^2 + x^3 + x^4 + \cdots)(p_0 + p_1 x + p_2 x^2 + \cdots) = p_0 x + (p_0 + p_1)x^2 + (p_0 + p_1 + p_2)x^3 + \cdots + \left(\sum_{i=0}^{n-1} p_i \right) x^n + \cdots.$$

Therefore, the generating function for $t(n)$ is

$$T(x) = (x + x^2 + x^3 + \cdots)P(x).$$

But the first term simplifies as

$$x + x^2 + x^3 + \cdots = x(1 + x + x^2 + x^3 + \cdots) = \frac{x}{1-x},$$

so we can rewrite $T(x)$ as

$$T(x) = \frac{x}{1-x}P(x) = \frac{x}{(1-x)(1-x)(1-x^2)(1-x^3)\cdots} = \frac{x}{1-x}\prod_{k=1}^{\infty}\frac{1}{1-x^k}.$$

Compare this with the generating function $S(x)$ we found before. They're the same! We see that $S(x) = T(x)$, so their coefficients are all equal, and in particular $s(n) = t(n)$ for all n. $\square$

Although this problem was difficult, once we decided to use generating functions, each individual step was a relatively straightforward algebraic manipulation. Writing the generating function was a matter of systematically converting the English-language problem statement into the appropriate algebraic objects. The final step was comparing the generating functions.

> **Concept:** If we want to show that the number of ways to count two different things, each of which depends on a positive integer n, is the same for all n, then often a convenient way to do this is to show that the generating functions for the two different things are the same.

Exercises

14.5.1 Write the generating function for the number of ways that we can form n cents using pennies (worth 1 cent), nickels (worth 5 cents), and dimes (worth 10 cents).

14.5.2 Write the generating function for the number of partitions of n into even parts.

14.5.3 Show that the number of partitions of n in which no part appears more than twice is equal to the number of partitions of n in which no part is a multiple of 3. **Hints:** 217

14.5.4★ The function

$$f(x) = \frac{1}{1 - x - x^2 - x^3 - x^4 - x^5 - x^6}$$

is the generating function for what sequence of numbers? In other words, what does the coefficient of x^n of $f(x)$ count? **Hints:** 203, 125

14.5.5★ Find the generating function for the number of noncongruent triangles with integral side lengths and perimeter n. (You'll need to use Problem 4.19.) **Hints:** 54, 341, 215, 346

14.6★ The Generating Function for the Fibonacci Numbers

Problems

Problem 14.17: Let
$$f(x) = x + x^2 + 2x^3 + 3x^4 + 5x^5 + 8x^6 + \cdots$$
be the generating function for the Fibonacci numbers, so that the coefficient of x^n is the nth Fibonacci number F_n. Prove that
$$f(x) = \frac{x}{1 - x - x^2}.$$

Problem 14.18: Use the generating function from the previous problem to find a closed-form formula for the nth Fibonacci number.

We can pretty easily write down an expression for the generating function for the Fibonacci numbers, as we see in the next problem.

Problem 14.17: Let
$$f(x) = x + x^2 + 2x^3 + 3x^4 + 5x^5 + 8x^6 + \cdots$$
be the generating function for the Fibonacci numbers, so that the coefficient of x^n is the nth Fibonacci number F_n. Prove that
$$f(x) = \frac{x}{1 - x - x^2}.$$

Solution for Problem 14.17: We want to check that
$$(x + x^2 + 2x^3 + 3x^4 + 5x^5 + 8x^6 + \cdots) = \frac{x}{1 - x - x^2}.$$

So let's multiply it out. Define
$$g(x) = (1 - x - x^2)(x + x^2 + 2x^3 + 3x^4 + 5x^5 + 8x^6 + \cdots).$$

Our goal is to show that $g(x) = x$.

We can immediately see that $g(x)$ has no constant term, that the coefficient of the x term of $g(x)$ will be 1, and that the x^2 term of $g(x)$ will have coefficient 0. For all higher degree terms, note that the x^k term of $g(x)$ (for $k > 2$) is
$$1(F_k x^k) - x(F_{k-1}x^{k-1}) - x^2(F_{k-2}x^{k-2}) = (F_k - F_{k-1} - F_{k-2})x^k.$$

But, by definition, $F_k = F_{k-1} + F_{k-2}$, so the above coefficient is zero! Hence, we have shown that

$$g(x) = (1 - x - x^2)(x + x^2 + 2x^3 + 3x^4 + 5x^5 + 8x^6 + \cdots) = x,$$

and thus the generating function for the Fibonacci numbers is $x/(1 - x - x^2)$. $\square$

This generating function may seem like just an amusing curiosity, but it does have some uses; in particular, it gives us another way to find the closed-form formula for the Fibonacci numbers.

> **Problem 14.18:** Use the generating function from the previous problem to find a closed-form formula for the nth Fibonacci number.

Solution for Problem 14.18: We really don't know how to deal with a quadratic term in the denominator. But we do have some confidence with our ability to deal with linear terms. So let's start by factoring the denominator into linear factors. By the quadratic formula, the roots of $x^2 + x - 1$ are

$$\frac{-1 \pm \sqrt{5}}{2}.$$

Let's call these roots $\phi_1 = (-1 + \sqrt{5})/2$ and $\phi_2 = (-1 - \sqrt{5})/2$. Then we can rewrite our generating function as

$$\frac{x}{1 - x - x^2} = \frac{-x}{x^2 + x - 1} = \frac{-x}{(\phi_1 - x)(\phi_2 - x)}.$$

(Note that the numerator is now $-x$, so that the sign of the expression is correct.)

We're still not terribly happy with the way this looks. What we'd really like is to break up this function into the form

$$x\left(\frac{A}{\phi_1 - x} + \frac{B}{\phi_2 - x}\right),$$

where A and B are some constants. This is called a **partial fraction decomposition**. To determine what A and B should be, we place the sum of the fractions over a common denominator:

$$\frac{A}{\phi_1 - x} + \frac{B}{\phi_2 - x} = \frac{A(\phi_2 - x) + B(\phi_1 - x)}{(\phi_1 - x)(\phi_2 - x)}.$$

Since we want this to match our original generating function, we must have

$$A(\phi_2 - x) + B(\phi_1 - x) = -1.$$

This gives us a system of linear equations in A and B:

$$\phi_2 A + \phi_1 B = -1,$$
$$-A - B = 0.$$

By the second equation, we see that $B = -A$, and plugging this into the first equation gives us

$$(\phi_2 - \phi_1)A = -1,$$

So $A = 1/(\phi_1 - \phi_2)$. Remembering the definitions of ϕ_1 and ϕ_2, we see that $\phi_1 - \phi_2 = \sqrt{5}$, so $A = 1/\sqrt{5}$.

Therefore, our generating function for the Fibonacci numbers can be written as

$$\frac{x}{\sqrt{5}}\left(\frac{1}{\phi_1 - x} - \frac{1}{\phi_2 - x}\right).$$

For any constant α, note that

$$\frac{1}{\alpha - x} = \frac{1/\alpha}{1 - x/\alpha}$$

$$= \frac{1}{\alpha}\left(1 + \frac{x}{\alpha} + \frac{x^2}{\alpha^2} + \cdots\right)$$

$$= \frac{1}{\alpha} + \frac{1}{\alpha^2}x + \frac{1}{\alpha^3}x^2 + \cdots.$$

In particular, the coefficient of x^n in $1/(\alpha - x)$ is $1/\alpha^{n+1}$. Therefore, the coefficient of x^n in our generating function, which is the nth Fibonacci number, is

$$\frac{1}{\sqrt{5}}\left(\left(\frac{1}{\phi_1}\right)^n - \left(\frac{1}{\phi_2}\right)^n\right).$$

With a little algebraic manipulation, we can simplify this to

$$\frac{1}{\sqrt{5}}\left(\left(\frac{1 + \sqrt{5}}{2}\right)^n - \left(\frac{1 - \sqrt{5}}{2}\right)^n\right),$$

which is exactly Binet's Formula (as we first saw in Section 9.4). □

14.7 Summary

➤ Generating functions allow us to use algebraic techniques to solve a variety of counting problems.

➤ A generating function is an polynomial of the form

$$f(x) = a_0 + a_1 x + a_2 x^2 + \cdots.$$

We think of the coefficient a_k of x^k as counting something that depends on a parameter k.

➤ For simple problems, generating functions give us a convenient way of keeping track of a lot of information at once. Another way to think of this is that generating functions allow us to do casework by doing all of the cases at once.

➤ Often the key step in using generating functions is finding the most efficient algebraic way to compute the coefficient(s) that you need.

➤ One of the most commonly-used generating functions is for distributions of indistinguishable items. This function is $\dfrac{1}{(1 - x)^n}$, and the coefficient of x^k in $\dfrac{1}{(1 - x)^n}$ is $\dbinom{n - 1 + k}{k}$, which counts the number of ways to distribute k items into n boxes (where some box(es) may remain empty).

➤ Many distribution problems with ugly-looking casework can be solved much more simply by using generating functions.

➤ Generating functions are very valuable for problems involving partitions. The generating function for the number of partitions of a positive integer n is

$$(1 + x + x^2 + \cdots)(1 + x^2 + x^4 + \cdots)(1 + x^3 + x^6 + \cdots)\cdots = \frac{1}{(1-x)(1-x^2)(1-x^3)\cdots} = \prod_{k=1}^{\infty} \frac{1}{1-x^k}.$$

➤ The generating function for the Fibonacci numbers is

$$f(x) = \frac{x}{1 - x - x^2},$$

and we can use this as another method for proving Binet's Formula.

REVIEW PROBLEMS

14.19 Find the coefficient of x^{10} in the expansion of each of the following:

(a) $\dfrac{1}{1-x}$

(b) $\dfrac{1}{1-x^2}$

(c) $\dfrac{1}{(1-x)^2}$

(d) $(1 + x^5)^5$

(e) $(1 + x + x^2 + x^3 + x^4 + x^5 + x^6)^6$

(f) $\dfrac{1}{(1-x^2)^3}$

14.20 Find the number of ways to place 25 people into three rooms with at least 1 person in each room. (Solve using generating functions.)

14.21 How many ways are there to distribute eight identical toys among four children if the first child gets at least two toys? (Solve using generating functions.)

14.22 Find the generating function for the number of ways to give n cents change using U.S. pennies, Canadian pennies, British pennies, U.S. nickels, and Canadian nickels. (Assume each "penny" is worth 1 cent and each "nickel" is worth 5 cents, regardless of the country of origin.)

14.23 Three of my friends and I are going to split the bill for dinner. Sara and I will each contribute an odd number of dollars, while Krishna contributes a number of dollars that is a multiple of 3. Shyster will either contribute nothing, or steal one or two dollars. In how many ways can we pay a $30 bill?

14.24 I have a bowl with 31 candies. Ten of the candies are indistinguishable and the other 21 are all different (and different from the first 10). In how many ways can I choose 10 of the candies?

14.25 Superman has super-strength and can carry any number of boulders, but insists on carrying an odd number. Batman can carry up to 40 boulders. Mighty Mouse can only carry up to 2 boulders. Batman or Mighty Mouse might go empty-handed.

(a) How many ways can the three distribute exactly 37 boulders to carry?

(b) How many ways can the three distribute exactly 87 boulders to carry?

14.26 How many solutions in positive integers are there to the equation $y_1 + y_2 + y_3 + y_4 = 30$ such that no y_i is greater than 12?

14.27 The expression
$$(x + y + z)^{2006} + (x - y - z)^{2006}$$
is simplified by expanding it and combining like terms. How many terms are in the simplified expression? *(Source: AMC)*

Challenge Problems

14.28 Compute
$$\sum_{k=0}^{\infty} k \left(\frac{1}{3}\right)^k$$
using generating functions. **Hints:** 48, 234

14.29 Define $Q(n, k)$ to be the coefficient of x^k in the expansion of $(1 + x + x^2 + x^3)^n$. Prove that

$$Q(n, k) = \sum_{j=0}^{k} \binom{n}{j}\binom{n}{k - 2j},$$

where $\binom{n}{r} = 0$ if $r < 0$. *(Source: Putnam)* **Hints:** 21

14.30★ We can generalize the Binomial Theorem to negative powers. Specifically, for any positive integer n, we can write
$$(x + y)^{-n} = a_0 x^{-n} + a_1 x^{-(n+1)} y + a_2 x^{-(n+2)} y^2 + \cdots,$$
for some coefficients a_k, provided that $\left|\frac{y}{x}\right| < 1$.

(a) Prove that $a_k = (-1)^k \binom{n + k - 1}{k}$. **Hints:** 77, 50

(b) Explain why it makes sense to write $a_k = \binom{-n}{k}$. **Hints:** 18

14.31★ Let k be a positive integer. Prove that there are exactly k ordered pairs (x, y) of nonnegative integers that satisfy *any one* of the following equations:

$$x + 3y = 2k - 1,$$
$$3x + 5y = 2k - 3,$$
$$\vdots$$
$$(2k - 1)x + (2k + 1)y = 1.$$

(For example, if $k = 2$, then there are two equations $x + 3y = 3$ and $3x + 5y = 1$. The 2 solutions are $(3, 0)$ and $(0, 1)$; notice that both satisfy the first equation.) *(Source: Mandelbrot)* **Hints:** 184, 86, 111

14.32★

(a) Let
$$C(x) = 1 + x + 2x^2 + 5x^3 + 14x^4 + \cdots$$

be the generating function for the Catalan numbers, so that the coefficient of x^n is the nth Catalan number. Prove that
$$C(x) = x(C(x))^2 + 1.$$

Hints: 230

(b)★ Prove that
$$C(x) = \frac{1 - \sqrt{1 - 4x}}{2x}.$$

Hints: 100

(c)★ Use the generating function to find a closed-form formula for the nth Catalan number. **Hints:** 324

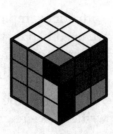

At what point on the graph do "must" and "cannot" meet? Yet I must, but I cannot! – Ro-Man, Robot Monster

CHAPTER 15

Graph Theory

15.1 Introduction

A **graph** is a visual tool that is very useful for solving a large number of problems. More specifically, a graph depicts connections or relationships between objects.

For example, we can use a graph to show the six New England states, and which of them border each other:

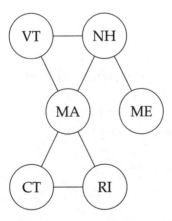

Whenever we have a set and we also have relationships between some pairs of items in the set, we can represent the set and its relationships as a graph. Drawing a graph is usually a better way for us to imagine and work with the relationships than simply listing them. For instance, which is easier to work with, the graph of the New England states above, or the list of bordering states shown below?

{(VT, NH), (NH, ME), (MA, VT), (NH, MA), (MA, CT), (MA, RI), (CT, RI)}

The graph lets us see complex relationships a lot more easily than a simple list of pairs does. To continue with our example, we can see instantly from the graph that to get from Maine (ME) to Connecticut (CT), we need to pass through New Hampshire (NH) and Massachusetts (MA), and that there's no route that passes through fewer states. Getting this information from the list of states that border each other is a lot more difficult, even though the list contains the same information.

Although the theory of graphs is vast enough to fill an entire book (and many such books have been written), we're going to cover just the basics of graph theory. The goal is to give you another tool that you can put into your problem-solving toolbox, ready for you to grab when you might need it. Many problems that can be solved with graph theory can also be solved without it, but as we see in our "map" of New England above, graphs can often provide a convenient tool for organizing information.

15.2 Definitions

Before we can start solving problems using graphs, we need to learn a little bit of terminology.

A **graph** is a set of **vertices** together with a set of **edges** connecting the vertices. Graphs are usually displayed visually with dots representing the vertices, and lines or arcs representing the edges. For example, the figure below depicts a graph with 5 vertices and 7 edges.

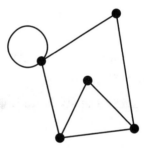

Notice that one of the edges in the above picture connects a vertex with itself. Such an edge is called a **loop**.

What defines a graph is how many vertices it has and which edges are present. The exact geometric configuration of vertices and edges in a diagram of the graph is not important. In particular, when drawing a graph, we can place the vertices and edges wherever is convenient. So, for instance, the following three diagrams all represent the same graph:

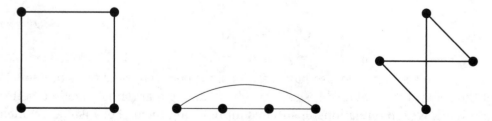

When drawing a graph, we don't mind if edges appear to intersect in our picture (as in the example graph on the far right above). We only consider edges to connect pairs of vertices, so the "intersection points" of edges (other than vertices) are irrelevant.

The number of vertices in a graph is called the **order** of the graph. (We usually only work with graphs with a finite number of vertices, so "order" makes sense. However, there is a more general theory for infinite graphs.) The number of edges that are connected to a given vertex is called the **degree** of that vertex. If a vertex has a loop, that loop counts as 2 towards the vertex's degree. The degree of vertex V is denoted $\deg(V)$.

Generally, we only allow at most one edge between any pair of vertices. However, there are some types of graphs that break this rule. If a graph allows multiple edges between vertices, we call it a **multigraph**.

Sometimes, we will think of the edges as "one-way," meaning that each edge has a "start" vertex and an "end" vertex. In this case, we call the graph **directed**, and we draw little arrows on the edges to indicate which way the edges point:

We can use this type of graph to represent relationships among *ordered* pairs of elements of a set. For example, if we have a set of teams in a basketball league, we could draw a graph in which each team is a vertex, and a directed edge from team A to team B represents team A defeating team B in a basketball game. We could also think of our state diagrams from Chapter 13 as examples of directed graphs: the vertices are states, and the edges are transitions between states. Directed graphs are typically allowed to have up to 2 edges between any pair of vertices, with 1 edge running in each direction.

As we said at the beginning, we could fill an entire book with graph theory. In order to keep the amount of material a bit more manageable, we're going to restrict our attention to graphs that don't have any of the more exotic properties that we've described above. Specifically:

> **Important:** When we use the word "graph," we will mean a graph with undirected
> ⚠️ edges, no loops, and at most one edge between any pair of vertices,
> unless we specifically say otherwise.

Some books call these types of graphs **simple graphs**. More exotic types of graphs—such as directed graphs, graphs with loops, and/or multigraphs—are important and are very useful in problem solving. But in this book, we'll focus solely on simple graphs. We'll still be able to use graphs as a powerful problem-solving tool, without necessarily having all the extra bells and whistles.

> **WARNING!!** Not all books or contests follow the convention we just stated. Make
> ☢ sure that you are clear on what the default definition of a "graph" is.

15.3 Basic Properties of Graphs

Problems

Problem 15.1: The island nation of Aopslandia has 10 cities, and a network of roads connecting the cities. Roads only intersect in cities.

(a) If each city has 4 roads leading out of it, then how many total roads are there?

(b) Suppose six of the cities have 6 roads leading out of them, and the other four cities have 5 roads. How many total roads are there?

(c) Is it possible for five of the cities to have 5 roads and for the other five to have 4 roads? Why or why not?

Problem 15.2: Suppose that a graph has n vertices and that there is an edge between every pair of distinct vertices. How many edges are in the graph?

Problem 15.3: Random Airlines flies to 21 cities, and from each city, they plan to have nonstop service to 7 others. (Nonstop service works in both directions, so that if the airline flies nonstop from A to B, then they also fly nonstop from B to A.) Prove that this flight schedule is impossible.

Problem 15.4: Prove that, given any 6 people, there are 3 of them that all know each other, or 3 of them such that no two of the 3 know each other. (Assume that "knowing" is a symmetric relationship, meaning that if person A knows person B, then person B also knows person A.)

Problem 15.5: In a tournament involving 20 players, there are 14 games (each game is between two players). Given that every player plays in at least one game, show that we can find 6 games involving 12 distinct players.

(a) Represent the tournament as a graph with 20 vertices (the players) and 14 edges (the games). In words, describe what we are trying to prove about this graph.

(b) Determine an algorithm for finding the 6 edges that we are looking for.
(Source: USAMO)

Problem 15.6: Random Airlines is going to try again. Now they are only going to service 10 airports. How many flights must they have such that among any 5 airports, there are at least two pairs of airports that have nonstop flights? (The same airport can be part of both flights.)

(a) Initially, forget about trying to minimize the number of flights. Try to find *any* example that works.

(b) Is there a way that you can delete flights from your example in (a) to create an example with fewer flights?

(c) The smallest example is 12 flights. Can you find it?

(d) Show that 11 flights is insufficient.

We'll explore some of the important basic properties of graphs through a series of progressively more difficult problems.

One of the most basic and most useful properties of graphs is the relationship between the number of vertices and the number of edges. The key fact, of course, is that each edge connects 2 vertices.

Problem 15.1: The island nation of Aopslandia has 10 cities, and a network of roads connecting the cities. Roads only intersect in cities.

(a) If each city has 4 roads leading out of it, then how many total roads are there?

(b) Suppose six of the cities have 6 roads leading out of them, and the other four cities have 5 roads. How many total roads are there?

(c) Is it possible for five of the cities to have 5 roads and for the other five to have 4 roads? Why or why not?

Solution for Problem 15.1: For all of these parts, we'll think of Aopslandia as a graph, where the cities are the vertices and the roads are the edges. Generally, whenever we are presented with a problem in which a finite number of objects are connected in pairs in some way, we might want to think about representing the problem using a graph. In particular, the word "network" in the problem strongly suggests using a graph.

Concept: Graph theory is a useful tool for problems in which a finite number of objects are connected or paired up in some fashion. The objects become the vertices, and the connections become the edges.

(a) "Each city has 4 roads leading out of it" means that each of the 10 vertices in our graph has degree 4, meaning that each vertex has 4 edges connected to it. Since each of the 10 vertices has 4 edges, this makes a count of $10 \times 4 = 40$ total edges in the graph; however, this counts each edge twice, once for the vertex at each end of the edge. So there are $40/2 = 20$ edges in the graph.

This result can easily be generalized:

Important: If a graph has n vertices and each vertex has degree d, then the graph has $nd/2$ edges. More generally, if the degrees of the vertices of a graph sum to s, then the graph has $s/2$ edges.

(b) This is not much harder than part (a). The 6 vertices of degree 6 contribute $6 \times 6 = 36$ edges, and the 4 vertices of degree 5 contribute $4 \times 5 = 20$ edges. So we have an initial count of $36 + 20 = 56$ edges, but this counts each edge twice (once for each endpoint), and thus there are $56/2 = 28$ edges in the graph.

(c) If 5 vertices had degree 5 and the other 5 vertices had degree 4, this would give an initial count of $(5 \times 5) + (5 \times 4) = 25 + 20 = 45$ edges. But this should count each edge twice, which would give a total of $45/2$ edges in the graph. However, the number of edges must be an integer. So such a graph is not possible. □

Extra! *Perhaps the most surprising thing about mathematics is that it is so surprising. The rules which* ➥➥➥➥ *we make up at the beginning seem ordinary and inevitable, but it is impossible to foresee their consequences.* – Edward Titchmarsh

---

Problem 15.2: Suppose that a graph has n vertices and that there is an edge between every pair of distinct vertices. How many edges are in the graph?

Solution for Problem 15.2: If a graph with n vertices has every possible edge, then each vertex has degree $n - 1$. So there are a total of $n(n - 1)/2$ edges, using the same logic as in the previous problem.

Alternatively, we can see that there is a 1-1 correspondence

$$\{\text{edges}\} \quad \leftrightarrow \quad \{\text{pairs of distinct vertices}\},$$

and since there are $\binom{n}{2}$ pairs of distinct vertices, we may conclude that there are $\binom{n}{2}$ edges. $\square$

A graph in which there is an edge connecting every pair of distinct vertices is called a **complete graph**. The complete graph with n vertices is often denoted by K_n. Below are two representations of K_4:

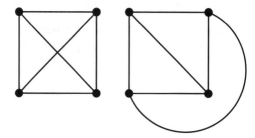

The left picture has the advantage that we show all the edges as straight lines, and the right picture has the advantage that the edges, as drawn, only intersect at vertices. As we've mentioned, both of these pictures represent the same graph, so most of the time it doesn't matter which one we draw.

Problem 15.3: Random Airlines flies to 21 cities, and from each city, they plan to have nonstop service to 7 others. (Nonstop service works in both directions, so that if the airline flies nonstop from A to B, then they also fly nonstop from B to A.) Prove that this flight schedule is impossible.

Solution for Problem 15.3: As expected, we'll model this problem with a graph, where the cities are the vertices and the nonstop routes are the edges. Then we have a graph with 21 vertices, each of which has degree 7. But this means that there are $(21)(7)/2 = 73.5$ edges, which is not possible since the number of edges must be an integer. So the flight schedule described is impossible. $\square$

There is a more general statement that can be made:

> **Important:** In any graph, the number of vertices with odd degree must be even.

Hopefully after Problems 15.1(c) and 15.3, you have a pretty good idea why this is true. You'll be asked to formally prove this statement in the Exercises.

Problem 15.4: Prove that, given any 6 people, there are 3 of them that all know each other, or 3 of them such that no two of the 3 know each other. (Assume that "knowing" is a symmetric relationship, meaning that if person A knows person B, then person B also knows person A.)

Solution for Problem 15.4:   The clarification inside the parentheses—that "knowing" is a symmetric relationship—is a pretty strong clue to model this problem with a graph. We have 6 vertices corresponding to the 6 people, and we draw an edge between any two people if they know each other.

If 3 people all know each other, then there will be edges between all 3 pairs of the 3 corresponding vertices, forming a triangle. A more technical way to say this is that there will be a **subgraph** of our graph that is a copy of K_3, the complete graph with 3 vertices.

If 3 people are such that no two of them know each other, then there will be 3 vertices in our graph for which there are no edges between any pair of them. In more technical language, we would say that there is an **empty** subgraph with 3 vertices: "empty" means that there are no edges.

So how can we go about proving that one of these situations must occur?

As we have seen, we can often get information about a graph by looking at the degrees of its vertices. So let's pick a vertex—call it A—and examine its degree.

If the degree of A is 3 or more, then A is connected to at least 3 other vertices: call three of them B, C, and D. If any two of B, C, and D are connected, then we can take those two, together with A, to form a triangle. But if none of them are connected, then the threesome B, C, and D form the other type of triple that we're looking for: a set of 3 people, none of whom know each other. Thus, if $\deg(A) \geq 3$, we're guaranteed to have the threesome that we are looking for.

Now suppose that the degree of A is less than 3. Since A can be connected to at most 2 of the other 5 vertices, we conclude that there are at least 3 vertices—call three of them E, F, and G—that are *not* connected to A. We can now just do the reverse of the above argument. If any two of E, F, and G are not connected, then those two along with A form the unconnected triple we seek, as they represent 3 people, none of whom know each other. On the other hand, if E, F, and G are all connected, then they form a triangle.

So we're done! □

Note that we could have solved Problem 15.4 without first representing the problem as a graph. However, thinking of the people as vertices in a graph, and the "knowing" relationships as edges of the graph, gave us a nice way to visualize the problem. We didn't even need to draw the graph—just thinking about the problem in terms of a graph gave us the path to the solution.

> **Concept:**   The usefulness of graphs is that they allow us to visualize relationships among objects.

Problems involving tournaments often lend themselves to graph theory approaches. Here is a more difficult example:

> **Problem 15.5:**   In a tournament involving 20 players, there are 14 games (each game is between two players). Given that every player plays in at least one game, show that we can find 6 games involving 12 distinct players. *(Source: USAMO)*

Solution for Problem 15.5: Naturally, we can represent the tournament with a graph consisting of 20 vertices (the players) and 14 edges (the games). We are also told that the degree of each vertex is at least

1; this just means that there are no isolated single vertices that are not connected to the rest of the graph. We are trying to prove that we can find 6 different edges that (pairwise) have no vertices in common.

We might notice that 14 edges doesn't seem like a whole lot of edges for a graph with 20 vertices. Indeed, 14 edges means that the sum of the degrees of the vertices is only 28 (since each edge gets counted twice). Moreover, this means that at most 8 vertices can have degree of more than 1; conversely, at least 12 vertices have degree of exactly 1.

Here's a bad way to finish:

> **Bogus Solution:** We know that at least 12 vertices have degree exactly 1, so there are $12/2 = 6$ edges connecting them. Take the 6 edges connecting those 12 vertices. Since each vertex has degree 1, the edges cannot have any vertices in common (otherwise such a vertex in common would have degree 2 or more).

The problem with this reasoning is that we don't know that those 12 vertices are necessarily all connected to each other. The diagram at the right shows that they may not be. Note that all but 2 of the vertices have degree 1. If we were to select the "wrong" set of 12 of them, as indicated by the circled vertices in the diagram at right, then we would not get the 6 edges that we need, as not all of the circled vertices are connected to each other in pairs.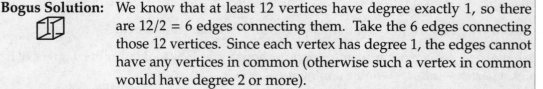

We must be a bit more careful in our reasoning. As we stated earlier, the total degree of all the vertices combined is 28. Since each vertex must have degree at least 1, this means that there are at most 8 "extra" degrees. Suppose that we remove any edge from any vertex that has degree greater than 1. We repeat this step as necessary until every vertex has degree 1 or less. Since we start with at most 8 "extra" degrees, we'll need to remove at most 8 edges, and we'll be left with a graph in which there are at least 6 edges, and each vertex has degree at most 1. But then we're done—we just select any 6 edges that are remaining. No two of these edges can have a vertex in common (since each vertex has degree at most 1), so they represent 6 games with no players in common; in other words, they must involve 12 distinct players, 2 for each game. □

> **Problem 15.6:** Random Airlines is going to try again. Now they are only going to service 10 airports. How many flights must they have such that among any 5 airports, there are at least two pairs of airports that have nonstop flights (the same airport can be part of both flights)?

Solution for Problem 15.6: We're starting with a graph with 10 vertices. We want to populate it with as few edges as possible such that for any subset of 5 vertices, the subgraph consisting of those vertices contains at least 2 edges.

One idea is to first look for *any* example that satisfies the criteria, and then try to modify it to reduce the number of edges. Obviously we can take the trivial example K_{10}, where we draw *every* possible edge, but this has $\binom{10}{2} = 45$ edges; surely we can do better than that.

We can remove a lot of edges by simply getting rid of all the edges to one of the vertices. This leaves us with K_9 together with a disconnected vertex. If we choose any 5 vertices, at least 4 of them must be in the K_9, so we will get at least $\binom{4}{2} = 6$ edges among them. We can repeat this process and disconnect a second vertex, leaving us with K_8 together with two disconnected vertices. Now, if we choose any 5 vertices, at least 3 of them must be in the K_8, so we will get at least $\binom{3}{2} = 3$ edges among them. This graph has a total of $\binom{8}{2} = 28$ edges.

We suspect that we can do even better, since our above example (K_8 plus 2 extra vertices) guarantees us 3 edges for any subset of 5 vertices, and the problem statement only requires 2 edges. But we fail if we try to split off a third vertex, leaving us with K_7 plus 3 disconnected vertices. In this case, if we pick 2 vertices in K_7 plus the 3 disconnected vertices, we'll only have one edge. In the diagram at right, which is K_7 plus 3 disconnected vertices, the five circled vertices have only 1 edge among them (shown in bold).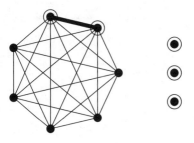

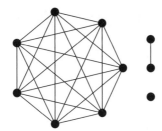 However, if we draw an edge between two of our disconnected vertices, we'll have a copy of K_7 and a copy of K_2 together with an extra vertex, as shown to the left. Now if we pick any 5 vertices, either at least 3 of them will be in the K_7 (giving us 3 edges among them), or 2 will be in the K_7 (giving us 1 edge) and the others will be the 3 outside K_7 (giving us a second edge). In either case, we get the required 2 edges. This graph has $\binom{7}{2} + 1 = 22$ edges, our best yet!

This last example gives us an idea for how to improve even more. If we divide our original 10 vertices in three groups, and draw every possible edge in each group, then by the Pigeonhole Principle, either we will have 3 vertices in one group (and thus edges between all 3 pairs of them), or else we will have 2 vertices in each of two groups (and thus an edge between each pair within a group).

Now we need to figure out how to divide the 10 vertices into 3 groups so that we have the fewest number of edges. Intuitively, it seems as if the fewest edges will result when we make the groups as evenly-sized as possible. Therefore, two good candidates to consider are dividing the 10 vertices into groups of $4, 3, 3$ or into groups of $4, 4, 2$, as in the following graphs:

The $4, 3, 3$ grouping has $6 + 3 + 3 = 12$ edges, and the $4, 4, 2$ grouping has $6 + 6 + 1 = 13$ edges. So we prefer the $4, 3, 3$ grouping, since it has fewer edges. But can we do any better?

After a little more experimentation, you might become somewhat convinced that 12 edges is the best that we can do. Now we need to prove it. Specifically, what we need to prove is that in any graph with 10 vertices and 11 (or fewer) edges, we can always choose 5 vertices with at most 1 edge among them.

> **Concept:** In order to prove that n is the minimum possible integer number of something in a problem, we need to show two things: that n is possible, and that $n - 1$ (or less) is impossible.

We'll use a technique sometimes called the **extremal principle**, which is often very handy in graph theory problems. The extremal principle can be loosely described as "consider the most extreme case or item first." In this problem, since we want to find a subgraph that has as few edges as possible, let's work backwards by eliminating vertices that have the most edges.

> **Concept:** When solving a problem, it often pays to focus on the object with the most of something (or the least of something). This is sometimes referred to as the **extremal principle**.

Let's recall the goal: we have a graph with 10 vertices and 11 (or fewer) edges. We want to prove that we are able to find 5 vertices with at most 1 edge among them. We will do this via an algorithm that systematically removes vertices from the graph until we are left with the 5 that we want.

At the start, if we have fewer than 11 edges, then we arbitrarily add edges to the graph until we have 11 edges. Adding edges cannot hurt us: if the edge that we end up with at the end of the algorithm is one of our added edges, this means that in the original graph, we have found 5 vertices with no edges at all among them.

As we have done in several previous problems, we can start by converting the number of edges into the total degrees of the vertices: since there are 11 edges, the sum of the degrees of the vertices is 22. Therefore, by the Pigeonhole Principle, there must be some vertex that has degree at least 3. Using our extremal principle philosophy, we'll suppose that this vertex is not going to be one of the 5 vertices that has at most 1 edge among them—this makes sense, since we're looking for vertices with few edges, and this one has lots of edges.

So we'll go ahead and delete this vertex from consideration. This will also delete at least 3 edges from consideration. Thus, we're now reduced to considering a graph with 9 vertices (since we've eliminated 1 vertex) and at most 8 edges (since we've eliminated at least 3 edges). If we have fewer then 8 edges, then as we discussed at the beginning of the algorithm, we add edges to the graph so that we have 8 edges total. Similarly, at any step of the algorithm, we may end up with fewer than the maximum possible number of edges, in which case we add edges as necessary.

Now we repeat the process:

- 9 vertices and 8 edges means that we must have a vertex with degree at least 2. So we remove it and the 2 edges attached to it from consideration, and now we're down to only 8 vertices and 6 edges.

- Remove another vertex of degree 2 and its 2 edges. We now have 7 vertices and 4 edges.

- Again remove a vertex of degree 2 and its 2 edges. We now have 6 vertices and 2 edges.

- To finish, just remove any vertex with an edge, and the attached edge. We now have 5 vertices and only 1 edge, which is what we were looking for! (Recall that if this edge is an edge that we added along the way, then these 5 vertices had no edges in the original graph.)

The diagram at the top of the next page shows an example of this algorithm in action. At each step, we remove the circled vertex by changing it from solid black to white, and we remove the attached edges by changing them from solid to dashed. The 5 vertices remaining when we're done are 5 vertices of the original graph with only 1 edge among them.

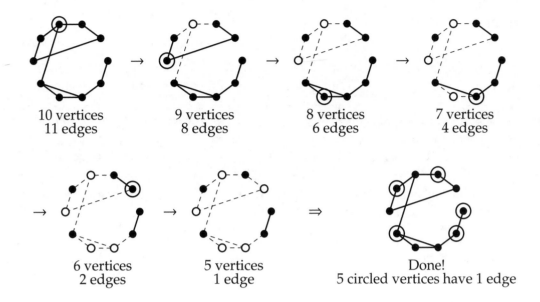

10 vertices	9 vertices	8 vertices	7 vertices
11 edges	8 edges	6 edges	4 edges

6 vertices	5 vertices	Done!
2 edges	1 edge	5 circled vertices have 1 edge

In summary, we have exhibited a graph with 10 vertices and 12 edges such that every subset of 5 vertices has at least 2 edges, and we have also proved that a graph with 11 or fewer edges must have a subset of 5 vertices with at most 1 edge. Therefore, 12 is the minimum number of edges in our graph, and thus 12 is the minimum number of flights that the airline must schedule. □

Exercises

15.3.1 Prove that any graph must have an even number of vertices of odd degree.

15.3.2 Prove that any graph must have two vertices with the same degree. **Hints:** 68

15.3.3 Random Airlines is now trying yet another schedule. New York is served by nonstop flights to 39 other cities. San Diego is only served by a nonstop flight to Los Angeles and no other flights. Every other city in the network (including Los Angeles) is served by nonstop flights to 22 other cities. Prove that it is possible to fly from New York to San Diego by making connections enroute. **Hints:** 153

15.3.4 Suppose that the nation of Graphdom has 13 cities, and that each city is connected to 6 others via a direct train line. Show that it is possible to get from any city to any other city, either directly or with at most one connection. **Hints:** 127

15.3.5 Is it possible to draw:

(a) 7 line segments in the plane, each of which intersects exactly 3 others?

(b) 6 line segments in the plane, each of which intersects exactly 4 others?

15.3.6★ Suppose that in the freshman class at my university, if I choose any 4 students, at least one of the four knows all of the other three. (Assume that "knowing" is symmetric: if A knows B, then B knows A.) Prove that there must be a student that knows everybody in the class. **Hints:** 29

15.3.7★ Suppose that a graph G has an even number of vertices. Vertex V is said to be a *neighbor* of vertex W if there is an edge between V and W. Show that there must exist two distinct vertices of G that have an even number of neighbors in common. **Hints:** 182, 232, 253

15.4 Cycles and Paths

Before we get to the problems in this section, let's define a few more graph theory concepts. Now that you've had some experience working with graphs, some of these concepts will be common sense.

> **Definition:** A **path** between two vertices A and B is a (finite) sequence of edges $e_1, e_2, \ldots, e_k$ such that e_1 connects A with some vertex V_1, then e_2 connects V_1 with some vertex V_2, and so on, until edge e_k connects vertex V_{k-1} with B. The positive integer k is called the **length** of the path.

A path is just common sense: it's a way to travel from A to B along the graph. The number of edges that we use to get from A to B is the length of the path.

There is not universal consensus as to whether a path is allowed to travel along the same edge twice. For the purposes of this book, we will allow a path to be any sequence of adjoining edges, including possibly the same edge twice in a row (which would be a "u-turn" in the path). We will call a path with no repeated edges a **simple path**.

Often, there will be lots of paths between two vertices A and B, but a path that has the smallest possible length is called a **minimal path** from A to B, and the length of a minimal path is called the **distance** between A and B. There might be more than one minimal path from A to B, but the distance (if it exists) is uniquely determined. Also note that a minimal path between A and B must be a simple path, because if it had any repeated edges, we could systematically remove them to get a shorter path (we will let you explain this in the Exercises at the end of the section). We will denote the distance between A and B by $d(A, B)$.

> **Definition:** A simple path is called a **cycle** if it starts and ends at the same vertex.

We also have a term for a graph that only has "one piece," in the sense that every vertex can be reached from any other vertex:

> **Definition:** A graph is called **connected** if, for every pair of distinct vertices A and B, there exists a path from A to B.

> **Definition:** A connected graph with no cycles is called a **tree**.

Problems

> **Problem 15.7:** Prove that a graph is a tree if and only if for any two distinct vertices A and B, there is a unique simple path from A to B.

> **Problem 15.8:**
> (a) Prove that a connected graph with n vertices must have at least $n - 1$ edges.
>
> (b) Prove that a tree with n vertices must have exactly $n - 1$ edges.
>
> (c) Prove that a tree (with at least 2 vertices) must contain at least 2 vertices of degree 1. (Such a vertex is called a **leaf**.)

Problem 15.9: During a certain lecture, each of 5 mathematicians fell asleep exactly twice. For each pair of these mathematicians, there was some moment when both were sleeping simultaneously. Prove that, at some moment, some three of them were sleeping simultaneously. *(Source: USAMO)*

Problem 15.10: Assume that a connected graph G has no cycles of odd length. Show that the vertices of G can be divided into two groups, S and T, such that every edge of G joins a vertex in S and a vertex in T.

First, we explore exactly what a tree is. Trees are one of the most important special types of graphs (with many applications to computer science in particular), so it is useful to understand them well. We'll start with an alternative characterization of trees:

Problem 15.7: Prove that a graph is a tree if and only if for any two distinct vertices A and B, there is a unique simple path from A to B.

Solution for Problem 15.7: First, suppose that our graph is a tree, and let any two distinct vertices A and B be given. We know that there must be a simple path from A to B, because by definition all trees are connected. Now we must show that this path is unique.

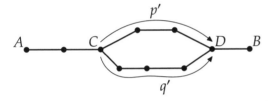

Suppose that there are two different simple paths p and q from A to B; we will show that this leads to a contradiction. The paths p and q might have some edges in common as they leave A, so let C be the first vertex after which the paths p and q differ (in other words, the edge of p that leaves C is different from the edge of q that leaves C). Note that we could have $C = A$. Also let D be the vertex where the paths first come back together (in other words, the edge of p that enters D is different from the edge of q that enters D); again note that we could have $D = B$. Let p' be the part of p that travels from C to D, and let q' be the part of q that travels from C to D. By the way we've constructed them, the paths p' and q' have no vertices or edges in common other than C and D.

Let r be the path the follows p' from C to D, then follows q' backwards from D back to C. Then r is a cycle that starts and ends at C. This is a contradiction, since we know that a tree has no cycles. Therefore, there cannot be two different simple paths from A to B; in other words, the simple path from A to B must be unique.

Conversely, suppose we have a graph such that for any two distinct vertices A and B, we have a unique simple path from A to B. Then the graph is connected (by definition), and it cannot have any cycles: if it did have a cycle starting and ending at A, then we could "cut" the cycle at any point B, and we'd have two different simple paths from A to B, as shown in the picture to the right. Therefore, the graph has no cycles, and thus is a tree. □

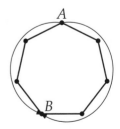

Another nice property of trees is that the number of edges in a tree is always one fewer than the number of vertices, as we see in the next problem.

Problem 15.8:

(a) Prove that a connected graph with n vertices must have at least $n - 1$ edges.

(b) Prove that a tree with n vertices must have exactly $n - 1$ edges.

(c) Prove that a tree (with at least 2 vertices) must contain at least 2 vertices of degree 1. (Such a vertex is called a **leaf**.)

Solution for Problem 15.8: For parts (a) and (b), we have statements that we wish to prove for all positive integers n, so let's use induction.

Base case: Suppose that $n = 1$; that is, we have a graph with 1 vertex. Then naturally it has 0 edges. It is connected and has at least 0 edges, which establishes the base case for part (a). Also, it is a tree and has exactly 0 edges, which establishes the base case for part (b).

Inductive step: Suppose that the statements are true for all graphs with fewer than k vertices, for some positive integer k, and consider a graph with exactly k vertices.

(a) If every vertex has degree at least 2, then the graph has at least $2k/2 = k$ edges, and we're done. So assume that some vertex V has degree 1. Delete the vertex V and its corresponding edge from the graph. The remaining graph has $k - 1$ vertices. The remaining graph is also connected, since no simple path not starting or ending at V passes through V, thus no simple paths connecting the remaining vertices have been removed from the graph. So, by the inductive hypothesis, the remaining graph has at least $k - 2$ edges. Therefore the original graph has at least $k - 1$ edges.

(b) We first prove that every tree has a vertex of degree 1, by proving that any graph in which the degree of every vertex is at least 2 must contain a cycle. Suppose we have a graph in which the degree of each vertex is at least 2. Pick any vertex A, and construct a path starting at A by picking any unused edge and following it to the next vertex. At each subsequent vertex, extend the path if possible by adding an edge that we have not yet traversed. Every time we visit a vertex for the first time, we will always be able to extend our path, since the condition that the degree is at least 2 guarantees that we will always have an unused edge the first time we arrive at a vertex. But this means that our path will stop only when we visit a vertex for a second time. The subset of the path between the two times that we visit this vertex is a cycle.

Therefore, a tree must have some vertex V of degree 1. We now use a similar argument as in part (a): remove V and its adjacent edge from the tree. What remains is still a tree, since it is still connected and still has no cycles, and this tree has $k - 1$ vertices. Therefore it has $k - 2$ edges by the inductive hypothesis, and the original tree has $k - 1$ edges.

(c) You can draw a few examples to convince yourself that the statement is true. Since we want to show that every tree has at least 2 leaves, we prove that the only alternatives, 0 or 1 leaves, are impossible.

We proved in part (b) that a tree cannot have 0 leaves. Suppose a tree with n vertices has exactly 1 leaf. Then every vertex other than the leaf has degree at least 2, so the total degree of all the vertices is at least $2(n - 1) + 1 = 2n - 1$, and hence the number of edges is at least $(2n - 1)/2 = n - \frac{1}{2}$. But part (b) tells us that there are exactly $n - 1$ edges, so this is a contradiction. Thus there must be at least 2 leaves.

$\square$

There's not much else that we can say about the number of leaves. Note that we can get a tree with exactly 2 leaves by drawing all the vertices in a "line," as in the picture on the left below. We can get a tree with n vertices, $n - 1$ of which are leaves, by drawing all the leaves connected to a central "hub," as in the picture on the right below.

We use the name **leaf** to describe any vertex that has degree 1, even if the graph is not a tree.

Let's try a hard problem involving some of the concepts that we've just introduced.

Problem 15.9: During a certain lecture, each of 5 mathematicians fell asleep exactly twice. For each pair of these mathematicians, there was some moment when both were sleeping simultaneously. Prove that, at some moment, some three of them were sleeping simultaneously. *(Source: USAMO)*

Solution for Problem 15.9: There doesn't seem to be a lot of information in the problem statement. For instance, we don't know how long each mathematician slept. How can we possibly approach this problem?

One key to converting a relatively abstract problem such as this into a graph-theoretic representation is to identify the objects that will become the vertices, and the relationships between the objects that will become the edges. It might be tempting to simply have a graph with 5 vertices, one for each mathematician, but that doesn't really give us a way to encode the information "each... fell asleep exactly twice" into the graph.

Instead, we need to have 10 objects: one for each nap. In other words, if we call our mathematicians A, B, C, D, E, then mathematician A has naps A_1 and A_2, mathematician B has naps B_1 and B_2, and so on. Then, we can draw an edge between two naps if they overlap, meaning that the corresponding mathematicians are asleep at the same time during their respective naps.

Now we have a graph with 10 vertices. How many edges does it have? We're told that each pair of mathematicians are simultaneously asleep at least once. Since there are $\binom{5}{2} = 10$ pairs, we conclude that our graph must have at least 10 edges.

We'd like to conclude, using Problem 15.8, that the graph must contain a cycle. However, there is one technicality we have to worry about: Problem 15.8 requires that the graph be *connected*, and the graph that we've constructed for this problem might not be connected. For example, if the lecture runs from 9:00 to 10:00, and all the professors take their first nap between 9:00 and 9:20 and their second nap between 9:40 and 10:00, then there will be no edges connecting any of the early naps with any of the late naps, so the graph will not be connected.

We can get around this difficulty by looking at the **connected components** of the graph. Intuitively, a connected component of a graph is a vertex together with all the other vertices that are connected to it via paths, and all of the edges between any pair of those vertices. Any graph can be split into its connected components; each component is connected, and there are no edges between any two vertices in different components. (If the graph is already connected to begin with, then it has only one connected component, namely the graph itself.) For instance, the graph below has 4 connected components, with 5, 2, 1, and 3 vertices, respectively, in the components.

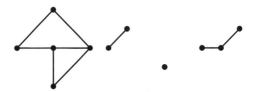

Since our original graph of naps has at least as many edges as vertices, and every edge of the graph is in exactly one connected component, we can apply the Pigeonhole Principle and conclude that at least one of the connected components must have at least as many edges as vertices. Therefore, by Problem 15.8, this connected component cannot be a tree, and thus must contain a cycle.

> **Important:** If a graph contains at least as many edges as vertices, then the graph must contain a cycle.

Now we'd like to show that three of the naps in the cycle must all simultaneously overlap. Just to have a frame of reference, suppose that one of the naps in the cycle lasts from 1:00 to 2:00. (Note that the times are arbitrary; we will say more about this assumption after the solution.) We'll look at all the other naps relative to this one.

If the second nap in the cycle is a "subnap" of the first nap, meaning that it occurs entirely between 1:00 and 2:00, then we're done: the third nap in the cycle must overlap with the second nap, and hence with the first nap as well. Similarly, if the second nap is a "supernap" of the first nap, meaning that it starts before 1:00 and ends after 2:00, then we're also done: the last nap of the cycle must overlap with the first nap, and hence with the second nap as well.

So we only need to consider the case where the second nap extends either before 1:00 or after 2:00, but not both. Since it doesn't matter which (the analysis in the two cases is exactly the same), let's suppose it extends beyond 2:00. This means in particular that the two naps share 2:00. Just to simplify the argument a bit, suppose that the second nap starts at 1:30 and ends at 2:30 (the exact times do not matter for the proof; all that matters is that the nap starts between 1:00 and 2:00 and ends after 2:00).

Here's a diagram with the first two naps plotted on the real line:

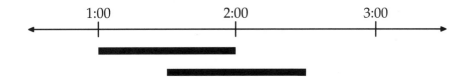

If the next nap in the cycle contains some time between 1:30 and 2:00, then we're done. But what if the next nap starts between 2:00 and 2:30? Our goal in this case is to show that there must be a nap later in the cycle that also contains 2:00. The key is that the *last* nap in the cycle must overlap the original 1:00–2:00 nap, in order to complete the cycle. Therefore, the subsequent nap times in the cycle cannot keep getting later and later past 2:00, because the last nap of the cycle must start before 2:00. Therefore, there must be some nap later in the cycle that starts before 2:00 but whose preceding nap starts after 2:00. Thus, this nap must end after 2:00, which means that it overlaps with the first two naps in the cycle. An example is shown in the diagram below:

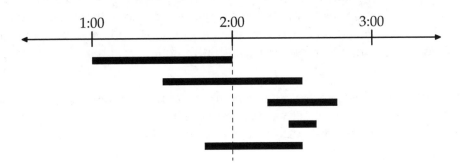

Therefore, three naps in the cycle must overlap at some point, and hence there must be three mathematicians asleep at the same time. □

Note that in the last part of the above solution, we simplified the argument a bit by choosing specific times for the first two naps. This was done to give us a convenient frame of reference to refer to in the rest of the proof. Choosing those times didn't in any way restrict the validity of our proof, since the specific times themselves were irrelevant to the argument.

> **Concept:** If you can choose specific numbers or values of unknown quantities in a proof, in a way that doesn't destroy the generality of the proof, then doing so often makes the rest of the proof easier to understand.

> **WARNING!!** ☢ The phrase "in a way that doesn't destroy the generality of the proof" in the above sentence is very important! You cannot select values that are insufficiently general for your argument.

> **Problem 15.10:** Assume that a connected graph G has no cycles of odd length. Show that the vertices of G can be divided into two groups, S and T, such that every edge of G joins a vertex in S and a vertex in T.

Solution for Problem 15.10: "Divided into two groups" and "no cycles of odd length" are clues. Our idea is to somehow divide the vertices into two groups called "odd" and "even." We just need to figure out how to make this division precise.

Pick any vertex V. Let S be the set of all vertices that are an even distance from V (including V itself, which is distance 0 from V), and let T be the set of all vertices that are an odd distance from V. Because the graph is connected, we know that every vertex must be in exactly one of S or T. We must now show that every edge connects a vertex in S and a vertex in T, by showing that there are no edges between vertices in S, and also no edges between vertices of T.

Suppose there is an edge e between two vertices A and B in S. Let $a = d(V, A)$ and $b = d(V, B)$; by construction both a and b are even. Let p be a path of length a from V to A, and let q be a path of length b from V to B. For now, suppose that p and q don't overlap anywhere along their length, and that they don't include the edge e between A and B. (As an Exercise, you'll see what happens when we don't make these simplifying assumptions.)

Notice what happens if we start at V, follow path p to A, then follow edge e to B, and finally follow path q (in reverse) back to V. We have constructed a cycle that starts and ends at A, and has length $a + b + 1$. But a and b are even, so $a + b + 1$ is odd, and we're not permitted to have a cycle of odd length in our graph. This is a contradiction, and therefore the edge e cannot exist.

If A and B are both in T, then we can repeat the exact same argument, except that this time a and b are odd. If an edge exists between A and B, then we again find a cycle of length $a + b + 1$, which is odd because a and b are both odd. This is a contradiction, and thus there cannot be an edge between A and B.

The only remaining issues are those simplifying assumptions that we made: that the paths p and q don't overlap, and that neither includes an edge between A and B. It turns out that making these assumptions does not invalidate our proof. You'll be asked to explain why in the Exercises. $\square$

Graphs of the type described in Problem 15.10 occur relatively frequently, so we give them a special name.

> **Definition:** A graph G is called **bipartite** if the vertices of G can be partitioned into two sets S and T, such that every edge of G connects a vertex in S with a vertex in T.

Note that, in particular, every tree is bipartite: trees have no cycles at all, so they clearly have no cycles of odd length.

Exercises

15.4.1 How many different trees are there with 3 vertices? With 4 vertices? With 5 vertices?

15.4.2 Suppose that graph G has 80 vertices and each vertex has degree at least 40. Prove that G must contain a cycle of length 4. **Hints:** 154

15.4.3 Explain why, if there is a path between two vertices A and B, then a minimal path between them must be a simple path.

15.4.4 Explain why, in Problem 15.10, we can make the assumptions that we made about the paths p and q: that they don't overlap and that they don't include the edge between A and B.

15.4.5 Show the converse of Problem 15.10: If a connected graph is bipartite, then it contains no cycle of odd length. **Hints:** 250

15.4.6 Generalize Problem 15.10, by showing that any graph G, not necessarily connected, is bipartite if and only if it contains no cycle of odd length. **Hints:** 204

15.4.7★ Suppose that a graph has 10 vertices. What is the maximum number of edges that it can have if it does not contain any triangles (cycles of length 3)? **Hints:** 306, 339

15.5 Planar Graphs

As we saw earlier, we can draw the graph K_4 (the graph with 4 vertices and all 6 possible edges) with the edges "crossing" in the middle, as in the left diagram of Figure 15.1 below, or we can draw it so that none of the edges cross, as in the right diagram of Figure 15.1 below.

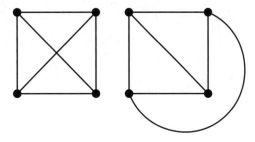

Figure 15.1: Two depictions of K_4

Although the left diagram is more compact, the right diagram has the advantage that there are no "phantom" vertices where edges cross at a non-vertex point. Graphs like K_4, in which we can draw all of the edges with no non-vertex crossing points, have some special nice properties. First, let's name them:

> **Definition:** A connected graph that can be drawn on the plane so that none of its edges intersect at a non-vertex point is called **planar**.

Note the phrase "can be drawn" in the definition of planar. If we drew K_4 as in the left picture of Figure 15.1, we might not think it was planar. But because we *can* draw it in the plane with no edges crossing, as in the right diagram of Figure 15.1, we see that it is in fact planar. To make the distinction precise, we say that a drawing of a planar graph in which no edges cross is called a **planar representation** of the graph.

We have one more definition that pertains to planar representations of graphs:

> **Definition:** A **face** of a planar representation of a graph is a region of the plane with no edges in its interior and whose boundary is a union of edges of the graph. (Note that we consider the region of the plane outside of the entire graph as a face; its "boundary" consists of the edges on the "outside" of the graph.)

Intuitively, a face is just an empty space in the plane that is surrounded by edges. For example, in the figure at right, the different solidly-colored regions are faces. The "exterior" of the graph (in white) is also a face, so the graph shown at right has 4 faces.

Problem 15.11: Find a relationship among the number of vertices V, the number of edges E, and the number of faces F in a planar representation of a connected planar graph.

Problem 15.12: Show that if a graph is connected, planar, and has more than one edge, then $E \leq 3V - 6$.

Problem 15.13: Prove that K_5 is not planar.

Problem 15.14: There are 25 points inside a rectangle, no 3 of which are collinear. Segments are drawn between them and the four corners of the rectangle, such that the rectangle is partitioned into triangles. (All 25 points must be a vertex of some triangle.) How many triangles do we have?

Problem 15.15: There are 3 houses on Lake Street, and 3 wells in a field behind the houses. Can we construct paths such that there is one path from each house to each well, and such that none of the paths cross?

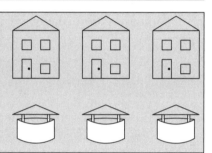

Often it's hard to tell whether a graph has a planar representation or not. It would be nice to know some additional properties that planar graphs must satisfy. One very useful property is that connected planar graphs have a nice relationship among the numbers of vertices, edges, and faces in the graph.

Problem 15.11: Find a relationship between the among of vertices V, the number of edges E, and the number of faces F in a planar representation of a connected planar graph.

Solution for Problem 15.11: Since we really have no idea where to begin, let's just draw a few simple examples of planar graphs, count up V, E, and F for each of them, and see if there's any pattern.

Graph	V	E	F
Tree	n	$n-1$	1
(triangle)	3	3	2
(graph)	4	6	4
(graph)	6	7	3

It may not be terribly easily to see the pattern, but if we add V and F, we always get 2 more than E.

So let's conjecture that the relationship is

$$V + F = E + 2.$$

We already see that this formula is true for all of the examples in our table above. Another simple example in which the formula holds is the graph that consists of n vertices arranged in a cycle. This graph has n edges and 2 faces (the "inside" and the "outside"), so $V = n$, $E = n$, $F = 2$, and the equation holds.

How can we prove the formula works for all connected planar graphs? As we often do in graph theory, we use induction. In previous problems (such as Problem 15.8), we used induction on the number of vertices. However, if you try that here, you will find some considerable technical difficulties in completing the proof. So instead, we'll try induction on the number of edges.

If the graph has only 1 edge, then it must be a tree with 2 vertices, and we already know that the formula is true for trees. This proves the base case of the induction.

Now suppose that the formula works for connected planar graphs with m edges, and consider a connected planar graph with V vertices, $E = m + 1$ edges, and F faces. We already know that the formula works for trees, so assume that our graph is not a tree. Then by definition it must contain a cycle, and if we choose an edge (call it e) in any cycle and remove it, the remaining graph will have the same number of vertices, 1 fewer edge, and 1 fewer face, as shown in the example below:

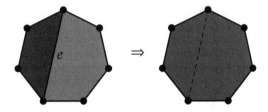

A potential problem is that the new graph might not be connected. The graph will be disconnected only if there are two vertices A and B, whose path between them (in the original graph) contained the edge e that we removed. But if this is the case, we can use the cycle containing e as a "detour" to reconnect the path between A and B.

Therefore, the graph with e removed is still connected, so by the inductive hypothesis it satisfies the formula

$$V + (F - 1) = (E - 1) + 2.$$

Adding 1 to both sides gives $V + F = E + 2$ as desired, completing the proof. $\square$

Important: A planar representation of a connected graph with V vertices, E edges, and F faces satisfies the formula

$$V + F = E + 2.$$

This is called **Euler's formula**. An important consequence is that any planar representation of a connected graph with V vertices and E edges must have $E - V + 2$ faces; in other words, the number of faces in a planar representation of a connected graph is independent of how we draw it.

Note that "Euler" is pronounced "oiler," not "yuler."

One nice consequence of Euler's formula is that it gives a way to "test" a graph to see if it might be planar. Unfortunately, until we know whether a graph is planar or not, we have no direct way of determining F: we can only find F if we already know that the graph is planar. So we need some sort of criterion for planarity that does not involve F.

Problem 15.12: Show that if a graph is connected, planar, and has more than one edge, then $E \leq 3V - 6$.

Solution for Problem 15.12: Call an edge *normal* if it borders two different faces, otherwise call the edge *special*. For example, the graph shown at right has two special edges (shown in bold). We count the total number of edges bordering any face by counting each normal edge once and each special edge twice. For example, the shaded face in the graph at right is bordered by 10 edges, and the white exterior face is bordered by 6 edges.

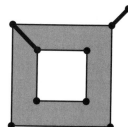

Counting in this manner, we see that every face must be bordered by at least 3 edges (unless there is only 1 edge in the entire graph). In addition, if we sum the number of border edges over all of the faces in the graph, then we will count each edge exactly twice: we'll count every normal edge once for each of the two faces that it borders, and we'll count every special edge twice for the single face that it borders. This means that $3F \leq 2E$, or $F \leq \frac{2}{3}E$. If we apply this inequality to Euler's formula, we see that in any connected planar graph with more than one edge,

$$E + 2 = V + F \leq V + \frac{2}{3}E \quad \Rightarrow \quad E \leq 3V - 6.$$

$\square$

This inequality can be used to test a graph to see if is non-planar.

Important: If a graph is connected, planar, and has more than one edge, then $E \leq 3V - 6$. Therefore, if a connected graph with more than one edge satisfies $E > 3V - 6$, then the graph is not planar.

Problem 15.13: Prove that K_5 is not planar.

Solution for Problem 15.13: K_5 has 5 vertices and $\binom{5}{2} = 10$ edges. But $10 > 3(5) - 6 = 9$, so K_5 cannot be planar. $\square$

The same argument shows that K_n is not planar for all $n \geq 5$, since $\binom{n}{2} > 3n - 6$ for all $n \geq 5$. (We'll leave the details as an Exercise.)

Problem 15.14: There are 25 points inside a rectangle, no 3 of which are collinear. Segments are drawn between them and the four corners of the rectangle, such that the rectangle is partitioned into triangles. (All 25 points must be a vertex of some triangle.) How many triangles do we have?

Solution for Problem 15.14: Let T be the number of triangles. Since every triangle that we create is a face, and since the exterior of the rectangle is also a face, we have $T + 1$ faces. Counting the edges by face gives us $3T + 4 = 2E$, since each triangle has 3 edges surrounding it, and the exterior face has the 4 sides of the rectangle surrounding it. Finally, we have 29 vertices (the 25 points inside plus the 4 corners of the rectangle). So plugging in $V = 29$, $E = \frac{3}{2}T + 2$, and $F = T + 1$ into Euler's formula, we get

$$29 + (T + 1) = \left(\frac{3}{2}T + 2\right) + 2.$$

Solving this gives $T = 52$. □

Problem 15.15: There are 3 houses on Lake Street, and 3 wells in a field behind the houses. Can we construct paths such that there is one path from each house to each well, and such that none of the paths cross?

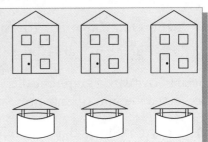

Solution for Problem 15.15: If we could construct all of the paths as required, then we would have a planar graph with 6 vertices and $3 \times 3 = 9$ edges. We note that $9 \leq 3(6) - 6 = 12$, so it passes our planarity test. Remember: this doesn't mean it's planar; it merely means we're going to have to work a bit harder.

Notice that our graph is bipartite: every edge runs between a house and a well. This means that there are no odd cycles. How can we use this extra bit of information?

The fact that there are no odd cycles means that if the graph is planar, then every face is bordered by at least 4 edges. (In other words, there are no triangular faces.) So we can strengthen our inequality a bit, using $4F \leq 2E$, or $F \leq \frac{1}{2}E$. When we plug this into Euler's formula, we get

$$E + 2 = V + F \leq V + \frac{1}{2}E \quad \Rightarrow \quad E \leq 2V - 4.$$

So now we know that if a bipartite graph (with more than one edge) satisfies $E > 2V - 4$, then it cannot be planar. And indeed, in our example, we have $V = 6$ and $E = 9$, and $9 > 2(6) - 4 = 8$. So the graph cannot be planar, and thus we cannot construct the paths as specified in the problem. □

Exercises

15.5.1 Is the graph on the left below planar? Why or why not?

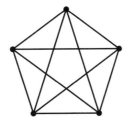

15.5.2 Is the graph on the right above planar? Why or why not?

15.5.3   Show that K_n is not planar for any $n \geq 5$.

15.5.4   Show that if G is a planar graph, then G must contain a vertex with degree less than 6. **Hints:** 159

15.5.5★   Consider the graph at right, where the vertices represent squares on a 4×4 checkerboard, and two squares are connected by an edge if a knight can move between them (that is, if the squares are 2 apart in one direction, horizontal or vertical, and 1 apart in the perpendicular direction).

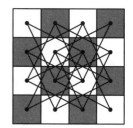

(a)   Is this graph bipartite? **Hints:** 78

(b)   Is this graph planar? **Hints:** 31

15.5.6★   A planar graph is called **Platonic** if all vertices have the same degree d, where $d \geq 3$, and all faces have the same number of boundary edges b.

(a)   Use your knowledge of graphs and Euler's formula to find equations for V, E, and F in terms of d and b.

(b)   Use the equations from part (a) to show that $(d - 2)(b - 2) < 4$.

(c)   Find all possible values of d and b. **Hints:** 299, 220

15.6   Eulerian and Hamiltonian Paths

This section concerns special types of paths that in some sense travel an entire graph. There are two different ways in which we can interpret the word "entire," giving us two different definitions.

Definitions:

- An **Eulerian path** of a graph is a path that contains every edge of the graph exactly once. If an Eulerian path is also a cycle, then it is called an **Eulerian cycle**.

- A **Hamiltonian path** of a graph is a path that contains every vertex of the graph exactly once. A **Hamiltonian cycle** is a cycle that contains every vertex (except the starting and ending vertex) exactly once. (The start/end vertex is included twice.)

Problems

Problem 15.16: For each of the following graphs, determine if it has an Eulerian path or cycle. If it does, show it; if not, explain why not.

(a) 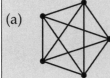   (b)    (c)

Problem 15.17:

(a) Show that a connected graph has an Eulerian path if and only if it has at most 2 vertices with odd degree.

(b) Show that a connected graph has an Eulerian cycle if and only if it has no vertices with odd degree.

Problem 15.18: A $1 \times 1 \times 1$ wire frame cube is to be made by gluing together pieces of wire. The wire can be bent to form corners of the cube. If exactly 12 units of wire is used to make the frame, what is the fewest number of pieces of wire that can be used to make the frame? *(Source: MATHCOUNTS)*

Problem 15.19: 7 players are in a round-robin tennis tournament in which each player plays a match against every other player exactly once. There is only one court, and we would like to schedule the tournament so that each match (after the first) has one player in common with the previous match.

(a) Is this scheduling possible?

(b) What if there were only 6 players?

Before looking at the general theory of Eulerian paths and cycles, let's look at some examples.

Problem 15.16: For each of the following graphs, determine if it has an Eulerian path or cycle. If it does, show it; if not, explain why not.

(a) (b) (c)

Solution for Problem 15.16:

(a) There are many ways in which we can draw an Eulerian cycle on this graph. If we label the points as shown at the right, then one such path is

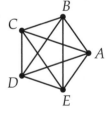

$$A \to B \to C \to D \to E \to A \to C \to E \to B \to D \to A.$$

(b) This one is a bit trickier. We can construct an Eulerian path as follows:

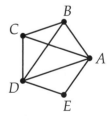

$$C \to D \to B \to C \to A \to E \to D \to A \to B.$$

Try as you might, you cannot make an Eulerian cycle on this graph. Also, after some experimentation, you might notice that all Eulerian paths must start at B and end at C, or else start at C and end at B. Why should this be the case?

As we often do with graphs, we analyze further by looking at the degrees of the graph's vertices. Note that B and C each have degree 3, whereas E has degree 2 and A and D each have degree 4. In an Eulerian cycle, the cycle must enter and leave each vertex an equal number of times, so the degree of each vertex must be even. (In the graph in part (a), each vertex has degree 4, so an Eulerian cycle is possible.) In an Eulerian path that is not a cycle, the path must also enter and leave each vertex an equal number of times, *except* for the start and end vertices: the start vertex will have one extra "leaving," and the end vertex will have one extra "entering." Therefore, these vertices must have odd degree.

In general, a graph with exactly 2 vertices of odd degree will always have an Eulerian path, starting at one of the odd-degree vertices and ending at the other one. We'll prove this general statement in the next problem.

(c) A bit of experimentation will probably convince you that no Eulerian path is possible. For example, if you try to start

$$A \to C \to D \to B \to E \to A \to B \to C,$$

then there's no way to get to that last edge between D and E.

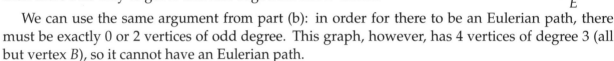

We can use the same argument from part (b): in order for there to be an Eulerian path, there must be exactly 0 or 2 vertices of odd degree. This graph, however, has 4 vertices of degree 3 (all but vertex B), so it cannot have an Eulerian path.

□

Now that we've seen a few examples, let's prove the general result that our examples led us to.

Problem 15.17:
(a) Show that a connected graph has an Eulerian path if and only if it has at most 2 vertices with odd degree.

(b) Show that a connected graph has an Eulerian cycle if and only if it has no vertices with odd degree.

Solution for Problem 15.17: One idea is to take a constructive approach. We ask ourselves: how would we construct such a path or cycle? Playing with a few simple examples might help here.

In the previous problem, we saw that any vertices with odd degree, if they exist, must come at the start or at the end of the path. Recall the explanation: each vertex (other than the starting and ending vertices) must be entered some number of times and then left the same number of times. So the total number of edges on the path at each vertex (except for the start and the end) must be even. But we have to use every edge! So every vertex, except the start and end vertices, must have even degree. For the start vertex, we'll be leaving it one more time (at the start) than we will be arriving at it, unless the path happens to be a cycle, so it's necessary to have an extra "odd" edge. The same thing is true at the end vertex, except in reverse: we'll be arriving at it one more time than we will be leaving from it (again unless the path is a cycle), so it's necessary for its degree to be odd.

This shows that a graph with more than 2 odd-degree vertices cannot have an Eulerian path, but how do we know that a connected graph with 2 or fewer odd-degree vertices does have one?

First of all, a graph cannot have exactly 1 vertex of odd degree, because any graph must have an even number of vertices of odd degree. (Go back and review Problem 15.3 if you do not recall why this fact is true.) So we only need to consider graphs that have either 0 or 2 odd-degree vertices. If the graph has 2 vertices of odd degree, then we are looking for an Eulerian path that starts at one of the odd-degree vertices and ends at the other one. If the graph has no vertices of odd degree, then we are looking for an Eulerian cycle.

We'll prove both parts of our statement by induction on the number of edges. For the base case, if the graph has 2 vertices with a single edge between them, then that edge itself is an Eulerian path, and the graph has 2 odd-degree vertices, so we're fine.

For the inductive step, assume that every graph with fewer than k edges satisfies the problem statement, and consider a graph with k edges and 2 or fewer vertices of odd degree. First, let's consider a graph that has exactly 2 vertices of odd degree. Our strategy is to remove an edge, use the inductive hypothesis to get an Eulerian path (or cycle) on the new smaller graph, and then reattach the edge to get an Eulerian path on the entire graph. There are 2 cases, depending on the type of edge that we can remove.

Case 1: the graph has a leaf.

If one of the vertices of odd degree is a leaf (call it L), then we can remove the edge connecting L to the rest of the graph. (This is an example of the extremal principle: remove an edge with minimum degree.) If L was connected to the other odd vertex (call it M), then the remaining graph has no vertices of odd degree, and thus by the inductive hypothesis it has an Eulerian cycle. We can then reattach the edge, and we get an Eulerian path that starts at L, follows its edge to M, and then follows the smaller graph's Eulerian cycle around the graph, ending back at M.

On the other hand, if L was connected to a vertex of even degree (call it M), then after removing the edge connected to L, the vertex M is of odd degree. Therefore, by the inductive hypothesis there is an Eulerian path starting at M and ending at the other vertex of odd degree (call it N). Then, we have an Eulerian path on the original graph starting at L, following the edge to M, and then following the smaller graph's Eulerian path to N.

Case 2: the graph does not have a leaf.

We have taken care of the case in which the graph has a leaf. If the graph does not have any leaves, then suppose A is a vertex of odd degree at least 3, and delete any edge from A to some other vertex B. There are now two subcases:

Case 2a: Removing the edge from A to B leaves a connected graph. Then the new graph has either 0 or 2 vertices of odd degree, and thus by the inductive hypothesis has an Eulerian path. If B has odd degree in the new graph, then the path runs from B to the other odd-degree vertex (call it C), and then we can extend this path to the original larger graph by starting at A, following the edge to B, and then continuing on the smaller graph's Eulerian path to C. On the other hand, if B has even degree in the new graph, then it had odd degree in the original larger graph, and thus A and B were the two odd-degree vertices in the original graph. Thus the new graph has an Eulerian cycle, and we can construct an Eulerian path in the original graph by starting at A, following the edge to B, and then following the Eulerian cycle of the smaller graph back to B.

Case 2b: Removing the edge from A to B leaves a disconnected graph. Then the new graph consists of two connected components. The component connected to A (call this graph J) must have only vertices of even degree: A is now of even degree, so if J has 2 vertices of odd degree, then the original graph had 3 (or more) vertices of odd degree, a contradiction. Therefore, by the inductive hypothesis, graph J has an Eulerian cycle. The component connected to B (call this graph K) must have either 0 or 2 vertices of odd degree, depending on whether B had odd or even degree in the original graph. In either case, by the inductive hypothesis there is an Eulerian path (or cycle) starting at B and ending somewhere (either the other vertex of odd degree, or back at B). We can construct an Eulerian path on the original larger graph, by starting at A, following J's cycle around back to A, then traversing the edge to B, and following K's path (or cycle) to its ending point. We can see this in the diagram below.

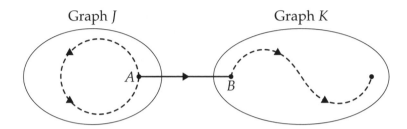

In either case, we have constructed an Eulerian path starting at vertex A and ending at the other vertex with odd degree.

Finally, if the original graph has 0 vertices with odd degree, then every vertex has degree at least 2, so there are at least as many edges as vertices. This means that the graph must contain a cycle. Simply remove any edge that is part of a cycle; note that the edge's removal does not disconnect the graph. This will leave two vertices A and B with odd degree, so there is an Eulerian path from A to B with this edge removed. Then we just add back the edge from B to A, giving us an Eulerian cycle on the original graph. $\square$

As we have just seen, there is a fairly easy criterion to determine whether or not a graph has an Eulerian path or cycle, and in fact, the proof gives an algorithmic way of constructing the path or cycle if we so desire. Unfortunately, it is not so simple to determine if a graph has a Hamiltonian path or cycle. We will present one criterion in the Exercises. For now, let's look at a couple additional problems involving Eulerian paths and cycles.

> **Problem 15.18:** A $1 \times 1 \times 1$ wire frame cube is to be made by gluing together pieces of wire. The wire can be bent to form corners of the cube. If exactly 12 units of wire is used to make the frame, what is the fewest number of pieces of wire that can be used to make the frame? *(Source: MATHCOUNTS)*

Solution for Problem 15.18: We can model this problem by letting the corners of the cube be vertices of a graph, so that the edges of the cube become edges of the graph. The graph has 8 vertices and 12 edges, as shown below; the left picture is perhaps the most "natural" depiction of the cube, but the right picture has the advantage of being a planar representation.

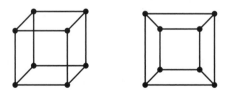

Note that each piece of wire will correspond to a path that does not repeat any edges. Each vertex has degree 3, so there is no Eulerian path, meaning that we cannot use a single piece of wire. Moreover, since each vertex has odd degree, we will need a piece of wire to begin or end at each vertex, since each time a wire passes through a vertex in the middle of the wire piece, it uses up 2 edges: one "entering" edge and one "leaving" edge. Therefore, we will need at least $8/2 = 4$ pieces of wire to construct the cube. It is easy to come up with a way to construct the cube using 4 pieces of wire; we will leave it as a task for you to come up with one. □

Problem 15.19: 7 players are in a round-robin tennis tournament in which each player plays a match against every other player exactly once. There is only one court, and we would like to schedule the tournament so that each match (after the first) has one player in common with the previous match.

(a) Is this scheduling possible?

(b) What if there were only 6 players?

Solution for Problem 15.19:

(a) Tournaments are natural to model using graphs. In this problem, we have 7 vertices, corresponding to the 7 players. There is an edge between any two players corresponding to the required match between those players. Therefore, our graph is K_7.

   Finding a schedule that satisfies the condition of the problem is equivalent to finding an Eulerian path on our graph: two consecutive edges in the path share a vertex that corresponds to the player competing in the corresponding consecutive matches. The Eulerian condition means that every match will occur. In K_7, every vertex has degree 6, so there exists an Eulerian cycle, and thus the schedule is possible.

(b) With only 6 players, our graph is K_6. Now each vertex has degree 5; in particular, each vertex has odd degree. So there is no Eulerian path, and hence such a tournament schedule is not possible.

□

║║║	**Exercises**	▶

15.6.1 Recall the 4×4 checkerboard knights-move graph from Problem 15.5.5, shown again at right.

(a) Does this graph have an Eulerian path or cycle?

(b) Does this graph have a Hamiltonian path or cycle? **Hints:** 347

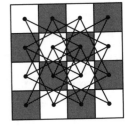

15.6.2 Generalize Problem 15.18: show that if a connected graph has $2n$ vertices of odd degree, then there are n disjoint paths that include all of the edges. (Two paths are *disjoint* if they have no edges in common.)

15.6.3 Show that the graph at right has a Hamiltonian path but does not have a Hamiltonian cycle. **Hints:** 229

15.6.4★ An n-dimensional cube (for $n \geq 2$) is a graph with 2^n vertices that are in 1-1 correspondence with n-tuples of 0's and 1's. Two vertices are connected by an edge if and only if they differ in exactly one coordinate. (The graph of the 3-dimensional cube is show at right.) Show that an n-dimensional cube has a Hamiltonian cycle. **Hints: 84**

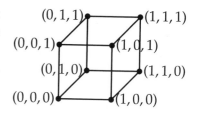

15.7 Summary

➤ For our purposes, "graph" means a simple graph with no loops, no directed edges, and at most one edge between any pair of vertices.

➤ Whenever we have a set and we also have relationships between some pairs of items in the set, we can represent the set and its relationships as a graph. The elements of the set are the vertices of the graph, and the relationships between pairs of elements are the edges.

➤ It is often easier to visualize complex relationships using a graph.

➤ If a graph has n vertices and each vertex has degree d, then the graph has $nd/2$ edges. More generally, if the degrees of the vertices of a graph sum to s, then the graph has $s/2$ edges.

➤ In any graph, the number of vertices with odd degree must be even.

➤ A **path** between two vertices A and B is a (finite) sequence of edges $e_1, e_2, \ldots, e_k$ such that edge e_1 connects A with some vertex V_1, edge e_2 connects V_1 with some vertex V_2, and so on, until edge e_k connects vertex V_{k-1} with B. The positive integer k is called the **length** of the path. A path with no repeated edges is called a **simple path**. A simple path that starts and ends at the same vertex is called a **cycle**.

➤ A graph is called **connected** if, for every pair of distinct vertices A and B, there exists a path from A to B.

➤ A connected graph with no cycles is called a **tree**.

➤ A connected graph with n vertices has at least $n-1$ edges. A tree with n vertices has exactly $n-1$ edges, and contains at least 2 vertices of degree 1 (called **leaves**). A graph with at least as many edges as vertices must contain a cycle.

➤ A graph G is called **bipartite** if the vertices of G can be partitioned into two sets S and T, such that every edge of G connects a vertex in S with a vertex in T. A graph is bipartite if and only if it has no cycle of odd length.

➤ A connected graph that can be drawn on the plane so that none of its edges intersect at a non-vertex point is called **planar**.

➤ Euler's formula for connected planar graphs: $V + F = E + 2$, where V is the number of vertices, E is the number of edges, and F is the number of faces.

➤ If a graph is planar, connected, and has more than 1 edge, then $E \leq 3V - 6$. Therefore, if a connected graph with more than 1 edge satisfies $E > 3V - 6$, then the graph is not planar.

➤ An **Eulerian path** of a graph is a path that contains every edge of the graph exactly once. If an Eulerian path is also a cycle, then it is called an **Eulerian cycle**. A **Hamiltonian path** of a graph is a path that contains every vertex of the graph exactly once. A **Hamiltonian cycle** is a cycle that contains every vertex (except the starting and ending point) exactly once.

➤ A connected graph has an Eulerian path if and only if it has at most 2 vertices with odd degree. A connected graph has an Eulerian cycle if and only if it has no vertices with odd degree.

We also saw the following general problem-solving concepts:

> **Concept:** In order to prove that n is the minimum possible integer number of something in a problem, we need to show two things: that n is possible, and that $n - 1$ (or less) is impossible.

> **Concept:** Often when solving a problem, it pays to focus on the object with the most of something (or the least of something). This is sometimes referred to as the **extremal principle**.

REVIEW PROBLEMS

15.20

(a) Can a country in which 3 roads lead out of every city have exactly 50 cities? Why or why not?

(b) Can a country in which 3 roads lead out of every city have exactly 35 cities? Why or why not?

15.21 17 people are at a party. For any pair of people present, they are friends, enemies, or don't know each other. Prove that there must be a group of 3 people all of whom are mutual friends, enemies, or strangers.

15.22 In the school ping-pong tournament only 6 students entered. The organizers decide to have exactly 7 games, and also mandate that out of any 3 players, at least two must play each other. Prove that:

(a) There exists a student who plays against at least 3 other students.

(b) There exist three people who each play against the other two.

(Source: Romania)

15.23 A graph G has n vertices, with degrees $d_1, d_2, \ldots, d_n$. How many paths of length 2 are there?

15.24 Show that distance on a graph satisfies the Triangle Inequality: if A, B, and C are vertices and are all connected to each other, then

$$d(A, C) \leq d(A, B) + d(B, C).$$

15.25   A connected planar graph G has 10 faces, and every vertex of G has degree 4. Find the number of vertices of G.

15.26   In a connected planar graph, every vertex has degree 3, and every face is bordered by 5 or 6 edges. How many faces are bordered by 5 edges?

15.27   At a conference, everybody has at least 1 friend among the people attending. It is known that among any group of at least 3 people, there are never exactly two pairs of friends. Prove that there is a person that is friends with everyone at the conference.

15.28   Is it possible for a knight to travel around a standard 8×8 checkerboard such that it makes every possible move exactly once? (We consider a move to be completed if it is made in either direction.)

Challenge Problems

15.29   In the ham radio club, there are 78 members, and each member is friends with at least 52 other members. Prove that there are 4 members with the same number of friends. **Hints:** 318, 247

15.30   Suppose there is a group of people in which each person has at most 3 enemies, and enemies are mutual (i.e. if person A hates person B, then person B also hates person A). Prove that it is possible to split the people into two groups so that each person has at most one enemy in his or her group. **Hints:** 28

15.31   A city has the following property: if any set of citizens were to meet, the number of introductions necessary to acquaint everyone would be less than the number of citizens present. Prove that the city may be partitioned into two groups of people, such that each person knows everyone else in their group. **Hints:** 221

15.32   At a homecoming dance, no boy dances with every girl, but each girl dances with at least one boy. Prove that there are two couples gb and $g'b'$ who dance such that g doesn't dance with b' and g' doesn't dance with b. *(Source: Putnam)* **Hints:** 152

15.33★   Prove that if G is a connected graph with n vertices, and each vertex of G has degree at least $n/2$, then G has a Hamiltonian cycle. (This is known as **Dirac's Theorem.**) **Hints:** 22

15.34★   A domino consists of two squares, each of which is marked with between 0 and 6 dots (inclusive). The complete set of 28 dominos is shown at right. Is it possible to arrange all 28 dominos in a circle, such that adjacent halves of neighboring dominos have the same number of dots? **Hints:** 88, 349

15.35★   In the country of Eulandia, any two cities are connected either by rail or by bus (but not both). Prove that for any city, we can choose a type of transportation (rail or bus) such that every other city in Eulandia can be reached via the chosen type of transportation with at most 1 transfer. **Hints:** 135

Every problem has in it the seeds of its own solution. If you don't have any problems, you don't get any seeds.

– Norman Vincent Peale

CHAPTER **16**

Challenge Problems

We finish the book with a collection of challenging problems. They are arranged roughly in increasing order of difficulty; the starred ones at the end are particularly difficult.

Enjoy!

Challenge Problems

16.1 A manufacturer of airplane parts makes a certain engine that has a probability p of failing on any given flight. There are two planes that can be made with this sort of engine, one that has 3 engines and one that has 5. A plane crashes if more than half its engines fail. For what values of p do the two plane models have the same probability of crashing? *(Source: HMMT)*

16.2 If we roll n standard 6-sided dice, the probability of rolling a sum of 2004 is nonzero and is exactly the same as getting a sum of X. What are the possible values of X? *(Source: AMC)*

16.3 Each of two boxes contains both black and white marbles, and the total number of marbles in the two boxes is 25. One marble is taken out of each box randomly. The probability that both marbles are black is $\frac{27}{50}$. Find the probability that both marbles are white. *(Source: AIME)*

16.4 Can two six-sided dice be numbered in such a way that if the dice are fairly weighted, there are 12 equally likely possible sums of the two numbers shown on a roll of the dice? **Hints:** 173

16.5 A deck contains 52 cards of 4 different suits, but not necessarily 13 cards of each suit. You are told the number of cards in each suit. You pick a card from the deck, guess its suit, and set it aside; you repeat this until there are no more cards. Show that if you always guess a suit having no fewer remaining cards than any other suit, then you will guess correctly at least 13 times. **Hints:** 30

16.6 There are 1000 rooms in a row along a long corridor. Initially, the first room contains 1000 people and the remaining rooms are empty. Each minute, the following happens: for each room containing more than one person, someone in that room decides it is too crowded and moves to the next room. All these movements are simultaneous (so nobody moves more than once within a minute). After one hour, how many different rooms will have people in them? *(Source: HMMT)* **Hints:** 279

16.7 I've created a game called Cit-Cat-Cut-Eot. It works like this: A $4 \times 4 \times 4$ board is used. Each player takes turns putting an X or an O in the board until one of the players gets 4 in a row, either along a column or row or diagonally; such a group of 4 marks is called a "winning set." Count the number of possible winning sets. **Hints:** 44

16.8 Alice has two bags. Each bag has 4 slips of paper with the numbers 1 through 4 on them. Betty also has two bags, each with 4 slips of paper with positive integers on them. They decide to play a game whereby each girl pulls a slip from each of her own bags, records the sum of the numbers, then returns each slip to the bag it came from. The numbers in Betty's bags are not 1 through 4 in each bag, but the expected distribution of her sums is the same as Alice's. What are the numbers in Betty's bags? **Hints:** 312

16.9 Let $f(n)$ equal the number of ways n can be written as the sum of 1's and 2's, taking order into account. Let $g(n)$ equal the number of ways n can be written as the sum of integers greater than 1, again taking order into account. For example,

$$f(3) = 3, \text{since } 3 = 1 + 2 = 2 + 1 = 1 + 1 + 1,$$

and

$$g(5) = 3, \text{since } 5 = 5 = 2 + 3 = 3 + 2.$$

Prove that $f(k) = g(k + 2)$ for all positive integers k. **Hints:** 322

16.10 An unfair coin has a $\frac{2}{3}$ probability of turning up heads. If this coin is tossed 50 times, what is the probability that the total number of heads is even? *(Source: AMC)* **Hints:** 298

16.11 How many nonempty subsets S of $\{1, 2, 3, \ldots, 15\}$ have the following two properties?
(1) No two consecutive integers belong to S.
(2) If S contains k elements, then S contains no number less than k.
(Source: AMC) **Hints:** 243, 308

16.12 Let n be a positive integer, and let Pushover be a game played by two players, standing squarely facing each other, pushing each other, where the first person to lose balance loses. At a Pushover tournament, 2^{n+1} competitors, numbered 1 through 2^{n+1} clockwise, stand in a circle. They are equals in Pushover: whenever two of them face off, each has a 50% probability of victory. The tournament unfolds in $n + 1$ rounds. In each round, the referee randomly chooses one of the surviving players, and the players pair off going clockwise, starting from the chosen one. Each pair faces off in Pushover, and the losers leave the circle. What is the probability that players 1 and 2^n face each other in the last round? (Express your answer in terms of n.) *(Source: HMMT)* **Hints:** 179, 92

16.13 Five pirates find a cache of 500 gold coins. They decide that the shortest pirate will serve as the *bursar* and determine a distribution of the coins however he sees fit, and then they all will vote. If at least half of the pirates (including the bursar) agree on the distribution, it will be accepted; otherwise, the bursar will walk the plank, the next shortest pirate will become the new bursar, and the process will continue. Assume that each pirate is super-intelligent and always acts so as to maximize his wealth, and also that each pirate is vindictive: he would like to see someone walk the plank, as long as it does not hurt him financially. How many coins will the shortest pirate get? *(Source: Mandelbrot)* **Hints:** 277

16.14 At a summer math program a contest is held for n teams. Each team, composed of n individuals, is given the same n problems to work on. Suppose that on each team, there is one person who knows how to solve all n problems, another who knows how to solve $n - 1$ of the problems, and so on down to one member who only knows how to solve a single problem. One person is chosen from each team at random. These selected individuals are randomly ordered and asked one at a time to select a problem that has not already been taken. What is the probability that every team successfully presents a solution? *(Source: Mandelbrot)* **Hints:** 240, 208

16.15 In a tournament each player played exactly one game against each of the other players. In each game the winner was awarded 1 point, the loser got 0 points, and each of the two players earned 1/2 point if the game was a tie. After the completion of the tournament, it was found that exactly half of the points earned by each player were earned against the ten players with the least number of points. (In particular, each of the ten lowest scoring players earned half of her/his points against the other nine of the ten.) What was the total number of players in the tournament? *(Source: AIME)* **Hints:** 11

16.16 A carnival game is set up so that a ball put into play has an equally likely chance of landing in any of 60 different slots at the base. The operator of the game allows you to choose a certain number of balls and put them all into play. If every ball lands in a separate slot, you receive $1 for each ball played; otherwise, you win nothing. How many balls should you choose to play in order to maximize your expected winnings? *(Source: Mandelbrot)* **Hints:** 207

16.17★ At a certain college, there are 10 clubs and some number of students. For any two different students, there is some club such that exactly one of the two belongs to that club. For any three different students, there is some club such that either exactly one or all three belong to that club. What is the largest possible number of students? *(Source: HMMT)* **Hints:** 292, 332, 157, 146

16.18★ Prove that

$$\sum_{k=0}^{n} \binom{n+k}{k}\left(\frac{1}{2}\right)^k = 2^n.$$

Hints: 49, 143

16.19★ Eight coins are arranged in a circle heads up. A move consists of flipping over two adjacent coins. How many different sequences of six moves leave the coins alternating heads up and tails up? *(Source: HMMT)* **Hints:** 272

16.20★ A stack of 2000 cards is labeled with the integers from 1 to 2000, with different integers on different cards. The cards in the stack are not in numerical order. The top card is removed from the stack and placed on the table, and the next card in the stack is moved to the bottom of the stack. The new top card is removed from the stack and placed on the table, to the right of the card already there, and the next card in the stack is moved to the bottom of the stack. This process—placing the top card to the right of the cards already on the table and moving the next card in the stack to the bottom of the stack—is repeated until all cards are on the table. It is found that, reading from left to right, the labels on the cards are now in ascending order: $1, 2, 3, \ldots, 1999, 2000$. In the original stack of cards, how many cards were above the card labeled 1999? *(Source: AIME)* **Hints:** 118, 74, 66

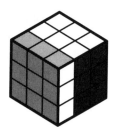

References

In addition to the books cited below, there are links to many useful websites on the book's links page at

`http://www.artofproblemsolving.com/BookLinks/IntermCounting/links.php`

1. B. Averbach and O. Chein, *Problem Solving Through Recreational Mathematics*, Dover Publications, Inc., 2000.

2. A. Engel, *Problem-Solving Strategies*, Springer-Verlag, 1998.

3. D. Fomin, S. Genkin, and I. Itenberg, *Mathematical Circles (Russian Experience)*, American Mathematical Society, 1996.

4. L. Larson, *Problem-Solving Through Problems*, Springer-Verlag, 1983.

5. S. Lehoczky and R. Rusczyk, *the Art of Problem Solving, Volume 1: the Basics*, 7th edition, AoPS Incorporated, 2006.

6. S. Lehoczky and R. Rusczyk, *the Art of Problem Solving, Volume 2: and Beyond*, 7th edition, AoPS Incorporated, 2006.

7. R. Morash, *Bridge to Abstract Mathematics*, Random House, 1987.

8. I. Niven, *Mathematics of Choice*, Mathematical Association of America, 1965.

9. D. Patrick, *Introduction to Counting & Probability*, AoPS Incorporated, 2005.

10. F. Roberts, *Applied Combinatorics*, Prentice-Hall, Inc., 1984.

11. R. Stanley, *Enumerative Combinatorics* (Volumes 1 & 2), Cambridge University Press, 1997, 1999. `http://www-math.mit.edu/~rstan/ec/`

12. A. Tucker, *Applied Combinatorics*, 2nd edition, John Wiley & Sons, Inc., 1984.

13. P. Zeitz, *The Art and Craft of Problem Solving*, 2nd edition, John Wiley & Sons, Inc., 2007.

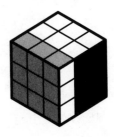

Hints to Selected Problems

1. For the winning positions, examine the 3 cases separately: A, B both odd, A odd and B even, and A even and B odd.

2. Note that $x_1 + x_2 + \cdots + x_n + x_{n+1} = x_1 + x_2 + \cdots + (x_n + x_{n+1})$.

3. Compute the expected number of fixed points of a random permutation, in two different ways.

4. Write p and q in terms of c.

5. The number 100 is not important. The same solution would work for any regular polygon with an even number of sides.

6. Think of the 5 people as dividing the 15 empty chairs into 6 groups. Which of these groups must be positive and which are allowed to be 0?

7. There are 4 possible strategies. For each strategy, compute the probability that the contestant wins the car.

8. Do casework on the non-double-good arrangements based on the point where they go bad (that is, the point where we first see too many ")"s).

9. Try to use smaller silver matrices to construct larger ones.

10. For any n, how many of $\{n, n+1, n+2, n+3, n+4\}$ can appear in the subset?

11. Count the total number of points won by all players.

12. Add a 5th variable d and solve $w + x + y + z + d = 20$.

13. First count how many ways to stack 8 identical coins, then count how many ways to assign 4 of them to be gold and 4 of them to be silver.

14. Show that the set $\{1, 2, \ldots, n\}$ can be partitioned into sets of size $\left\lfloor \dfrac{n+1}{2} \right\rfloor, \left\lfloor \dfrac{n+2}{4} \right\rfloor$, etc.

15. It may help to add balls labeled "0" and "45" (which can never be drawn).

16. The smallest losing position is $(1,2)$. This means that any position of the form $(1,n)$, $(2,n)$, or $(m, m + 1)$ is winning. What is the next smallest losing position?

17. It's not enough to just say that the Pigeonhole Principle doesn't work. You have to construct a counterexample.

18. We write $\binom{n}{k} = \frac{n(n-1)(n-2)\cdots(n-k+1)}{k!}$. Try a similar thing for $\binom{-n}{k}$.

19. The alternating signs and binomial coefficients sure look a lot like PIE!

20. Use Binet's formula and the formula that you found in part (a).

21. It may help to factor $(1 + x + x^2 + x^3)$.

22. Arrange the n vertices in any order. Show that if an edge is missing, then you can rearrange the vertices so that there are fewer missing edges.

23. Put each term of the product over a common denominator.

24. Proceed by careful casework.

25. Let $x\%$ denote the percentage of households that own both items. Use PIE to determine what percentage owns at least one item (in terms of x). What can you conclude?

26. Make the nonnegative integers correspond with the even positive integers, and the negative integers correspond with the odd positive integers.

27. First figure out the probability that the player going first in any individual game wins that game.

28. Divide the people into two groups in the way that minimizes the number of enemies within each group. Then show that this means there are no enemies in a group.

29. First show that any person does not know at most 2 other people.

30. Every time you guess wrong, the number of cards in the largest remaining suit stays the same.

31. Look for a subgraph that you know is not planar.

32. How can we use $f(1) = 30$ and $f(-1) = 12$ to give equations?

33. Count the number of 10-element arithmetic sequences by determining all possible (a, d) such that

$$\{a, a + d, \ldots, a + 9d\} \subset S.$$

34. How many questions on the test can be true?

35. Note that $144 = F_{12}$.

36. Write $f(x) = a_6 x^6 + a_5 x^5 + \cdots + a_1 x + a_0$, and plug in $x = 1$ and $x = -1$.

37. How much, in terms of expected value, does Henry's last step (choosing a coin and, if it is a tail, flipping it) add to the total number of heads?

38. When we go from n^2 people from n countries to $(n + 1)^2$ people from $n + 1$ countries, we need to add $n + 1$ people from the new country, plus 1 extra person from each of the original n countries. Pair up new country people with original country people, and add them to the table.

39. How can you convert this to a "standard" distributions problem?

40. Prove the base case $|x + y| \leq |x| + |y|$ by squaring both sides.

41. Why does the observation in part (a) allow you to use induction?

42. The RHS looks a lot like a PIE expression.

43. You can restrict to looking at lines that are parallel to one side of the square.

44. Use casework based on the number of coordinate planes that the winning line is parallel to.

45. Count the number of sequences of 10 flips with no two consecutive heads.

46. There's nothing particularly special about the numbers 1000 and 2003. It's actually simpler to think of the more general question: if a number greater than b is currently displayed, what is the probability that b will appear?

47. Count separately those with $d < 50$ and $d \geq 50$.

48. Which generating function looks like $\sum_{k=0}^{\infty} kx^k$?

49. The coefficients $\binom{n}{0}$, $\binom{n+1}{1}$, etc. should remind you of the Hockey Stick identity.

50. Prove by induction on n.

51. Compute the first several Lucas numbers and compare them to the Fibonacci numbers.

52. The 4 Aces divide the other 48 cards into 5 groups. How does this help?

53. You'll have to do a bit of algebra to compute the number of "balls" for use in the Pigeonhole Principle.

54. Try to relate this problem to a problem involving partitions.

55. If you counted the paths directly, you'll notice that your answer is a Catalan number. Can you find a Catalan recurrence for this problem?

56. For the losing positions, you'll need to do some casework for when one (or both) piles have 3 or fewer chips.

57. Iterate the recursion, meaning apply the recursion to itself so that you can write a_n in terms of a_{n-2}. What can you conclude?

58. Write a statement that is equivalent to $p \wedge (q \vee r)$.

59. It may be helpful to first compute the number of sequences with no consecutive 0's.

60. Look at the first spot in the second row in which there is no coin.

61. Count the number of ways to have each outcome (in terms of n), and set them equal. Then use algebra to solve for n.

62. If a number is both a perfect square and a perfect cube, then it must be a perfect sixth power.

63. Consider permutations of $1, 2, \ldots, n + 1$. (This should be suggested by the $(n + 1)!$ term.) The LHS is casework.

64. All outcomes are equally likely, so just count them.

65. You can simplify the calculation a bit by imagining that each die has only 3 equally-likely outcomes: ⚀ or ⚅, ⚁ or ⚄, and ⚂ or ⚃.

66. How do you use your analysis of the 2048-card problem to get the answer to the 2000-card problem? What role do those extra 48 cards play?

67. If we can divide the points into 25 categories, then one of the categories must contain at least 3 points.

68. Use the Pigeonhole Principle.

69. Count how many times each element appears in a geometric mean.

70. If the last coin is heads, then an even number of the previous $n - 1$ coins have to be heads. If the last coin is tails, then an odd number of the previous $n - 1$ coins have to be heads. Use this to set up a recurrence for the probability.

71. First show that there must be a car with enough gas to reach the next car.

72. Focus on where the partial sum first becomes 0, and set up a Catalan-style recurrence.

73. Take a look at the picture below, showing a typical triangle:

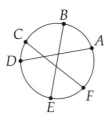

What can you conclude?

74. Consider the problem for a stack of 2048 cards.

75. Reformulate the coloring condition as: none of the odd numbers can have a color in common with any of the even numbers, and vice versa. So some subset of the colors can be assigned to the odd numbers, and a disjoint subset of the colors can be assigned to the even numbers.

76. Think of partitioning, in two different ways, all of the correctly solved problems on all of the tests. One partition is where each term corresponds to a student. Another partition is where each term corresponds to a question.

77. Try $(x + y)^{-1} = \frac{1}{x+y}$ first.

78. The colors of the checkerboard are a very powerful clue.

79. Construct a simple example.

80. Note that condition is equivalent to being able to choose members x, y, z (not necessarily distinct) from the same country such that $x + y = z$.

81. Do casework based on the last letter of the word.

82. There are only four 2×2 squares that might be white, one for each corner of the 3×3 grid.

83. The only states that you should need are the starting state and the states where the final sequences of flips are T, TT, HHHH, and HHHHH.

84. Prove by induction on n.

85. Counting the number of successful outcomes is a distribution problem.

86. Write a separate generating function for the number of solutions to each equation (in terms of k). How do you then get the generating function for the solutions to any of the equations?

87. If 6 students get handouts, then 9 students don't get one. But no more than 2 consecutive students can be without a handout. Determine the possible configurations of the "handout-less" students.

88. Make a graph where the dominos are the edges.

89. In how many ways can we color the first row?

90. You can count the paths directly using complementary counting and PIE.

91. One solution is to write and solve a recursion with the "initial" conditions $a_0 = a_{28} = 1$.

92. How does the original problem relate to the more symmetric version of the problem suggested in the previous hint?

93. Remember that the Binomial Theorem works for complex numbers too!

94. Note that we have to place 8 1's, such that each row and each column contains two 1's.

95. First draw the two 7-pointed stars. Why are there only two of them?

96. Use complementary counting.

97. Plug $x = \sqrt[3]{1} = \frac{-1+i\sqrt{3}}{2}$ into the Binomial Theorem.

98. Once you find a lower bound for $\#(X \cap Y \cap Z)$, you still need to show examples of sets X, Y, and Z that achieve this bound.

99. Show that a_n satisfies the recurrence $a_n = a_{n-1} + 4a_{n-2} - 4a_{n-3}$.

100. Solve the equation from part (a) using the quadratic formula.

101. It's enough to show that the interval $[0, 1]$ is uncountable.

102. Let x and y denote the quantities given by the first two bullet points. Write equations for the rest of the bullet points in terms of x and y.

103. The two situations are very different. There is a relatively straightforward description of $\mathcal{P}(S \cap T)$. However, if you play with some simple examples, you should convince yourself that there is no similar description for $\mathcal{P}(S \cup T)$.

104. Compute a couple more: S_5, S_6, etc. This should make the pattern more clear.

105. Count the outcomes using casework based on the number showing on the largest die.

106. There are 2 cases. One case is if there are 5 numbers, each with a different remainder. If this case does not occur, then what can you conclude?

107. We can always arbitrarily color the first row. Use casework from there.

108. Order the 330 members of the first country as $a_1 < a_2 < \cdots < a_{330}$. The differences $a_{330} - a_i$ cannot be in the first country without satisfying the condition. This gives a list of 329 members which must be in the other 5 countries. Use Pigeonhole again. Repeat this several times.

109. Place the purple balls first, then the green balls.

110. Find a 1-1 correspondence between this problem and a block-walking problem on the half-grid.

111. You may have to experiment with small values of k in order to guess at correct generating function, which you can then prove by induction.

112. To show that a set is even, we often try to break it up into two halves that are in 1-1 correspondence.

113. Try to construct the largest possible subset with the desired property.

114. Look for a counting argument.

115. If a number n is in B, then $125 - n$ cannot be in B.

116. Just focus initially on the probability that 1000 shows up.

117. Compute p first, then q, then r, then s.

118. 2000 is an ugly number to work with. Since every other card gets laid on the table as you progress through the deck, what might be a much nicer number to start with?

119. You can also solve the problem (or check your answer) by using a Venn diagram.

120. Don't try to find a formula for the losing positions; instead, try to find a recursive procedure for generating all the losing positions.

121. There are fewer B's, so it's easier to deal with them first.

122. You want to split the polygon into two smaller polygons, in order to use induction.

123. Split the big square into 25 little squares of side length $\frac{1}{5}$.

124. The first five odd Catalan numbers are C_0, C_1, C_3, C_7, and C_{15}.

125. Consider rolling an arbitrary number of distinguishable dice.

126. Show that any student must answer at least 20 pairs of questions in the same way (either both correctly or both incorrectly). So the number of "balls" is $20n$.

127. For any two vertices V and W, look at the cities connected to V and the cities connected to W. How many cities are accounted for?

128. Use a path-reversing argument similar to that which we used to prove the formula for the Catalan numbers.

129. Constructively count the number of pairs of subsets (A, B) with $B \subseteq A$. Do the same for pairs (A, B) with $B \subseteq (S \setminus A)$.

130. Think of X as white stones with weights $x_1, x_2, \ldots, x_m$ and Y as black stones with weights $y_1, y_2, \ldots, y_n$. We are trying to find subsets of stones of each color with the same weight.

131. If each player has an odd sum, how many odd tiles must each player select?

132. Try an example where $n = 3$.

133. Try to pair up the subsets in some way that makes the sums easy to count.

134. Use casework, depending on whether the two pairs have a common element or not.

135. Prove by induction on the number of cities.

136. It may help to note that the generating function for a single fair die factors: $(x + x^2 + x^3 + x^4 + x^5 + x^6) = x(1 + x)(1 + x + x^2)(1 - x + x^2)$.

137. The "boxes" for the Pigeonhole Principle are possible pairs of questions answered correctly and possible pairs of questions answers incorrectly.

138. You'll need to find the GCD's of all the subsets of $\{10^{10}, 15^7, 18^{11}\}$.

139. Note that $p \uparrow q$ is the same as $\neg((\neg p) \vee (\neg q))$. How do we represent $\neg p$ using the Sheffer stroke?

140. The first step is figuring out what the question is asking!

141. Remember that all of your variables must be integers.

142. Imagine climbing up and then down an n-step staircase.

143. Look for a block-walking argument on Pascal's Triangle.

144. Find a 1-1 correspondence between $[0, 1]$ and $\mathcal{P}(\mathbb{N})$.

145. Note that $\mathcal{P}(\emptyset) \neq \emptyset$. What is the only subset of $\emptyset$?

146. The answer is 513. It's not a coincidence that this is $2^9 + 1$.

147. Break the count into cases: sequences that end with a head and sequences that end with a tail.

148. Try to count the non-double-good arrangements.

149. Show that the process described in the text is reversible.

150. Look at any one particular point of the path. What is the probability that a random path has a turn at that point?

151. If we color the first 2006 rows, in how many ways can we finish by coloring the 2007th row?

152. Use the extremal principle: start with the boy who dances with the most girls.

153. The numbers 39 and 22 are not particularly important; all that matters is that 39 is odd and 22 is even.

154. Finding a cycle of length 4 is equivalent to finding two vertices that have two neighbors in common.

155. Note that for any integer a, no two consecutive terms of the sequence $a, 2a, 4a, 8a, \ldots$ can be in the subset. How does that help you construct the largest possible subset?

156. Prove by induction.

157. If you're still stuck, the next hint will tell you the answer. Once you know the answer, you might be better able to see a way to prove it.

158. There are two cases: x and y are both even, or x and y are both odd.

159. If every vertex has degree at least 6, then $V \le E/3$. Use this to show a contradiction.

160. Use a couple of previously-used identities to help simplify this sum.

161. You can save yourself some algebra by ignoring terms in the generating function that cannot possibly contribute to the x^{20} coefficient.

162. The first step is that one country, by Pigeonhole, must have at least 330 members.

163. Use some of the identities that we've already proven (either in the text or in previous exercises).

164. You might suspect that it cannot be done. How can you prove it?

165. The characteristic equation is a cubic. Does it factor?

166. Pick a triangle to start with, and work your way around the decagon.

167. Rewrite the inequality with the N term on one side and all the $S(\cdots)$ terms on the other side. Interpret the $S(\cdots)$ terms as a PIE expression.

168. To set up the recursion, you may have to consider a more general version of the problem.

169. How is the problem related to $\phi(1000)$?

170. Use a "greedy" algorithm. Construct subsets of X and Y, where in each step we add an element to the subset with the smaller total.

171. First choose which kids will receive which type of candy.

172. Use the pattern to make a conjecture, and then prove your conjecture by induction.

173. Note that 12 equally likely sums means that each sum must be rolled in 3 different ways.

174. Experiment with some small values of n and look for a pattern.

175. Use an argument similar to Problem 5.10 to complete the proof.

176. Show that the condition is equivalent to there being 0, 1, 9, or 10 true questions on the test.

177. If a circle has diameter d, what is the probability that a randomly chosen line will intersect it? Using this, what is the expected number of circles that a randomly chosen line intersects?

178. Losing \$1 corresponds to the x^{-1} term in the generating function.

179. First, find the probability that players 1 and $2^n + 1$ face off in the final round. This should be easier because of the symmetry involved.

180. Use PIE to count the number of seatings in which at least one pair of countrymen are seated in the same row.

181. Color it first without worrying about the rotational symmetry, and then divide to correct for the symmetry.

182. Prove by contradiction: assume that every pair of vertices in G has an odd number of neighbors in common.

183. For any integer n, use the prime factorization of n to compute the number of divisors. How would you count the number of odd divisors? What can you conclude from this?

184. First note that no (x, y) can be a solution to more than one of the equations.

185. What is the expected number of monochromatic 10-element arithmetic sequences in a random coloring of S?

186. Try to construct a pair of unfair dice such that the generating function for their sum matches the corresponding generating function for a pair of fair dice.

187. Try it with a smaller classroom, say 3×5.

188. Let S equal the desired sum. Compare S, $S/3$, and $S/9$.

189. Use the Hockey Stick identity to sum the numerators.

190. Observe that once we decide which rows and which columns contain empty desks, we've determined which desks are occupied.

191. Don't read random hints!

192. Fix one of the three points, and compute everything in terms of the positions of the other two relative to the first point.

193. Use PIE to get bounds on $\#(X \cup Y)$ and $\#(X \cup Y \cup Z)$.

194. To simplify things, treat the maple trees and the oak trees as being the same.

195. Let A_i be the number of deals in which player i receives a pair (for $1 \le i \le 4$). Use PIE to count $\#(A_1 \cup A_2 \cup A_3 \cup A_4)$.

196. Do casework based on what combination of the last 2 houses get mail.

197. Compute the first few a_n. A pattern should emerge.

198. Show that once you place the purple balls and two of the green balls, the positions of the other two green balls are fixed.

199. Use casework, depending on which stair is the last stair stepped on in both directions.

200. Think about symmetry.

201. Break up the movement of the ants into cycles of the form $A_1 \to A_2 \to \cdots \to A_d \to A_1$; that is, A_1 moves to A_2's spot, A_2 moves to A_3's spot, and so on, up to A_d moves to A_1's spot. Count the different ways in which this can happen.

202. Compute the expected number of pairs of judges that agree on a randomly-chosen contestant, first by looking at a judge's point-of-view, then by looking at a contestant's point-of-view.

203. That term in the denominator has x, x^2, up through x^6. What do you know of that has possible outcomes 1, 2, up through 6?

204. Show that each connected component is bipartite. Why does this mean that the whole graph is bipartite?

205. Note that whichever team wins the last game must win the series.

206. Take advantage of the symmetry in the problem.

207. Write a formula for the expected win if you choose n balls. How does this expected win change as you go from n to $n + 1$?

208. The Hockey Stick identity will be useful, as will the identity $\binom{n}{k}\binom{k}{r} = \binom{n}{r}\binom{n-r}{k-r}$.

209. What does 9^k count?

210. Compute the probability that a drawing has no two balls consecutive.

211. Count the number of internal nodes on either side of the root node.

212. This calls for clever substitution into the Binomial Theorem.

213. Write a cubic polynomial with a as one of its roots.

214. Set up a state diagram, noting that only the last consecutive sequence of heads or tails is important.

215. We can't have any partition: we must have a partition that satisfies the Triangle inequality.

216. Start with the prime factorization of $20!$.

217. Use a procedure similar to Problem 14.14.

218. Consider the cases separately when n is even and when n is odd.

219. What does equation (3.7.6) imply if $B = 10$?

220. There are 5 graphs altogether.

221. Consider the graph where two vertices are connected if the corresponding people *don't* know each other.

222. Use complementary counting: count the number of ways in which at least one class is left empty.

223. Remember that every nonzero rational number is of the form p/q, where p, q are relatively prime integers. How does this help?

224. Look at your examples from parts (a) and (b). Notice anything?

225. Break up into cases depending on whether the pentagon has adjacent sides on the n-gon or not.

226. Use the Pigeonhole Principle, where the boxes are the pairs of members in the club.

227. If we color k squares in any given row, then these squares determine $\binom{k}{2}$ columns. We can't have the same pair of columns colored in two different rows.

228. The hard step is proving that a polygon must have an interior diagonal. Try to construct one.

229. Notice there are 5 "outer" vertices and 5 "inner" vertices. The outer vertices must appear on a Hamiltonian cycle in groups.

230. Compute $(C(x))^2$ using the recursive formula for the Catalan numbers.

231. Set up a recurrence for the number of spacy subsets of $\{1, 2, \ldots, n\}$, based on whether n is an element of the subset or not.

232. First show that if we assume every pair of vertices has an odd number of neighbors in common, then every vertex has even degree.

233. Count the number of ways to place k distinguishable items into 9 distinguishable boxes, such that at least one box is empty.

234. Try the generating function $\frac{x}{(1-x)^2}$.

235. For each $1 \le n \le 15$, count the number of subsets that has n as an anchor.

236. Draw a Ferrers diagram.

237. Experiment with small values of n. Does this lead to a conjecture as to what values of n have silver matrices?

238. If a pentagon has 3 sides on the original n-gon, then at least 2 of these sides must be adjacent. If a pentagon has 4 sides on the original n-gon, then they must all be adjacent.

239. All LHS terms are of the form $2^k\binom{n}{k}$. This suggests forming k-person committees in which every committee member has 2 choices (of something).

240. There's no real magic here—you just have to grind through the algebra. But you should get a nice, simple-looking answer at the end.

241. Add an extra number to the end of the sequence so that the sum is 0.

242. Try to construct some simple examples to better see what's going on.

243. Count using casework based on k.

244. The worst-case scenario seems to be when students enroll in exactly 3 of the 6 courses. Try to construct an example in which each student enrolls in exactly 3 courses.

245. As usual, you need to show that every winning position has some move that leads to a losing position, and that all moves from a losing position lead to winning positions.

246. Compare non-double-good arrangements of n "("s and $2n$ ")"s with arrangements of $n-1$ "("s and $2n+1$ ")"s. (This is similar to the "path-reversing" argument used to prove the formula for the Catalan numbers.)

247. It may initially appear that you don't have enough data to use the Pigeonhole Principle. You'll need to use another fact about degrees of vertices in a graph.

248. Set up a recurrence and look for a pattern.

249. Start at 1000 and work backwards to determine the winning and losing positions.

250. Try to construct a cycle of odd length. What happens?

251. Break into cases, depending on whether a boy or a girl is sitting in the first seat.

252. Uh-oh, the characteristic equation has complex roots. Don't worry about it: keep plowing ahead with the solution.

253. For any vertex V, count the neighbors of the neighbors of V.

254. Draw the "possible" region (subject to the given condition) on the Cartesian plane. What is the "successful" region?

255. Write out expressions for a_n and a_{n-1} and use algebra to try to convert the recurrence into a linear recurrence.

256. First imagine that you have an infinite supply of each candy, and count the number of ways to distribute 20 pieces to each sister. Then subtract the distributions in which one sister gets more than 10 of one type.

257. Compute P(the die is loaded | k ⚃'s are rolled). Then find k such that this probability is at least 0.9.

258. Try constructing a list of concerts.

259. Multiply both sides by $k!$. The counting argument may be easier to see now.

260. Count the possible number of tournaments, and the number of tournaments with the desired property.

261. Next count the number of two-coin sequences (HH, HT, TH, and TT) that must be in the sequence.

262. What are the possible sizes of the circles of celebrities (from the previous hint)?

263. Try to show a 1-1 correspondence between solutions to this problem and solutions to the previous exercise.

264. If a_n is the number of integers whose digit sum is $2n$, then computing the first few terms of $\sqrt{a_1}, \sqrt{a_2}, \sqrt{a_3}, \ldots$ should show an interesting pattern.

265. The "boxes" for applying the Pigeonhole Principle are the diagonals of the chessboard.

266. Every parallelogram has sides parallel to 2 of the 3 sides of the triangle. Try to use the third side somehow to count the parallelograms with sides parallel to the other 2 sides.

267. Prove by induction that for all $k \geq 3$, we have $F_{k+1} < L_k < F_{k+2}$.

268. During the inductive step of the proof, consider separately those subsets of $\{1, 2, \ldots, n + 1\}$ that contain $n + 1$ as an element.

269. Use Binet's formula and the formula that you found in part (a).

270. Establish a 1-1 correspondence between subsets S with $10 \in S$ and subsets T with $10 \notin T$.

271. Do casework based on the number of A's in the word.

272. Do very careful casework.

273. Ending in the digits 0001 means leaving a remainder of 1 upon division by 10000. This sort of problem should look familiar!

274. We can add $1, 2, \ldots, 24$ to B without worry, because the corresponding number that sums to 125 is not in our original set $\{1, 2, \ldots, 100\}$.

275. Break up the event "A or B" into exclusive cases.

276. When constructing a subset $T \subseteq S$, how many choices do you have for each element $x \in S$?

277. Try it with 2 pirates first, then work your way backwards to all 5 pirates.

278. Break up the interval $[0, 1]$ into $[0, \frac{1}{n}], [\frac{1}{n}, \frac{2}{n}]$, etc. Let $\langle r \rangle$ be the fractional part of r (that is, $\langle r \rangle = r - \lfloor r \rfloor$). Use the Pigeonhole Principle.

279. Work out by hand what happens over the first several minutes. Do you see a pattern?

280. Imagine a dance class with n couples and 1 teacher. The RHS counts the number of n-person committees. Show that the LHS does as well.

281. We want to find the x^0 term (that is, the constant term) in the combined generating function for the 4 friends.

282. What conditions on A or B must hold if both $B \subseteq A$ and $B \subseteq (S \setminus A)$?

283. Try an example, say $n = 25$.

284. Focus on the squares along the diagonal.

285. Write the sequences just using 0's and 1's: 0 for an even term, 1 for an odd term. This may make the pattern easier to see.

286. This looks like a job for substitution into the Binomial Theorem! How can we make all but every third term cancel?

287. Given a winning position, find the left-most digit position that has an odd number of 1's. Take chips from that pile.

288. Set up a recurrence for the probability that Alfred wins the nth game.

289. In any set of $3^4 + 1 = 82$ 4-tuples of points, there must be two that are colored in the same way.

290. Let S_i be the set of 6-digit numbers with no digit i (where $i = 1, 2, 3$). Compute $\#(S_1 \cup S_2 \cup S_3)$ using PIE.

291. Show that silver matrices exist for $n = 2^k$ where k is any nonnegative integer.

292. Try it first for a fewer number of clubs, say 2 or 3 or 4. You might notice a pattern.

293. Note that a rectangle is determined by a choice of a pair of rows and a choice of a pair of columns.

294. Note that all of the flips before the last flip are irrelevant.

295. Compute Tina's generating function. Then use casework.

296. After multiplying both sides by $k!$, the LHS counts the number of *ordered* partitions of $\{1, 2, \ldots, n\}$ into k disjoint sets. This is the same as distributing the elements of $\{1, 2, \ldots, n\}$ into k distinguishable boxes so that no box is left empty. Use PIE to show that the right side counts this as well.

297. Suppose we picked the 6 numbers first, then tried to assign 3 of them to be a's and 3 of them to be b's. In how many ways could we do this?

298. Think of using the Binomial Theorem.

299. Note that $(d - 2)(b - 2)$ must be 1, 2, or 3. Determine the possible graphs for each of these.

300. Answer the question where the numbers are chosen from the set $\{1, 2, 3, 4, 5, 6\}$. How is this different from the problem where we choose our numbers from $\{1, 2, \ldots, 1000\}$?

301. Pick a student; call her Sally. Compute the number of ways to distribute the handouts so that Sally gets one.

302. Assemble the celebrities into circles, in which each celebrity shakes hands with his or her neighbors.

303. Start by experimenting with the small cases.

304. Prove by induction.

305. Think about drawing all 6 balls at once, and then arranging them.

306. Use the extremal principle: focus on the vertex of highest degree.

307. It may be easier to visualize coloring the vertices of a cube. Why is this the same problem?

308. Once k is fixed, think of counting the number of subsets with k elements as a distribution problem.

309. Let f, s, and x be the number of students in French, Spanish, and both, respectively. Write equations to represent the given facts, and solve for s.

310. Count the selfish subsets by casework, depending on whether or not n is in the subset.

311. Classify the subsets in terms of (a, d), where a is an element of the arithmetic sequence and d is the common difference.

312. Write the generating function for Alice's sums. Then try to factor it in a different way to get Betty's bags.

313. Rather than one player taking all his swings and then the other player taking all his swings, imagine that they are alternating swings.

314. Counting the colorings in which more than one 2×2 square is white must be done carefully—the number of these with two adjacent 2×2 white squares is different than the number of these with two diagonally-opposite 2×2 white squares.

315. Note that you only need prove the statement for $0 \le N \le 2^{n-1}$. For $N > 2^{n-1}$, just color as for $2^n - N$ and then reverse the colors.

316. For any element $x \in S$, and any subset $T \subseteq S$, either $x \in T$ or $x \notin T$.

317. Another solution is to write a system of 27 equations and sum them all.

318. Use the Pigeonhole Principle.

319. In any set of 4 points, there must be two of the same color.

320. Count the number of ways that a tournament can have an undefeated team or a winless team.

321. Do casework, based on the first urn chosen and the color of the ball chosen from that urn.

322. Compute a few examples. You should see a pattern.

323. What choices do you have to make to place the rooks?

324. Use the Binomial Theorem on $(1 - 4x)^{\frac{1}{2}}$.

325. You might think that because there are 52 cards and 4 Aces, the answer must be $52/4 = 13$. It's not. But thinking in this way is the right idea—look for a simple answer that involves symmetry.

326. First count the number of heads and tails.

327. Use geometry. Draw the "possible" and "successful" regions on the Cartesian plane.

328. Place the first tile so that it covers the upper-left square. What's left to do?

329. Show by induction that C_k is odd if and only if k is 1 less than a power of 2.

330. Focus on the outside edges. At least one must be missing (otherwise there'd be a loop). Use casework based on the position of the missing edge.

331. Pay particular attention to the corners of the board.

332. Note that no 2 students can have the exact same club memberships. There are 2^{10} possible different subsets of clubs that any student could be a member of, so 1024 is an upper bound for the possible number of students.

333. The block 101 shows up a lot. Rewrite the S_n's in terms of these blocks.

334. The LHS counts pairs of pairs of n objects.

335. Experiment with $n = 2$ and $n = 3$.

336. Prove the more general statement where 2002 is replaced by n.

337. If no pair has overlap greater than 1, and $B \geq 10$, what can we conclude?

338. Think of extending the sides of a parallelogram to intersect the 3rd side of the triangle.

339. Bipartite graphs don't have any triangles.

340. One "half" (as in the previous hint) is sets that contain their average. This doesn't quite work, but how can we fix it?

341. The side lengths of a triangle with perimeter n form a partition of n into 3 parts.

342. The previous hint doesn't quite work as-is. Instead, draw a larger triangle with side length $n + 1$ outside of the original triangle, and try the previous hint again.

343. List the first 10 or so Catalan numbers. This should lead you to a conjecture.

344. Work backwards from the end of the game. First determine what the position must be for a player not to be able to move. Then determine what positions have a move that leads to the immediate loss. And so on.

345. What is the expected number of handouts received by a randomly-chosen student?

346. Try to find a 1-1 correspondence between "Triangle inequality" partitions and some other type of partitions, for which it is easier to write a generating function.

347. Focus on the corner squares of the checkerboard.

348. Consider the games as paths on a grid from $(0, 0)$ to $(n - m, n + m)$. What line can't be crossed?

349. It may be helpful to use a graph with loops.

Index